MEDICAL
INTELLIGENCE
UNIT

Fibrogenesis: Cellular and Molecular Basis

Mohammed S. Razzaque, M.D., Ph.D.
Department of Oral and Developmental Biology
Harvard School of Dental Medicine
The Forsyth Institute
Boston, Massachusetts, U.S.A.

LANDES BIOSCIENCE / EUREKAH.COM
GEORGETOWN, TEXAS
U.S.A.

KLUWER ACADEMIC / PLENUM PUBLISHERS
NEW YORK, NEW YORK
U.S.A.

FIBROGENESIS: CELLULAR AND MOLECULAR BASIS

Medical Intelligence Unit

Eurekah.com / Landes Bioscience
Kluwer Academic / Plenum Publishers

Softcover re-print of the Hardcover 1st edition 2005

Kluwer Academic / Plenum Publishers, 233 Spring Street, New York, New York, U.S.A. 10013
http://www.wkap.nl

Please address all inquiries to the Publishers:
Landes Bioscience / Eurekah.com, 810 South Church Street, Georgetown, Texas, U.S.A. 78626
Phone: 512/ 863 7762; FAX: 512/ 863 0081
http://www.eurekah.com
http://www.landesbioscience.com

ISBN: 1-4614-9652-7

Selective Sweep edited by Dmitry Nurminsky, Landes / Kluwer dual imprint / Landes series: Molecular Biology Intelligence Unit

Fibrogenesis : cellular and molecular basis / [edited by] Mohammed S. Razzaque.
p. ; cm. -- (Medical intelligence unit)
Includes bibliographical references and index.
ISBN: 1-4614-9652-7
1. Fibrosis. I. Razzaque, Mohammed S. II. Series: Medical intelligence unit (Unnumbered : 2003)
[DNLM: 1. Fibrosis--pathology. 2. Molecular Biology. QZ 150 F443 2005]
RB150.F52F53 2005
616.7'7--dc22

2004023420

To my father, N.M.A. Razzaque Mojumder,
who taught me how to hold a pen, how to write and,
most importantly, nudged me to believe in my dreams.
Although he passed away over twenty years ago,
his teaching of doing the right things in life over
the easy things is ever enduring for me.

CONTENTS

EDITOR

Mohammed S. Razzaque, M.D., Ph.D.
Department of Oral and Developmental Biology
Harvard School of Dental Medicine
The Forsyth Institute
Boston, Massachusetts, U.S.A.
Chapters 1, 2, 13

CONTRIBUTORS

Wesam Ahmed
Department of Cancer Immunology and AIDS
Dana-Farber Cancer Institute
Harvard Medical School
Boston, Massachusetts, U.S.A.
Chapter 13

Abdallah Azouz
Department of Radiology
Beth Israel Deaconess Medical Center
Harvard Medical School
Boston, Massachusetts, U.S.A.
Chapter 1

Laurent Baud
Service d'Explorations Fonctionnelles Multidisciplinaires and INSERM
Hôpital Tenon
Paris, France
Chapter 4

Gavin J. Becker
Department of Nephrology
The Royal Melbourne Hospital and Department of Medicine
University of Melbourne
Melbourne, Victoria, Australia
Chapter 6

Agnes Bellocq
Service d'Explorations Fonctionnelles Multidisciplinaires and INSERM
Hôpital Tenon
Paris, France
Chapter 4

Moussa El-Hallak
Department of Radiology
Beth Israel Deaconess Medical Center
Harvard Medical School
Boston, Massachusetts, U.S.A.
Chapter 1

Fabio Fiordaliso
Department of Cardiovascular Research
Istituto di Ricerche Farmacologiche "Mario Negri"
Milan, Italy
Chapter 8

Bruno Fouqueray
Service d'Explorations Fonctionnelles Multidisciplinaires and INSERM
Hôpital Tenon
Paris, France
Chapter 4

Jan Galle
Department of Medicine
Division of Nephrology
Julius-Maximilians-University
Würzburg, Germany
Chapter 3

Charles Giardina
Department of Molecular and Cell Biology
University of Connecticut
Storrs, Connecticut, U.S.A.
Chapter 15

Naoki Hagimata
Research Institute for Diseases
of the Chest
Graduate School of Medical Sciences
Kyushu University
Fukuoka, Japan
Chapter 14

Nobuyuki Hara
Research Institute for Diseases
of the Chest
Graduate School of Medical Sciences
Kyushu University
Fukuoka, Japan
Chapter 14

Jean-Philippe Haymann
Service d'Explorations Fonctionnelles
Multidisciplinaires and INSERM
Hôpital Tenon
Paris, France
Chapter 4

Tim D. Hewitson
Department of Nephrology
The Royal Melbourne Hospital
and Department of Medicine
University of Melbourne
Melbourne, Victoria, Australia
Chapter 6

Andrea K. Hubbard
Department of Pharmaceutical Sciences
University of Connecticut
Storrs, Connecticut, U.S.A.
Chapter 15

Yutaka Inagaki
Department of Community Health
Tokai University School of Medicine
Bohseidai, Isehara, Kanagawa, Japan
Chapter 12

Norifumi Kawada
Department of Hepatology
Graduate School of Medicine
Osaka City University
Osaka, Japan
Chapter 10

Shozo Kusachi
Department of Medical Technology
Faculty of Health Science
Okayama University Medical School
Shikata-cho, Okayama, Japan
Chapter 7

Kazuyoshi Kuwano
Research Institute for Diseases
of the Chest
Graduate School of Medical Sciences
Kyushu University
Fukuoka, Japan
Chapter 14

Roberto Latini
Department of Cardiovascular Research
Istituto di Ricerche Farmacologiche
"Mario Negri"
Milan, Italy
Chapter 8

Hyun Soon Lee
Department of Pathology
Seoul National University College
of Medicine
Seoul, Korea
Chapter 5

Serge Masson
Department of Cardiovascular Research
Istituto di Ricerche Farmacologiche
"Mario Negri"
Milan, Italy
Chapter 8

Kouji Matsushima
Department of Molecular Preventive
Medicine
The University of Tokyo
Tokyo, Japan
Chapter 2

Sarah Mowbray
Department of Pharmaceutical Sciences
University of Connecticut
Storrs, Connecticut, U.S.A.
Chapter 15

Maki Niioka
Department of Community Health and Environmental Health
Tokai University School of Medicine
Bohseidai, Isehara, Kanagawa, Japan
Chapter 12

Yoshifumi Ninomiya
Department of Molecular Biology and Biochemistry
Okayama University Graduate School of Medicine
Okayama, Japan
Chapter 7

Isao Okazaki
Department of Community Health
Tokai University School of Medicine
Bohseidai, Isehara, Kanagawa, Japan
Chapter 12

Julie Peltier
Service d'Explorations Fonctionnelles Multidisciplinaires and INSERM
Hôpital Tenon
Paris, France
Chapter 4

Thomas Quaschning
Department of Medicine
Division of Nephrology
Julius-Maximilians-University
Würzburg, Germany
Chapter 3

Ladislas Robert
Laboratoire de Recherche Ophtalmique
Université Paris
Paris, France
Chapter 9

Monica Salio
Department of Cardiovascular Research
Istituto di Ricerche Farmacologiche "Mario Negri"
Milan, Italy
Chapter 8

Stefan Seibold
Department of Medicine
Division of Nephrology
Julius-Maximilians-University
Würzburg, Germany
Chapter 3

Ichiro Shimizu
Department of Digestive and Cardiovascular Medicine
Tokushima University School of Medicine
Tokushima, Japan
Chapter 11

Yoshihiko Sugioka
Department of Environmental Health
Tokai University School of Medicine
Bohseidai, Isehara, Kanagawa, Japan
Chapter 12

Takashi Taguchi
Department of Pathology
Nagasaki University Graduate School of Biomedical Sciences
Nagasaki, Japan
Chapters 1, 2, 13

Michael Thibodeau
Department of Pharmaceutical Sciences
University of Connecticut
Storrs, Connecticut, U.S.A.
Chapter 15

Takashi Wada
Division of Blood Purification
Department of Gastroenterology
and Nephrology
Kanazawa University Graduate School
of Medical Science
Kanazawa, Japan
Chapter 2

Tetsu Watanabe
Department of Environmental Health
Tokai University School of Medicine
Bohseidai, Isehara, Kanagawa, Japan
Chapter 12

Hitoshi Yokoyama
Division of Blood Purification
Department of Gastroenterology
and Nephrology
Kanazawa University Graduate School
of Medical Science
Kanazawa, Japan
Chapter 2

PREFACE

This book is written to provide in-depth and current information on the molecular basis of fibrotic diseases in various tissues and organs. Why do we need a new book on fibrosis? Several currently available books on fibrosis are mostly devoted to specific tissue or organ systems; in contrast, this book is especially designed to provide integral and comprehensive information on fibrogenesis. Individual chapters of this book present up-to-date information on the cellular and molecular basis of fibrosis or sclerosis in heart, lung, liver, kidney, and blood vessels. Chapters present the effects on fibrogenesis of inflammation, oxidative stress, low-density lipoprotein, ischemia, apoptosis, and aging. The chapter authors who are actively involved in clinical and basic research have outlined the emerging therapies that can eventually treat fibrotic diseases. The future therapeutic potentials of matrix-degrading enzymes on the reversibility of fibrotic disorders are also included. The wide range of topics that are covered in this book will provide the reader with a comprehensive and detailed understanding of fibrogenesis. Recent molecular studies have changed the commonly held view that the extracellular matrix proteins are biologically passive, i.e., they mostly act as glue between cellular elements. Extensive research on matrix biology in the last decade or so has identified numerous essential dynamic functions for the extracellular matrix proteins, ranging from fetal development to tissue repair. Extracellular matrix proteins play crucial roles in determining the microenvironments and functional activities of the surrounding cells. However, excessive accumulation of extracellular matrix proteins in chronic inflammatory, immunologic, or metabolic diseases can lead to functional impairment of the affected tissues and eventual end stage organ failure. Irreversible fibrotic diseases are a major clinical problem, with high morbidity and mortality. The clinical importance, and significance of the fibrotic diseases in various tissues and organs have led me to conceive of this book. Leading experts in the field of matrix biology have contributed chapters containing essential information required to have an in-depth understanding of the cellular and molecular basis of fibrogenesis. This book will be important reading for all those involved in basic and clinical research on matrix biology. My utmost hope is that, amongst the readers, some will be inspired to take up the challenge and excitement of research to further enhance our knowledge and understanding of the complex biomechanistic pathways of fibrotic diseases. Such intellectually challenging efforts will help in developing new therapeutic strategies to treat fibrotic diseases.

I extend my sincerest thanks and gratitude to each of the contributing authors for his or her time and efforts. I acknowledge Dr. R. Landes of Landes Bioscience for his cooperation in publishing this book. Finally, I wish to acknowledge my mentor, and teacher, Prof. T. Taguchi of Nagasaki University Graduate School of Biomedical Sciences for his help and guidance. Particularly, I would like to thank the Razzaque family (Rafi, Yuki, Zahid, Ayesha, Amina, Abida, Muhit, Newaz, Tahiti, Karima, Shahid, Diane, Lisa, and Amma) for their encouragement, support, and help during my many hours of writing and editing. I hope that scientists and clinicians who need a quick update on fibrogenesis in their research and practice will find this a useful reference book.

Mohammed S. Razzaque

ABBREVIATIONS

TNFα	tumor necrosis factorα
IFNγ	Interferonγ
IL-1β	Interleukin-1β
TGFβ	transforming growth factorβ
IL-12	Interleukin-12
IL-18	Interleukin-18
IL-9	Interleukin-9
LPS	lipopolysaccharide
ROS	reactive oxygen species
RNS	reactive nitrogen species
MMPs	matrix metalloproteinases
TIMPs	tissue inhibitors of matrix metalloproteinases
AM	alveolar macrophages
PMN	polymorphonuclear leukocytes
LDH	lactate dehydrogenase
BAL	bronchoalveolar lavage
R	receptor
TH	T helper cell
ECM	extracellular matrix
TUNEL	terminal deoxynucleotidyl transferase-mediated X-dUTP nick-end labeling
L-NAME	N(G)-nitro-L-arginine methyl ester
FasL	Fas ligand
MEK	MAPK/ERK kinase
ERK	extracellular signal regulated kinase
AP-1	activating protein 1
iNOS	inducible nitric oxide synthase
NAC	n-acetyl cysteine
TCR	T cell receptor
KO	knock out

Chapter 1

Tissue Scarring:

Lessons from Wound Healing

Mohammed S. Razzaque, Moussa El-Hallak, Abdallah Azouz and Takashi Taguchi

Abstract

Tissue scarring due to abnormal matrix remodeling is an important cause of organ failure, which is a leading cause of morbidity and mortality. Current studies have fo cused on determination of the molecular basis of controlled wound healing and uncontrolled tissue scarring. Scarless repair is a unique feature of fetal wounds in early gestation. Our understanding of the molecular basis of fetal response during wound healing may represent a paradigm to modulate incomplete and/or excessive healing (tissue scarring) to an ideal scarless healing. Once the fetal microenvironment that steers scarless wound healing is known, attempts to create a similar artificial environment to modulate abnormal tissue scarring could be accomplished. This brief review addresses the pathogenesis of wound healing and its relevance to tissue scarring.

Wound Healing

A wound is a discontinuity of tissue integrity, while healing is the complex and dynamic process of reconstituting that integrity. Superficial surgical incision causes injury to the lining epithelial cells, disrupts underneath basement membrane, and damages the surrounding connective tissue cells. Such damage induces inflammatory reactions, matrix deposition, and resolution, which are essential components of wound healing. Immediately after the injury, clotted blood containing fibrin and blood cells accumulates in the wound area. In addition, locally generated inflammatory mediators start to recruit neutrophils within 24 hours, with increased proliferating activity in basal epithelial cells. By day 3, most neutrophils are replaced with macrophages. Granulation tissue usually consists of proliferating fibroblasts embedded in the extracellular matrix (ECM) and new blood vessels, and becomes apparent on day 3. The repair process continues, and matrix components in the granulation tissues begin to bridge the wound, which is accompanied by re-epitheliazation. The macrophage has an important role in the healing process. Macrophages are predominant within the injured areas in around 48 to 72 hours after injury. They do not only act as phagocytic cells, but also produce various factors involved in the proliferation and differentiation of matrix-producing fibroblasts to synthesize matrix proteins. Altered macrophage population may result in abnormal healing with poor debridement. A pathogenic role for T cells has also been suggested during the healing process. Experimental studies have suggested that dendritic epidermal T cells (DETCs) that bear a $\gamma\delta$

Fibrogenesis: Cellular and Molecular Basis, edited by Mohammed S. Razzaque.

T-cell receptor (TCR) play an important role in wound healing. Compared with wild-type mice, a delayed wound closure and reduced epithelial hyperthickening has been observed in $TCR\delta^{-/-}$ mice, which lack γδ T cells, after a full-thickness skin wound.[1]

At about one week after the original wound, although the inflammatory features essentially disappear, fibroblast proliferation and matrix deposition continue. At the end of the first month, the wound is usually healed with scar tissue, covered with a layer of surface epithelium. Every so often, non-surgical wounds are irregular, and the extent of the injury to the surrounding cells and tissues is severe, which makes the healing process difficult; in such cases, healing is accomplished by second intention. Healing by second intention is a complex dynamic process with intensified inflammatory responses, granulation tissue formation, and wound contraction. Usually the pre-injury architecture is not recovered, and original structures are replaced with a scar tissue. The matrix remodeling may continue for a long time with individual variations, which is often influenced by the aging process.

Complex interactions among various cytokines and growth factors regulate the wound healing process. EGF (epidermal growth factor) helps in regeneration of epithelial cells and proliferation of fibroblasts,[2] while PDGF (platelet-derived growth factor) plays a crucial role in proliferation, migration and differentiation of fibroblasts during wound healing.[3] FGF (fibroblast growth factor) and VEGF (vascular endothelial growth factor) exert angiogenic effects on endothelial cells and subsequent neovascularization.[4,5] TGF-β (transforming growth factor-β) helps in granulation tissue formation and subsequent matrix remodeling during the healing process.[6] The replaced scar tissue is rarely as strong as the original tissue. In addition to the above-mentioned factors, certain systemic and local factors influence the complex healing process. For example, vitamin C deficiency impairs healing, by inhibiting collagen synthesis, while, steroid therapy, age, presence of local infections, circulatory state, and genetic background might influence the healing process. Keloid is a scar tissue that is formed as a result of abnormal healing due to excessive accumulation of ECM proteins. The predominant occurrence of keloid in black population is suggestive of a genetic predisposition.[7]

Recent studies have identified the signaling cascade involved in wound healing. For instance, Jun kinase (JNK) signaling pathway plays a crucial role in efficient wound healing in *Drosophila*.[8] The expression of alpha v beta 6 integrin has been reported to be induced in keratinocytes, which is associated with the fusion of the epithelium and initiation of granulation tissue formation during wound healing.[9,10]

Fetal Wound Healing

The wound repair process in adults is characterized by scar tissue formation, often associated with contracture. Such deformity might be associated with pathological consequences. In contrast to adult wounds, fetal wounds heal without leaving any histological features of scarring during early gestation. Understanding the fetal microenvironment during scarless wound healing will provide important information that might help in creating a similar artificial environment to modulate abnormal scarring. Although the pathomechanism of scarless fetal wound healing is not completely understood, studies have documented that inflammatory mediators, their effects on fetal dermal fibroblasts, subsequent downstream signaling cascade, and composition of matrix proteins are different than those of adult wound healing. Accelerated healing process, and a relative lack of an acute inflammatory response distinguishes fetal wound healing from adult wound healing. Increased monocytic infiltration with relatively less neutrophilic activity is associated with fetal wound healing.[11] Similarly, relatively less amount of interleukin (IL)-6, IL-8, TGF-β and FGF have been detected during scarless fetal wound healing, compared with the adult healing.[11-16] The exogenous addition of TGF-β1 in fetal wounds resulted in healing with scar tissue formation.[14,15] Likewise, lower level of IL-6 expression has been detected during fetal wound healing, while exogenous addition of IL-6 resulted in fetal wound healing with scar formation.[16] Furthermore, compared with adult wound healing process, relatively lower levels of bone morphogenetic proteins (BMPs), members of TGF-β superfamily, and BMP receptors have been detected during fetal wound healing.[17] On the other hand, the fetal wound contains higher amounts of hyaluronic acid, and consists of less tightly cross-linked ECM components. During the first 3 days, hyaluronic acid was detected at high levels in the adult wound, but was not detected by day 7. Adult wound healing is mostly achieved by accumulation of collagen, which is the main component of scar tissue. In contrast, the level of hyaluronic acid remains elevated for 21 days during fetal wound healing,[18] while collagen deposition is rapid and not excessive. Rapid upregulation of epidermal integrin receptors specific for fibronectin and other matrix proteins during human fetal wound healing might facilitate the migration of keratinocytes and subsequent re-epithelialization, thereby limiting inflammatory responses, and fibrogenic activities.[19] Fetal fibroblasts do not only have greater migratory abilities, but also have greater ability to induce formation of dermal appendages. Further studies are needed to determine the molecular mechanisms of relatively less differentiated cells in the fetal wound bed, and the reasons for less tightly cross-linked ECM accumulation during scarless fetal wound healing. The understanding of the molecular basis of fetal wound healing might help in designing therapeutic strategies to help avert tissue scarring in adults.

Recent studies have focused on the potentials of bone marrow derived stem cells in controlled tissue repair. In experimental studies, when mouse bone marrow cells from nondiabetic and diabetic mice were enriched with putative stem cells and injected under skin wounds of nondiabetic or type 2 diabetic $Lepr^{db}$ mice, exogenous nondiabetic bone marrow–derived cells increased vascularization and improved wound healing in $Lepr^{db}$ diabetic mice with little effect on nondiabetic controls.[20] In clinical set up, autologous corneal stem cell grafting in patients with unilateral chemical burns was successful in restoring vision, reducing irritation with regeneration of regular corneal epithelium replacing the conjunctival overgrowth.[21] Similarly, in patients with partial limbal stem cell deficiency, amniotic membrane transplantation was effective in restoring a stable corneal epithelium.[22] Preliminary studies are projecting a promising role for stem cells in regeneration and healing process. Ongoing studies will help us understand the therapeutic potentials of stem cells in modulating tissue scarring.

Tissue Scarring

Irreversible end-stage tissue scarring continues to be a leading cause of morbidity and mortality, and to date no specific therapeutic agents are available to treat scarring of various tissues in mostly irreversible progressive diseases. Inflammatory reactions, matrix deposition, and resolution are important pathological events of the wound healing (Fig. 1). In contrast to the regular wound healing, no true resolution occurs during tissue scarring. It is the prolonged activation of the genes encoding for fibrogenic molecules that distinguishes controlled wound repair from uncontrolled pathologic tissue scarring, where accumulation of matrix proteins often continues in spite of apparent resolution and/or disappearance of the initial triggering factors. Activated fibroblasts and myofibroblasts mostly produce increased levels of matrix proteins, which eventually replace the normal tissue with scar tissue, leading to loss of function of the affected tissues or organs.[23]

Various factors and/or disorders are associated with tissue scarring. For instance, certain physical or chemical injuries and immunological disorders are associated with cutaneous fibrosis such as keloids, hypertrophic scars and scleroderma. Even a simple incision during a routine surgery may heal with scarring with unanticipated pathologic consequences. Often abdominal surgeries are associated with fibrous intraperitoneal adhesion with resultant mechanical obstruction of the intestine. Post-surgical strictures are encountered in the anastomotic sites like

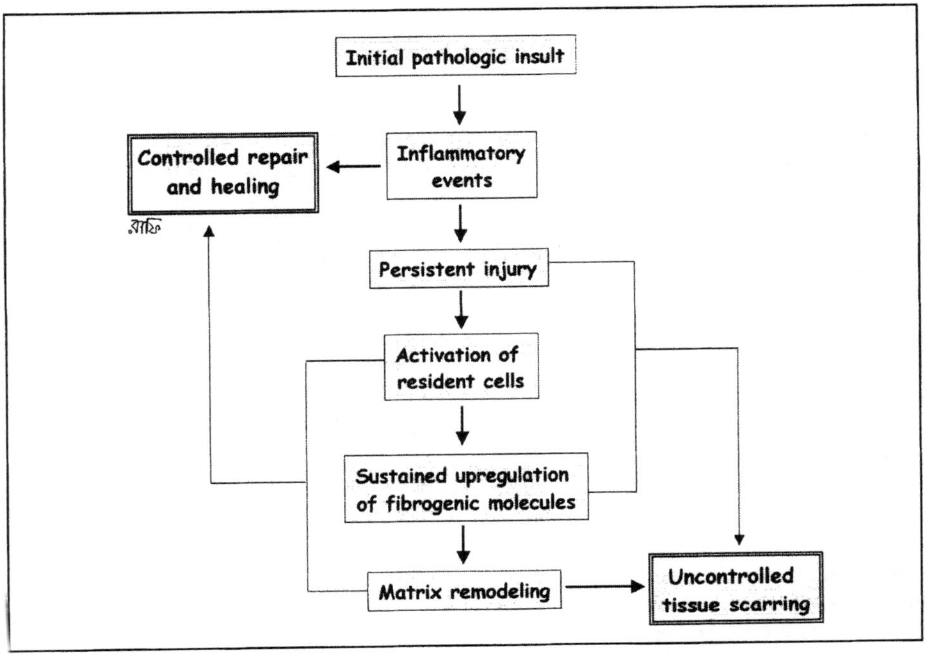

Figure 1. Simplified diagram showing various events during controlled wound healing and uncontrolled tissue scarring.

bowel, blood vessels, trachea, ureter, or bile duct, due to abnormal regulation of matrix metabolism. Similarly, metabolic diseases (diabetes mellitus), systemic diseases (hypertension), immunological diseases (glomerulonephritis), and even prolonged urinary protein leaking may lead to the development of renal scarring.[24,25] The clinical use of a significant number of useful drugs have been restricted for their tissue scarring effects, which includes bleomycin, cisplatin, cyclosporin, and gentamicin.[26-29] Prolonged alcohol consumption and chronic viral infections are usually associated with liver fibrosis,[30] while, toxic vapors, and certain drugs can induce pulmonary fibrosis. Although pulmonary fibrosis is triggered by diverse known and unknown factors including exposure to inorganic dusts or radiation, not all exposed individuals develop fibrosis, suggesting a possible genetic predisposition.[31,32] Identifying genetically susceptible population by genetic profiling might allow early detection, and early intervention.

The process and progression of tissue scarring can be broadly divided in to three phases and/or events: initial triggering events, leading to inflammatory events, which are usually followed by fibrotic events. These phases usually overlap, and are regulated by a wide range of phase-specific cytokines, chemokines, growth factors, and signaling molecules.[33] Details of the molecular and cellular interactions among some of these molecules during the progression of fibrosis/sclerosis in the lung, liver, heart, blood vessels, and kidney are described in the accompanying chapters. In general, there are similarities in tissue scarring in various organs, although the natures of the initial insults are diverse, and are not always detectable. In most fibrotic diseases, the initial insults trigger damage to the affected tissues or organs, which is accompanied by inflammatory reactions. Induction of various mitogenic and fibrogenic factors by inflammatory cells and activated resident cells results in activation of matrix-producing cells causing their proliferation, differentiation (to myofibroblasts), and eventual production of excessive matrix proteins, or scar tissue, in response to the injury.[23,33,34] Furthermore, the antioxidant defense system is thought to be impaired during fibrogenesis, and the pathologic role of reactive oxygen species (ROS) has been demonstrated in the early stage of various human and experimental fibrotic diseases. For instance, high levels of myeloperoxidase and low levels of glutathione, an important antioxidant, have been detected in the alveolar epithelial lining fluid of patients with idiopathic pulmonary fibrosis, suggesting the role of oxidant-mediated injury in pulmonary fibrosis.[35,36] Similarly, overproduction of ROS, and down-regulation in the expression of cellular antioxidants enzymes have been suggested in mediating renal injuries; the protective effects of antioxidants, including vitamin E and α-lipoic acid in modulating renal injuries provide evidence of oxidant-mediated injuries in the pathogenesis of scarring renal diseases.[37,38]

Accumulation of inflammatory cells, including neutrophils; lymphocytes, mast cells, monocytes, and macrophages is a consistent histological feature of the early stage of tissue scarring. Of these, macrophages are not only actively involved in the inflammatory events by producing high levels of IL-8 to facilitate neutrophil trafficking into the affected sites,[39] but also exert mitogenic and fibrogenic effects by secreting various cytokines, chemokines, and growth factors, including PDGF and TGF-β1. These factors in turn can act on fibroblasts and myofibroblasts to produce excessive amount of matrix proteins. Certain macrophage regulating molecules, including macrophage colony-stimulating factor (m-CSF) and macrophage migration inhibitory factor (MIF), by inducing increased accumulation and local proliferation of macrophages determine macrophage population in the injured tissues, which eventually develop tissue scarring.[40,41] Studies have convincingly demonstrated that the release and activation of certain transcriptional factors [e.g., activating protein (AP-1) and nuclear factor-kappaB (NF-κB)], chemokines [e.g., monocyte chemoattractant protein-1 (MCP-1), regulated on activation normal T-cell expression and secreted (RANTES)] cytokines (e.g., IL-1, IL-4, IL-8), and growth factors [e.g., PDGF, TGF-β1, tumor necrosis factor (TNF)-α], by infiltrating inflammatory cells could intensify both the inflammatory and fibrotic events in the affected organs.[42-44] Moreover, ECM-derived signals may exert chemotactic effects on a number of matrix-producing cells, which are involved in transcriptional activation of AP-1 and NF-κB to produce inflammatory cytokines, such as IL-1 and TNF-α and thereby exacerbate the fibroproliferative process.[45] In early stage of tissue scarring, inflammatory infiltration of lymphocytes and macrophages, with induction of adhesion molecules, such as vascular cell adhesion molecule 1 (VCAM-1) and intercellular adhesion molecule 1 (ICAM-1) by resident cells has been detected.[46] Both VCAM-1 and ICAM-1 are ligand for receptors expressed on memory-activated T cells and monocytes, and thus could create a microenvironment for the interaction of inflammatory cells and with resident cells. Change in the microenvironments due to altered composition of matrix proteins during tissue scarring may activate adjacent cells to produce factors that may influence both inflammatory responses and matrix remodeling. For instance, a type I collagen-rich microenvironment could activate matrix-producing hepatic stellate cells,[47] and these activated cells express increased levels of collagen receptors α1β1 and α2β1 integrins.[48] Some of these collagen receptors of integrins are also involved in the signal transduction, and may participate in transcriptional regulation of genes that might regulate proliferation, migration and adhesion of matrix-producing cells, and thus influence and contribute to tissue scarring. Heat shock protein 47 (HSP47) is a collagen-specific molecular chaperon, and is involved in the biosynthesis and secretion of procollagens.[49] Recent studies have documented a fibrogenic role for HSP47 in excessive accumulation of collagens during tissue scarring.[50-55]

An imbalance in the level of production and rate of degradation of matrix proteins could determine the extent of tissue scarring, in the involved tissues and organs. Accumulation of ECM is mostly achieved through the production collagenous and non-collagenous proteins. The degradation of ECM is mostly regulated by proteolytic enzymes, including MMPs (matrix metalloproteinases) and ADAMs (a disintegrin and metalloproteinase domain).[56,57] In addition, tissue inhibitors of metalloproteinases (TIMPs) also play an active role in matrix remodeling by neutralizing activities of MMPs. For instance, high expression levels of MMP-1, -2, and -9 have been detected during pulmonary fibrogenesis,[58] which was associated with increased levels of their inhibitory enzymes, TIMPs -1 and -2.[59] It is likely that the expression of TIMPs could neutralize the collagenolytic effects of MMPs, and thereby could facilitate matrix accumulation. Recent studies have suggested that TGF-β1 could induce a number of matrix remodeling molecules, including MMPs and plasminogen activator inhibitor-1 (PAI-1) during tissue scarring.[60,61] The reversibility of scarring in affected organs may be partly determined by the proteolytic activities of the relevant enzymes.

Conclusion

The basic science is meaningful, consequential and far-reaching when the hard, and often the difficult work of scientists is applied to minimize patient's sufferings, by clinical control of disease progression. The detailed understanding of the molecular mechanisms of scarless fetal wound healing might allow basic scientists and clinicians to influence the outcome of the scarring process, by preventing and/or delaying the progression of excessive tissue scarring. In this respect, it has been convincingly demonstrated that a similar type of injury elicits different responses by adult and fetal dermal fibroblasts, resulting in healing with scar formation in adults and healing without scar in fetus.[62] In this book, individual chapters do not only describe in detail the up-to-date information on fibrogenesis in various tissues and organs, but also discuss the implications of this information in developing new therapeutic strategies to treat progressive fibrotic diseases.

References

1. Jameson J, Ugarte K, Chen N et al. A role for skin gammadelta T cells in wound repair. Science 2002; 296:747-749.
2. Marikovsky M, Breuing K, Liu PY et al. Appearance of heparin-binding EGF-like growth factor in wound fluid as a response to injury. Proc Natl Acad Sci USA 1993; 90:3889-3893.
3. Seppa H, Grotendorst G, Seppa S et al. Platelet-derived growth factor in chemotactic for fibroblasts. J Cell Biol 1982; 92:584-588
4. Nissen NN, Polverini PJ, Koch AE et al. Vascular endothelial growth factor mediates angiogenic activity during the proliferative phase of wound healing. Am J Pathol 1998; 152:1445-1452
5. Tonnesen MG, Feng X, Clark RA. Angiogenesis in wound healing. J Invest Dermatol Symp Proc 2000, 5:40-46
6. Takehara K. Growth regulation of skin fibroblasts. J Dermatol Sci 2000; 24:S70-77.
7. Rockwell WB, Cohen IK, Ehrlich HP. Keloids and hypertrophic scars: A comprehensive review. Plast Reconstr Surg 1989; 84:827-837.
8. Reed BH, Wilk R, Lipshitz HD. Downregulation of Jun kinase signaling in the amnioserosa is essential for dorsal closure of the Drosophila embryo. Curr Biol 2001; 11:1098-108.
9. Breuss JM, Gallo J, DeLisser HM et al. Expression of the beta 6 integrin subunit in development, neoplasia and tissue repair suggests a role in epithelial remodeling. J Cell Sci 1995; 108:2241-51.
10. Haapasalmi K, Zhang K, Tonnesen M et al. Keratinocytes in human wounds express alpha v beta 6 integrin. J Invest Dermatol 1996; 106:42-8.
11. Liechty KW, Crombleholme TM, Cass DL et al. Diminished interleukin-8 (IL-8) production in the fetal wound healing response. J Surg Res 1998; 77: 80–84.
12. Dang CM, Beanes SR, Soo C et al. Decreased expression of fibroblast and keratinocyte growth factor isoforms and receptors during scarless repair. Plast Reconstr Surg 2003;111:1969-79.
13. Cowin AJ, Brosnan MP, Holmes TM et al. Endogenous inflammatory response to dermal wound healing in the fetal and adult mouse. Dev Dyn 1998; 212:385-93.
14. Lanning DA, Nwomeh BC, Montante SJ et al. TGF-beta1 alters the healing of cutaneous fetal excisional wounds. J Pediatr Surg 1999; 34:695-700.

15. Lin RY, Sullivan KM, Argenta PA et al. Exogenous transforming growth factor-beta amplifies its own expression and induces scar formation in a model of human fetal skin repair. Ann Surg 1995; 222:146-54.
16. Liechty KW, Adzick NS, Crombleholme TM. Diminished interleukin 6 (IL-6) production during scarless human fetal wound repair. Cytokine 2000; 12:671-6.
17. Hwang EA, Lee HB, Tark KC. Comparison of bone morphogenetic protein receptors expression in the fetal and adult skin. Yonsei Med J 2001; 42:581-6.
18. Longaker MT, Chiu ES, Adzick NS et al. Studies in fetal wound healing: V. A prolonged presence of hyaluronic acid characterizes fetal wound fluid. Ann Surg 1991; 213:292-296.
19. Larjava H, Salo T, Haapasalmi K et al. Expression of integrins and basement membrane components by wound keratinocytes. J Clin Invest 1993; 92:1425-35.
20. Stepanovic V, Awad O, Jiao C et al. Lepr^{db} diabetic mouse bone marrow cells inhibit skin wound vascularization but promote wound healing. Circ Res 2003; 92:1247-53.
21. Fagerholm P, Lisha G. Corneal stem cell grafting after chemical injury. Acta Ophthalmol Scand 1999; 77:165-9.
22. Anderson DF, Ellies P, Pires RT et al. Amniotic membrane transplantation for partial limbal stem cell deficiency. Br J Ophthalmol 2001; 85:567-75.
23. Moriyama T, Imai E. Role of myofibroblasts in progressive renal diseases. Contrib Nephrol 2003; 139:120-40.
24. Razzaque MS, Azouz A, Shinagawa T et al. Factors regulating the progression of hypertensive nephrosclerosis. Contrib Nephrol 2003; 139:173-86.
25. Matsuo S, Morita Y, Maruyama S et al. Proteinuria and tubulointerstitial injury: the causative factors for the progression of renal diseases. Contrib Nephrol 2003; 139:20-31.
26. Razzaque MS, Taguchi T. Role of glomerular epithelial cell-derived heat shock protein 47 in experimental lipid nephropathy. Kidney Int 1999; 71:S256-9.
27. Sund S, Forre O, Berg KJ et al. Morphological and functional renal effects of long-term low-dose cyclosporin A treatment in patients with rheumatoid arthritis. Clin Nephrol 1994; 41:33-40.
28. Cheng M, Razzaque MS, Nazneen A et al. Expression of the heat shock protein 47 in gentamicin-treated rat kidneys. Int J Exp Pathol 1998; 79:125-32.
29. Razzaque MS, Hossain MA, Kohno S et al. Bleomycin-induced pulmonary fibrosis in rat is associated with increased expression of collagen-binding heat shock protein (HSP) 47. Virchows Arch 1998; 432:455-60.
30. Marcellin P, Asselah T, Boyer N. Fibrosis and disease progression in hepatitis C. Hepatology 2002; 36:S47-56.
31. Zorzetto M, Ferrarotti I, Trisolini R et al. Complement receptor 1 gene polymorphisms are associated with idiopathic pulmonary fibrosis. Am J Respir Crit Care Med 2003; 168:330-4.
32. Chung MP, Monick MM, Hamzeh NY et al. Role of repeated lung injury and genetic background in bleomycin-induced fibrosis. Am J Respir Cell Mol Biol 2003; 29:375-80.
33. Razzaque MS, Taguchi T. Cellular and molecular events leading to renal tubulointerstitial fibrosis. Med Electron Microsc 2002; 35:68-80.
34. Razzaque MS, Taguchi T. Pulmonary fibrosis: Cellular and molecular events. Pathol Int 2003; 53:133-45.
35. Borok Z, Buhl R, Grimes GJ et al. Effect of glutathione aerosol on oxidant-antioxidant imbalance in idiopathic pulmonary fibrosis. Lancet 1991; 338:215-6.
36. Cantin AM, Fells GA, Hubbard RC et al. Antioxidant macromolecules in the epithelial lining fluid of the normal human lower respiratory tract. J Clin Invest 1990; 86:962-71.
37. Rohrmoser MM, Mayer G. Reactive oxygen species and glomerular injury. Kidney Blood Press Res 1996; 19:263-9.
38. Trachtman H, Chan JC, Chan W et al. Vitamin E ameliorates renal injury in an experimental model of immunoglobulin A nephropathy. Pediatr Res 1996; 40:620-6.
39. Pantelidis P, Southcott AM, Black CM, Du Bois RM. Up-regulation of IL-8 secretion by alveolar macrophages from patients with fibrosing alveolitis: a subpopulation analysis. Clin Exp Immunol 1997; 108:95-104.
40. Razzaque MS, Foster CS, Ahmed AR. Role of enhanced expression of m-CSF in conjunctiva affected by cicatricial pemphigoid. Invest Ophthalmol Vis Sci 2002; 43:2977-83.

41. Burger-Kentischer A, Goebel H, Seiler R et al. Expression of macrophage migration inhibitory factor in different stages of human atherosclerosis. Circulation 2002; 105:1561-6.
42. Inan MS, Razzaque MS, Taguchi T. Pathological significance of renal expression of NF-kappa B. Contrib Nephrol 2003;139:90-101.
43. Burton CJ, Combe C, Walls J et al. Secretion of chemokines and cytokines by human tubular epithelial cells in response to proteins. Nephrol Dial Transplant 1999; 14:2628-33.
44. Razzaque MS, Ahmed BS, Foster CS et al. Effects of IL-4 on conjunctival fibroblasts: possible role in ocular cicatricial pemphigoid. Invest Ophthalmol Vis Sci 2003; 44:3417-23.
45. Juliano RL, Haskill S. Signal transduction from the extracellular matrix. J Cell Biol 1993; 120:577-85.
46. Thorne SA, Abbot SE, Stevens CR et al. Modified low density lipoprotein and cytokines mediate monocyte adhesion to smooth muscle cells. Atherosclerosis 1996; 127:167-76.
47. Senoo H, Hata R. Extracellular matrix regulates and L-ascorbic acid 2-phosphate further modulates morphology, proliferation, and collagen synthesis of perisinusoidal stellate cells. Biochem Biophys Res Commun 1994; 200: 999–1006.
48. Carloni V, Romanelli RG, Pinzani M et al. Expression and function of integrin receptors for collagen and laminin in cultured human hepatic stellate cells. Gastroenterology 1996; 110: 1127-36.
49. Nagata K. Expression and function of heat shock protein 47: a collagen-specific molecular chaperone in the endoplasmic reticulum. Matrix Biol 1998; 16:379-86.
50. Razzaque MS, Foster CS, Ahmed AR. Role of collagen-binding heat shock protein 47 and transforming growth factor-beta1 in conjunctival scarring in ocular cicatricial pemphigoid. Invest Ophthalmol Vis Sci 2003; 44:1616-21.
51. Liu D, Razzaque MS, Cheng M et al. The renal expression of heat shock protein 47 and collagens in acute and chronic experimental diabetes in rats. Histochem J 2001; 33:621-8.
52. Razzaque MS, Taguchi T. The possible role of colligin/HSP47, a collagen-binding protein, in the pathogenesis of human and experimental fibrotic diseases. Histol Histopathol 1999; 14:1199-212.
53. Razzaque MS, Shimokawa I, Nazneen A et al. Life-long dietary restriction modulates the expression of collagens and collagen-binding heat shock protein 47 in aged Fischer 344 rat kidney. Histochem J 1999; 31:123-32.
54. Razzaque MS, Nazneen A, Taguchi T. Immunolocalization of collagen and collagen-binding heat shock protein 47 in fibrotic lung diseases. Mod Pathol 1998; 11:1183-8.
55. Razzaque MS, Kumatori A, Harada T et al. Coexpression of collagens and collagen-binding heat shock protein 47 in human diabetic nephropathy and IgA nephropathy. Nephron 1998; 80:434-43.
56. Zaoui P, Cantin JF, Alimardani-Bessette M et al. Role of metalloproteases and inhibitors in the occurrence and progression of diabetic renal lesions. Diabetes Metab 2000; 4:25-9.
57. Nagase H, Kashiwagi M. Aggrecanases and cartilage matrix degradation. Arthritis Res Ther 2003; 5:94-103.
58. Yaguchi T, Fukuda Y, Ishizaki M et al. Immunohistochemical and gelatin zymography studies for matrix metalloproteinases in bleomycin-induced pulmonary fibrosis. Pathol Int 1998; 48:954–63.
59. Madtes DK, Elston AL, Kaback LA et al. Selective induction of tissue inhibitor of metalloproteinase-1 in bleomycin-induce pulmonary fibrosis. Am J Respir Cell Mol Biol 2001; 24:599–607.
60. Jiang Z, Seo JY, Ha H et al. Reactive oxygen species mediate TGF-beta1-induced plasminogen activator inhibitor-1 upregulation in mesangial cells. Biochem Biophys Res Commun. 2003; 309:961-966.
61. Rechardt O, Elomaa O, Vaalamo M et al. Stromelysin-2 is upregulated during normal wound repair and is induced by cytokines. J Invest Dermatol 2000; 115:778-787
62. Mast BA, Diegelmann RF, Krummel TM et al. Scarless wound healing in the mammalian fetus. Surg Gynecol Obstet 1992; 174:441-51.

Chapter 2

Pathological Significance of Renal Expression of Proinflammatory Molecules

Takashi Wada, Mohammed S. Razzaque, Kouji Matsushima, Takashi Taguchi and Hitoshi Yokoyama

Abstract

Recent studies of cytokines, chemokines, adhesion molecules and growth factors have enhanced our understanding of molecular mechanisms of leukocyte trafficking and their activation in the inflammatory phase of various renal diseases. Interactions between infiltrated inflammatory cells and resident renal cells are actively involved in the pathogenesis of phase-specific renal disorders. Furthermore, a number of proinflammatory and fibrogenic cytokines, chemokines and growth factors exert their biological activities through their receptors expressed on resident renal cells, to induce inflammatory responses that eventually lead to the development of fibrosis in various renal diseases. Thus, measuring the levels of certain proinflammatory molecules might provide useful information about the inflammatory state of the diseased kidney and could have clinical importance and significance. The selective intervention of some of these molecules might have the therapeutic potential to modulate renal inflammatory responses, and thereby could alter disease progression. Despite the apparent redundancy, accumulating evidence supports this possibility. In this chapter, we will briefly summarize the specific roles of certain proinflammatory molecules in the pathogenesis of various human and experimental renal diseases.

Introduction

The interactions of activated infiltrated cells and resident renal cells are actively involved in the pathogenesis of renal inflammation. The trafficking of leukocytes from peripheral blood into the kidneys is an important inflammatory event that is encountered in the early phases of most renal diseases.[1] Proinflammatory molecules, through their interactions with receptors expressed on leukocytes and resident renal cells, can induce a number of cytokines, chemokines, adhesion molecules and growth factors, both by autocrine and/or paracrine loops, during acute and chronic phases of various renal diseases.[2] Some of these secreted molecules are not only involved in the inflammatory phase of the disease, but also contribute to the development of subsequent renal fibrosis.

Fibrogenesis: Cellular and Molecular Basis, edited by Mohammed S. Razzaque.

Participants in Renal Inflammation

Inflammatory Infiltrates

The main inflammatory cells actively involved in the early phase of various renal diseases are neutrophils, macrophages, and lymphocytes. One of the mechanisms of neutrophil and macrophage infiltration into diseased kidneys is via activation of adhesion molecules to induce the release of lysosomal enzymes and generation of superoxide anions to initiate inflammatory events and subsequent tissue damage.[3] In contrast, macrophages also have a scavenging role in the clearance of nonself and/or altered-self materials, including glycated proteins and oxidized lipoproteins.[4] Macrophages exert dual effects in renal injury, i.e., damaging and protective effects. Therefore, infiltrated and activated macrophages play a crucial role in the pathogenesis of both inflammatory and fibrotic phases of the disease process. Certain chemokines, including monocyte chemoattractant protein (MCP-1), are involved in the recruitment of macrophages in diseased kidneys. In addition, macrophage colony-stimulating factor (m-CSF) is actively involved in monocyte and macrophage survival, proliferation, and chemotaxis.[5] Overexpression of m-CSF by tubular epithelial cells is closely associated with the interstitial accumulation and proliferation of macrophages, as demonstrated in experimental anti-glomerular basement membrane (GBM) nephritis and unilateral ureteral obstruction models.[6] A correlation between overexpression of m-CSF and accumulation of macrophages has also been demonstrated in various human and experimental diseases, including glomerulonephritis.[7,8] Similarly, increased expression of macrophage migration inhibitory factor (MIF) has been shown to be involved in human and experimental models of tubulointerstitial injury, possibly by facilitating the interstitial accumulation of macrophages and lymphocytes.[9,10] Infiltration of inflammatory cells, including lymphocytes, into diseased kidneys, exerts inflammatory responses by secreting certain proinflammatory cytokines and chemokines.[11] In kidneys of autoimmune MRL-*Fas*^{lpr} mice, a dynamic interaction of macrophages and lymphocytes has been shown to be the result of nephritogenic cytokines.[11]

Resident Renal Cells

In addition to inflammatory infiltrates, there are at least two groups of resident renal cells that actively participate in the inflammatory phase of various renal diseases; these are glomerular cells (mesangial cells, endothelial cells and epithelial cells), and tubulointerstitial cells (tubular epithelial cells, interstitial cells and peritubular capillary endothelial cells). These intrarenal cells are capable of proliferation, differentiation, and synthesis of various proinflammatory cytokines, chemokines, and growth factors in response to various stimuli, which in turn augment inflammatory responses. In addition, activated and transformed resident renal cells regulate extracellular matrix (ECM) remodeling by inducing the synthesis of fibrogenic molecules. Recently, it has been suggested that injury to the renal microvasculature contributes to the progression of a number of renal diseases. For instance, Ohashi et al demonstrated that early angiogenic responses, such as peritubular capillary regression, were followed by progressive deletion of endothelial cells by apoptosis in experimental obstructive nephropathy.[12]

Using a bone marrow transplantation model, Imasawa et al[13] recently reported the potential of bone marrow-derived cells to differentiate into glomerular mesangial cells. In an experimental model of anti-Thy1 nephritis, Ito et al[14] reported an increased number of bone marrow-derived Thy1(+) cells constituting about 7 to 8% of glomerular cells. These reports are interesting, and suggestive of the presence of bone marrow-derived mesangial cells in nephritis. Further studies are needed to determine the pathogenic role of these marrow-derived mesangial cells in various renal diseases.

Major Mediators Involved in Renal Inflammation

Chemokines

The chemokine family is divided into four groups depending on conserved cysteine residues that form disulfide bonds in the chemokine tertiary structure.[15,16] The CXC subfamily contains a single nonconserved amino acid separating the first two cysteine residues. The CC and CXXXC subfamilies have either none or exactly three nonconserved amino acids between the cysteines, respectively. The C chemokine subfamily, is unique in that only two of the four cysteine residues (i.e., the first and third) are present. More than 44 chemokines and 18 chemokine receptors have been identified. Recent studies have documented that chemokines, through the binding of their cognate receptors, play an important role in the pathogenesis of various renal diseases by regulating leukocyte trafficking at inflammatory sites.

Interferon-inducible protein (IP)-10/CXCL10 is a mitogenic factor for mesangial cells and possibly acts via the cognate receptor, CXCR3.[17] The constitutive glomerular expression of CCR7 and its ligand, secondary lymphoid tissue chemokine (SLC/CCL21), by adjacent renal cells suggests the involvement of this specific chemokine/chemokine receptor interaction in regulating glomerular homeostasis and regeneration.[18] In addition, the regulation of interleukin (IL)-8/CXCL8 and MCP-1 (also known as monocyte chemotactic and activating factor [MCAF])/CCL2 are closely related to the urinary excretion of protein in experimental models[19,20] and human nephrotic syndrome;[21] the glomerular protein leakage is possibly due to increased permeability of the glomerular capillaries. A recent report described the expression of CCR4, CCR8, CCR9, CCR10, CXCR1, CXCR3, CXCR4, and CXCR5 in cultured podocytes; the expression of CXCR1, CXCR3, and CXCR5 was also detected in podocytes of human kidney sections.[22] It is likely that the release of oxygen radicals that accompanies the activation of CCRs and CXCRs may contribute to podocyte injury and the development of proteinuria.[22]

Recent studies have documented a direct link between locally and systemically produced chemokines and the infiltration and activation of leukocytes in the kidneys. The infiltration of Th1 T cells in the interstitium in human renal diseases is partly regulated upon activation of normal T cells that express and secrete (RANTES)/CCL5, through its interaction with its cognate receptors, CCR1 and CCR5.[23] RANTES was also upregulated in the kidneys of a murine lupus nephritis model, MRL-*Fas*^{lpr} mice, prior to renal injury, and increased with progression of the injury.[24] When tubular epithelial cells genetically modified to secrete RANTES were infused under the renal capsule, features of interstitial nephritis developed in MRL-*Fas*^{lpr} mice.[24] RANTES fostered the accumulation of $CD4^+$ cells. In addition, circulating components, including $CD4^+$ cells, were required to incite renal injury in MRL-*Fas*^{lpr} mice via both cellular and humoral immune responses.[11] Therefore, the manipulation of lymphocyte trafficking by blocking the bioactivities of certain chemokines could have a potential therapeutic effect in renal diseases.

Cytokines

The renal inflammatory response is a multistep and multifactorial process, mainly orchestrated by the cross-talk of cytokines. Among the cytokines, IL-1 and tumor necrosis factor (TNF)-α are the most extensively studied molecules, and have been found to play an important role in the inflammatory phase of various renal diseases. These two cytokines are capable of, (1) regulating proliferation of resident renal cells (mesangial cells, endothelial cells), (2) augmenting production of inflammatory mediators (cytokines, chemokines, prostaglandins, free radical oxygen and superoxide anions), and (3) facilitating recruitment of inflammatory cells into the injured kidneys via the expression of chemokines.[25] In addition, the upregulation of these proinflammatory cytokines is closely related to the degree of apoptosis during inflammation.

The two different subsets of CD4⁺ lymphocytes (Th1 and Th2 subsets) produce different groups of cytokines.[26] Th1 cytokines include IL-2, IL-12, IL-18 and interferon (IFN)-γ, while IL-4, IL-5, IL-10, and IL-13 represent Th2 cytokines. Th1 cytokines are capable of activating monocytes, lymphocytes and resident renal cells, which augment immunoinflammatory responses via cellular immunity. In contrast, Th2 cytokines directly activate B cells to switch on humoral immunity and induce human endothelial cell adhesiveness for T cells via generation of adhesion molecules.[27] In particular, Th1-predominant nephritogenic immune responses are associated with severe proliferative and crescentic glomerulonephritis. In contrast, Th2 predominance is presumed to contribute to minimal change nephrotic syndrome and membranous nephropathy. Interestingly, lupus nephritis and IgA nephropathy have aspects of both Th1 and Th2 predominance in a phase-specific manner.

The transition from neutrophil infiltration to mononuclear cell infiltration is an important feature of inflammation. Recently, a role for IL-6 and its soluble receptor has been suggested in the transition from neutrophil to monocyte recruitment during inflammation.[28] During acute inflammation, IL-6 might help in the resolution of the neutrophilic infiltrates. In contrast, during chronic inflammation, IL-6 might contribute to increased infiltration of mononuclear cells. These findings are suggestive of a new role for IL-6 in the inflammatory process.

Leukocyte Trafficking During Renal Inflammation

Cell-cell interaction has an enormous impact on renal inflammation. Leukocyte trafficking at the inflammatory site comprises two major events: first, the activation and firm adhesion of leukocytes on endothelial surfaces and, secondly, the diapedesis and transmigration through the endothelial cells into the inflammatory sites of the renal tissue (Fig. 1). Leukocyte adhesion molecules are responsible for these steps and/or events. Major molecules involved in leukocyte trafficking include the selectin family of molecules and their glycoprotein ligands, integrins and immunoglobulin-like leukocyte adhesion molecules.[29] Selectin-mediated rolling and integrin-mediated firm adhesion is involved in the binding of leukocytes to the endothelium. In addition, recent studies have shown that chemokines, through their cognate receptors, play an important role in these steps. Chemokines expressed on the surface of endothelial cells interact with their cognate receptors on specific leukocytes, a process that triggers activation of adhesion molecules and result in firm adhesion of leukocytes to the surface of endothelial cells. Once leukocytes migrate into the interstitium, chemokines and proinflammatory cytokines produced by both resident cells and infiltrated inflammatory cells exert a wide range of biological activities at the inflammatory sites. Selective expression of chemokine receptors and adhesion molecules on specific cell populations determines the specific types of infiltrating cells in inflamed kidneys. Thus, the interaction of infiltrating cells and endothelial cells orchestrated by chemokines and adhesion molecules regulates sequential migratory patterns of specific types of leukocytes in a multistep manner.

Chemokine Systems in the Kidney from Acute Injury to Renal Scarring: The Chemokine Cascade

In renal inflammation, the type of leukocytes that infiltrate into the kidneys depends on the type of insult and phase of the disease; neutrophils are the predominant cells in acute inflammation, while macrophages, lymphocytes and plasma cells comprise the dominant cell populations in chronic inflammation. Fibrogenic factors released by some of these chronic inflammatory cells are involved in subsequent renal fibrogenesis (Fig. 2). Given the presence of a biologically active chemokine amplification cascade in the kidney,[30] a switch from acute

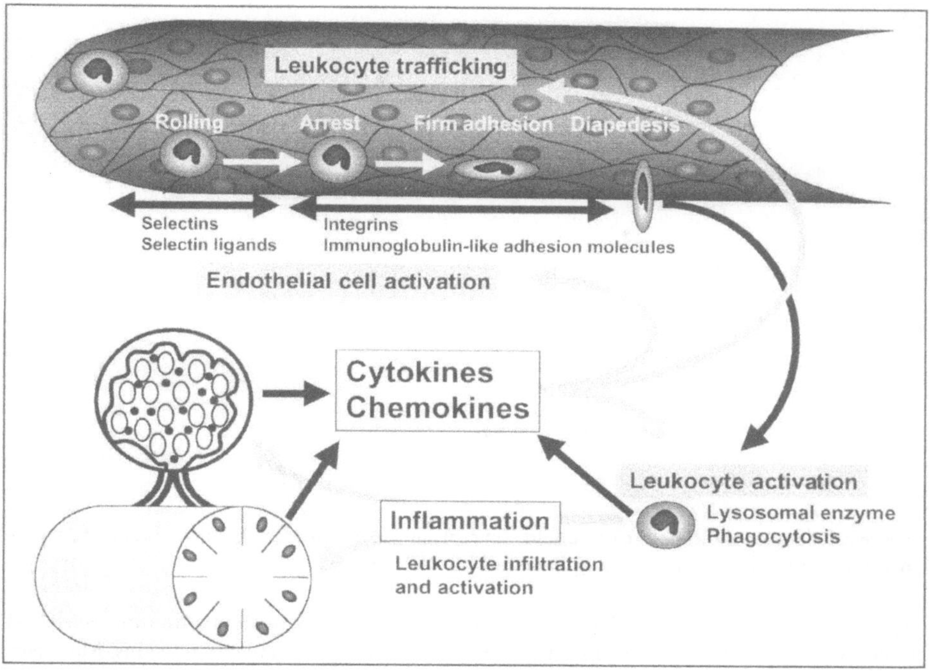

Figure 1. Proinflammatory molecules and kidney.

inflammation to chronic inflammation in various renal diseases may depend on the expression and bioactivity of specific chemokines. In human renal diseases, the presence of IL-8 in diseased kidneys reflects an acute disease stage, associated with neutrophil infiltration and hematuria[2] (Fig. 2). In contrast, the presence of elevated MCP-1 expression is suggestive of a chronic stage of disease, especially in tubulointerstitial lesions through the recruitment and activation of macrophages. Moreover, the measurement of urinary MCP-1 levels is a useful clinical tool for monitoring the disease activity of inflammatory renal disorders.[2] In addition, MCP-1 has been shown to mediate collagen deposition in an experimental model of nephritis, through its interaction with TGF-β[31] It is therefore likely that increased expression of MCP-1 in the chronic inflammatory stage of a particular disease may contribute to subsequent renal fibrosis, by regulating the synthesis of collagens. In support of this notion, the administration of anti-MCP-1 antibodies prevented the infiltration of leukocytes and development of renal fibrosis in a rat model of crescentic glomerulonephritis.[20] Thus, the positive amplification loop from CXC chemokines to CC chemokines (the "chemokine cascade") plays a crucial role, not only in early inflammatory events, but also in late fibrotic events of various renal diseases (Fig. 3).

The MCP-1/CCL2-TGF-β Axis: A Common Regulatory Pathway of Chronic Renal Inflammation Resulting in Renal Scarring

MCP-1 does not only regulate inflammatory events, but is also involved in progressive glomerular and interstitial damage resulting in renal scarring in various renal diseases. Recent studies suggest that MCP-1 plays an important role in the pathogenesis of metabolic renal disorders, such as diabetic nephropathy and noninflammatory nephrotic syndrome. Locally produced MCP-1, via recruiting and activating macrophages, contributes to the development

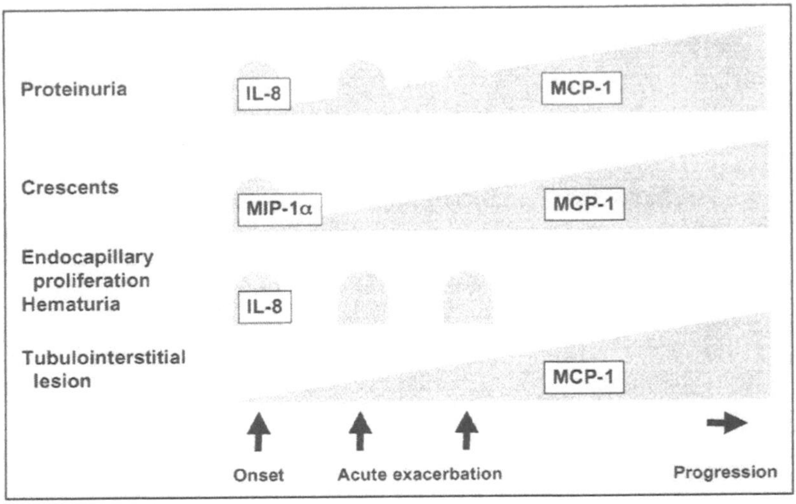

Figure 2. Relevance of the "chemokine cascade" in explaining the particular clinical symptoms and pathological changes in human renal diseases (modified from ref. 99).

of interstitial lesions in human diabetic nephropathy.[32] In diabetic nephropathy, urinary MCP-1 levels were significantly elevated in patients with advanced tubulointerstitial lesions. Moreover, urinary levels of MCP-1 showed a positive correlation with the increased infiltration of CD68-positive macrophages in the renal interstitium. Furthermore, studies using immunohistochemistry and in situ hybridization techniques showed increased number of MCP-1-expressing cells in the interstitium of kidneys affected by diabetes. These observations suggest that locally produced MCP-1 is involved in the development of tubulointerstitial lesions in diabetic nephropathy, possibly by regulating recruitment and activation of macrophages. Overexpression of MCP-1 and certain fibrogenic molecules, including platelet-derived growth factor (PDGF) and transforming growth factor beta (TGF-β), has been shown to be associated with interstitial accumulation of mononuclear cells and myofibroblastic transformation of resident cells in patients with membranous nephropathy.[33] Recently, an intrinsic regulatory loop, in which MCP-1 stimulates TGF-β production by resident glomerular cells, has been suggested in the absence of infiltrating immune competent cells.[34] Increased renal expression of TGF-β both at the mRNA and protein levels, was detected in isolated rat kidneys that were perfused with a polyclonal anti-thymocyte-1 antiserum and rat serum; this effect was attenuated by coperfusion with a neutralizing anti-MCP-1 antibody but was partly mimicked by perfusion with recombinant MCP-1 protein.[34] These results suggest an additional role for MCP-1 in fibrotic renal diseases, possibly by interacting with TGF-β.

Molecular Basis of Inflammation in Various Renal Diseases

Ischemia-Reperfusion Injury

Renal ischemia-reperfusion is usually encountered in renal transplantation, in patients with shock and circulatory collapse, or in renal artery stenosis. Ischemia-reperfusion injury in the kidney is pathologically characterized by tubular epithelial cell necrosis and/or apoptosis with marked inflammatory cell infiltration. Resident renal cell- and infiltrated inflammatory cell-secreted cytokines and chemokines, such as TNF-α, IL-1, IL-8, MCP-1 and RANTES play important roles in the induction and propagation of renal injury. Recently, a more specific

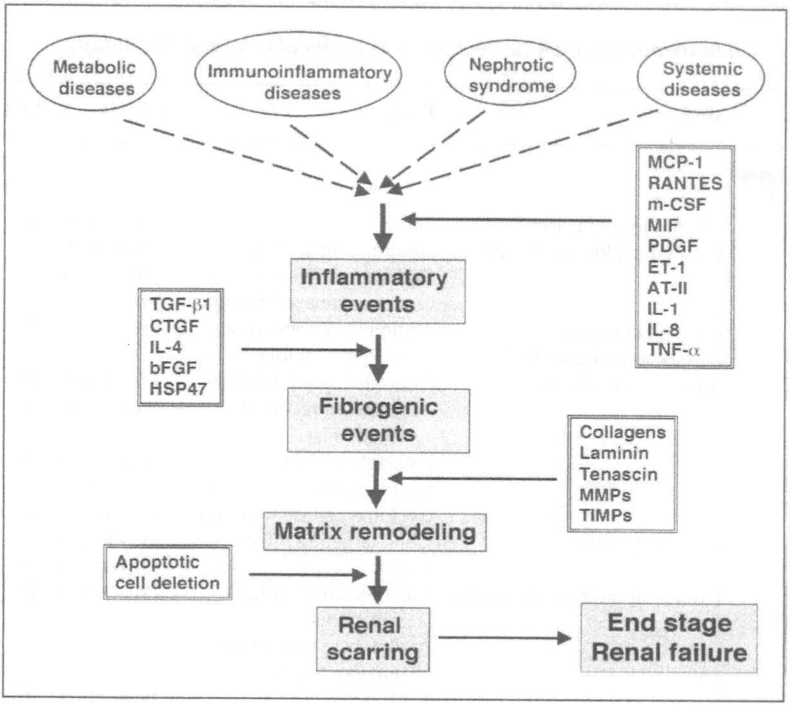

Figure 3. Schematic diagram of molecules involved in the multistep and multifactorial renal diseases that eventually lead to end stage renal failure. Not all molecules involved are included in this figure, in order to keep the diagram simple. [IL-1: Interleukin-1, IL-4: Interleukin-4, IL-8: Interleukin-8, TNF-α: Tumor necrosis factor-α, MCP-1: Monocyte chemoattractant protein, RANTES: Regulated upon activation, normal T cell expressed and secreted, m-CSF: Macrophage colony-stimulating factor, MIF: Macrophage migration inhibitory factor, PDGF: Platelet-derived growth factor, ET-1: Endothelin 1, ATII: Angiotensin II, TGF-β1: Transforming growth factor beta1, CTGF: Connective tissue growth factor, bFGF: Basic fibroblast growth factor, HSP47: Heat shock protein 47, MMP: Matrix metalloproteinase, TIMP: Tissue inhibitor of metalloproteinase].

role for chemokines in distinct steps of leukocyte extravasation has been demonstrated. Growth-related oncogene (GRO)/CXCL1 and fractalkine/CX3CL1 are thought to mediate the initial firm adhesion, whereas the chemokine MCP-1 is required for the later steps of leukocyte spreading and diapedesis.[35,36] Thus, these studies emphasize the significance of the interaction of endothelial cells with leukocytes in ischemia-reperfusion injury.

Various experimental models of ischemia-reperfusion injury have been reported (Table 1). Using a renal ischemia-reperfusion model, it was recently demonstrated that p38 mitogen-activated protein kinase (MAPK) plays a key role in cell infiltration and tubular necrosis by regulating the expression of IL-1, TNF-α, MCP-1 and RANTES.[37] In addition, glucocorticoids reduced ischemia-reperfusion injury in the kidneys by inhibiting the expression of JE/MCP-1.[38] Osteopontin not only promoted the accumulation of macrophages, but also exerted renoprotective effects in renal ischemia-reperfusion injury. Osteopontin inhibited the induction of inducible nitric oxide synthase and suppressed the synthesis of nitric oxide, resulting reduced peroxide levels in the cells, thus promoting the survival of cells exposed to hypoxia.[39] It has been demonstrated that the inhibition of neutrophil infiltration into ischemic kidneys during reperfusion by neutralizing the effects of chemokines (GRO and MIP-2)

Table 1. Chemokines/chemokine receptors as molecular targets in therapy

Targets	Models	Tools	Effects	Refs.
Chemokines				
1) MCP-1				
	Ischemia-reperfusion	7ND	Prevention	50
	Crescentic glomerulonephritis	Neutralizing antibodies	Prevention	20, 54-56
		Gene targeting mice	Prevention	58
		Cyclooxygenase inhibitor	Worse	69
	Immune complex glomerulonephritis	Angiotensin converting enzyme inhibitor	Prevention	73
	Anti-Thy-1 nephritis	Neutralizing antibodies	Prevention	66
		Angiotensin receptor 1 antagonist	Prevention	94
		Prostaglandin E1	Prevention	93
		Pentoxifylline	Prevention	68
		Cyclooxygenase inhibitor	Worse	69
	Lupus nephritis	Gene targeting mice	Prevention	75
		Bindarit	Prevention	79
	Unilateral ureteral obstruction	Angiotensin receptor 1 antagonist	Prevention	86
		Angiotensin converting enzyme inhibitor		
		Y-27632	Prevention	87
	Diabetic nephropathy	Angiotensin converting enzyme inhibitor	Prevention	81
2) MIP-1α				
	Crescentic glomerulonephritis	Neutralizing antibodies	Prevention	57
3) RANTES				
	Transplantation	Met-RANTES	Prevention	43, 48
	Anti-Thy-1 nephritis	AOP-RANTES	Prevention	67
4) IL-8				
	Immune complex glomerulonephritis	Neutralizing antibodies	Prevention	19
5) MIP-2				
	Ischemia-reperfusion	Neutralizing antibodies	Prevention	40
	Crescentic glomerulonephritis	Neutralizing antibodies	Prevention	60
6) GRO				
	Ischemia-reperfusion	Neutralizing antibodies	Prevention	40
7) CINC				
	Crescentic glomerulonephritis	Neutralizing antibodies	Prevention	57
Chemokine receptor				
1) CCR1				
	Transplantation	BX471	Prevention	49
		FTY720	Prevention	50
	Crescentic glomerulonephritis	Gene targeting mice	Worse	59
	Unilateral ureteral obstruction	BX471	Prevention	88

resulted in attenuation of the renal injuries.[40] Importantly, disruption of MCP-1/CCR2 signaling in CCR2 null mice could effectively attenuate renal ischemia reperfusion injury. Ongoing studies, using propagermanium, a selective inhibitor of MCP-1/CCR2 signaling, have shown relatively less inflammatory cell infiltration, and tubular necrosis in kidneys in the ischemia-reperfusion model (unpublished data). Using a gene therapy approach, expression of an amino-terminal deletion mutant of MCP-1 to inhibit MCP-1/CCR2 signaling resulted in decreased acute tubular necrosis and less infiltration of macrophages.[41]

Transplant Nephropathy

Renal transplantation is one of the most important clinical situations with exposure to ischemia-reperfusion injury. Recently, early nonspecific ischemic injury has been related to subsequent immunologic injuries in renal transplant rejection.[42] Rat models of transplantation have facilitated a functional study of the role of cytokines/chemokines in acute and chronic renal rejection.[43,44] In a rat model of acute renal transplant rejection, the expression of RANTES was upregulated, at the mRNA level, by as early as 6 hours, and this upregulation was again noted on day 3 to day 6.[45] Increased expression of RANTES in the early hours of engraftment may be related to ischemic injury, and could, in part, induce subsequent immunologic responses. In addition, macrophages and their secreted products play important roles in the eventual immune-mediated rejection process. Moreover, increased production of certain cytokines (IL-8, -10, -15), chemokines (RANTES, MIP-1, MCP-1) and hepatocyte growth factor (HGF) by infiltrating leukocytes, tubular epithelial cells, and endothelium may have a determinant role in renal transplant rejection.[44,46,47]

Recent studies have documented beneficial effects of blocking the bioactivities of certain chemokines in renal transplant rejection. For example, Met-RANTES, a chemokine receptor antagonist, not only reduced vascular and tubular damage in acute renal transplant rejection,[43] but also protected renal allografts from long-term deterioration.[48] In addition, a CCR1-specific nonpeptide antagonist, BX471, exhibited efficacy in a rabbit allograft rejection model.[49] FTY720 is a novel immunomodulator that acts by chemokine-dependent lymphocyte homing into secondary lymphoid organs leading to profound lymphocyte depletion in the blood.[50] Oral administration of FTY720 to cynomolgus monkeys with renal allotransplantation has been shown to prevent acute allograft rejection, with resultant rejection-free allograft survival.[50] The expression of MCP-1 in acute renal transplant rejection correlated with the number of infiltrating macrophages,[51] and elevated urinary MCP-1 excretion during rejection episodes, which diminished after successful treatment. However, the inhibitory impacts of MCP-1/CCR2 on allograft rejection need further comprehensive studies.

Crescentic Glomerulonephritis

Rapidly progressive crescentic glomerulonephritis is usually associated with clinical features of anemia and morphological features of tubulointerstitial nephritis that eventually lead to renal insufficiency. To monitor clinical symptoms, and to understand the disease activity of crescentic glomerulonephritis, it is essential to determine specific molecule(s) involved in the disease process. Increased urinary levels of MIP-1α have been selectively detected in patients with crescentic glomerulonephritis; MIP-1α was mostly undetectable in urine samples collected from healthy control subjects and in patients with renal diseases lacking crescent formation.[52] Urinary MIP-1α levels in patients with crescentic glomerulonephritis correlated well with the percentage of cellular crescents and the number of CD68-positive infiltrating cells, and CCR1- and CCR5-positive cells in the glomeruli.[23] Moreover, elevated urinary levels of MIP-1α and the number of CCR5-positive cells dramatically decreased during glucocorticoid therapy-induced convalescence. MIP-1α-positive cells were mainly detected in crescentic lesions. CCR1, and CCR5-positive cells, preferentially expressed on Th1 T cells,[53] were detected in diseased glomeruli and interstitium.[23] It is likely that MIP-1α plays a significant role in crescentic glomerulonephritis, by recruiting and activating macrophages and T cells. Measurement of urinary levels of MIP-1α appears to be clinically useful as its level correlates with disease activity of human crescentic glomerulonephritis.

Recent studies have documented the beneficial effects of neutralizing antibodies and specific chemokine/chemokine receptor antagonists in crescentic glomerulonephritis. Anti-MCP-1 or anti- MIP-1α antibodies or MCP-1 deficiency resulted in less glomerular accumulation of macrophages, reduced crescent formation, decreased interstitial damage, and most importantly, less proteinuria.[20,54-58] In contrast, aggravated renal dysfunction and increased proteinuria have been detected in CCR1-disrupted mice, compared to wild-type mice.[59]

The beneficial effects of blocking the bioactivities of CXC chemokines have been reported in crescentic glomerulonephritis. Neutralizing antibodies against cytokine-induced neutrophil chemoattractant (CINC) ameliorated the cell infiltration, including neutrophils, and reduced urinary protein excretion.[57] Similarly, a single dose of anti-MIP-2 antibody resulted in reduced neutrophil influx (40% at 4 hours) and periodic acid-Schiff-containing fibrin deposition (54% at 24 hours). These results suggest the crucial role of MIP-2 in recruiting neutrophils during glomerular inflammation.[60] However, the combination of anti-CINC and anti-MIP-1α antibodies did not show any additional beneficial effects.[57]

The viral macrophage inflammatory protein-II (vMIP-II) encoded by Kaposi's sarcoma-associated herpes virus is unique among all known chemokines in that vMIP-II shows a broad-spectrum interaction with both CC and CXC chemokine receptors, including CCR5 and CXCR4. vMIP-II patently inhibited MCP-1-, MIP-1α-, RANTES-, and fractalkine-induced chemotaxis of activated leukocytes isolated from nephritic glomeruli; it also reduced glomerular infiltration of leukocytes, and markedly attenuated proteinuria in the rat model of glomerulonephritis.[61]

Recently, modulation of the p38 MAPK pathway by using specific inhibitors was shown to have beneficial effects on the progression of crescentic glomerulonephritis. The MAPK signal transduction pathway is thought to be involved in the regulation of cell proliferation and apoptotic cell deletion in inflammatory diseases.[62] The activation of MAPK isoform p38, detected in mesangial cells, is closely associated with apoptosis, stress responses and acute and/or chronic inflammation.[62] Moreover, phosphorylation of p38 MAPK contributes to the activation of nuclear factor (NF)-κB and activating protein (AP)-1, which are essentially involved in inflammatory processes. FR167653 is a specific p38 MAPK pathway inhibitor. FR167653 markedly decreased IL-1β-induced phosphorylation of p38 MAPK in cultured rat mesangial cells.[63] In vivo administration of FR167653 reduced glomerular damage, including crescentic formation, proteinuria, and glomerulosclerosis in an experimental model of crescentic glomerulonephritis.[63,64] In addition, FR167653 markedly decreased renal expression of certain cytokines and chemokines. Hence, p38 MAPK might be a potential target for developing future therapeutic strategies against crescentic glomerulonephritis.

Acute/Experimental Nephritis

The acute phase of glomerulonephritis is morphologically characterized by infiltration of inflammatory cells into glomeruli and proliferation of mesangial cells. Renal parenchymal cells express chemokine receptors as well as their ligands in such conditions.[2] In vitro studies have demonstrated that proinflammatory stimuli, such as IL-1β, TNF-α, and IFN-γ, immune complexes, and certain growth factors, including PDGF and basic fibroblast growth factor (bFGF), are able to induce IL-8, MCP-1, IP-10, MIP-1α and RANTES from resident renal cells.[2] In turn, these stimuli induce the expression of CCR1 and CXCR3 on resident renal cells, especially mesangial cells.[18,65] These observations are suggestive of a positive feedback loop through cytokines/chemokines, which results in renal inflammation. Indeed, the CXC chemokines, MIP-2 and keratinocyte-derived cytokine (KC) are able to induce MCP-1 and RANTES expression in mesangial cells.[30] Autoinduction of MIP-2 and KC mRNA was also detected. This chemokine amplification mechanism is thought to contribute to the maintenance and chronic course of glomerular inflammation.

Anti-Thy1 antibody-induced nephritis is a well-studied rat model of mesangial proliferative glomerulonephritis, characterized by complement-dependent mesangiolysis, inflammatory reactions, and subsequent glomerulosclerosis. MCP-1 mediated infiltration of monocytes is suggested to play an important role in the progression of glomerular lesions in Thy-1 nephritis, and blocking the bioactivity of MCP-1 has been shown to reduce renal injury in this model of nephritis.[66] Similarly, a CCR5 chemokine receptor antagonist, AOP-RANTES, ameliorated monocyte/macrophage infiltration and improved glomerular pathology in an experimental model of nephritis.[67] It was concluded that the use of chemokine receptor antagonists might offer a new therapeutic option in treating inflammatory renal diseases. In addition, treatment with a clinically available nonselective inhibitor of cyclic 3',5'-nucleotide phosphodiesterase, pentoxifylline, resulted in reduced accumulation and proliferation of glomerular macrophages, suppression of activation and proliferation of mesangial cells, and proteinuria; all the above-mentioned changes were associated with decreased glomerular expression of MCP-1 and intercellular adhesion molecule-1 (ICAM-1) in the experimental model of nephritis at 2 hours and on day 1 of the study.[68] In contrast, the effect of cyclooxygenase (COX) inhibitors was evaluated in an anti-thymocyte antibody model and an anti-GBM model of glomerulonephritis.[69] These studies have suggested that COX products might serve as endogenous repressors of MCP-1 formation in the studied models of experimental nephritis. COX-1 and COX-2 products may regulate differently as selective COX-2 inhibitors exert less influence on chemokine expression.

It has been well documented that angiotensin II regulates the synthesis of certain proinflammatory cytokines and chemokines in the kidney. Angiotensin II plays an active role in inflammatory responses in renal diseases in concert with chemokine/cytokine expression, possibly by activating NF-κB.[70] Rats with experimental immune complex nephritis, treated with the angiotensin converting enzyme (ACE) inhibitor quinapril,[71] exhibited reduced expression of MCP-1.

NF-κB stays inactive in the cytoplasm, however upon activation by a wide range of factors, it translocates into the nucleus and regulates the expression of genes encoding cytokines, growth factors, oncogenes, transcription factors, and receptors involved in various pathological processes, including immunoinflammatory disorders. Activation of NF-κB leads to transcription of such genes as IL-1 and TNF-α, ICAM-1 and vascular cell adhesion molecule-1 (VCAM-1), MCP-1, m-CSF, inducible nitric oxide synthase (iNOS) and tissue factor. All these above-mentioned molecules have been shown to play important roles in the induction and propagation of various renal diseases.[72,73]

Lupus Nephritis

MCP-1 promotes autoimmune renal disease through the recruitment of macrophages and T cells, and the recruitment process is augmented by locally produced cytokines and/or chemokines in New Zealand Black x New Zealand White (NZB/W) F1 mice[74] and MRL-*Fas*^{lpr} mice.[75] In addition, modulation of the biological activities of MCP-1 dramatically reduced the recruitment of macrophages and T cells that not only reduced pathological alterations in the kidney, lung, skin, and lymph node, but also diminished proteinuria, and prolonged survival.[75]

Mononuclear cell infiltration has been demonstrated in the kidneys of MRL-*Fas^lpr^* mice, by weeks 10 to 12. At week 12, the expression of certain chemokines, including CCR1, CCR2, and CCR5 was upregulated in the mice kidneys, associated with morphological features of renal injuries and proteinuria. These results are in accord with the notion that chemokine-mediated leukocyte infiltration precedes proteinuria and renal damage in MRL-*Fas^lpr^* mice.[76] From the perspective of Th1/Th2 balance, $CCR4^+$, but not $CCR5^+$ T lymphocytes in peripheral blood, which represent Th2 cells, preferentially migrate into the kidneys of patients with lupus nephritis. It is likely that the disproportionate distribution of $CCR4^+$ T lymphocytes might play an important role in the development of subsequent renal injuries that are found in patients with lupus nephritis.[77] In addition, markedly enhanced expression of the B lymphocyte chemokine (BLC/CXCL13) has been detected in the thymus and kidneys of aged (NZB/W) F1 mice developing lupus nephritis. These observations suggest that myeloid dendritic cells in the target organs in aged (NZB/W) F1 mice may play a pivotal role in disrupting immune tolerance in the thymus and in recruiting autoantibody-producing B cells in the development of murine lupus.[78]

Recent studies have documented the beneficial effects of modulating cytokines/chemokines in lupus nephritis. Mice [(NZB/W) F1] treated with Bindarit (50 mg/kg/day, p.o.), a novel molecule devoid of immunosuppressive effects, resulted in a delayed onset of proteinuria and reduced impairment of renal function, and prolonged survival.[79] Similarly, daily oral administration of FR167653 (a selective inhibitor of p38 MAPK) decreased p38 MAPK phosphorylation in the kidneys, which resulted in reduced renal accumulation of macrophages and lymphocytes and an improvement of overall renal pathology, with prolonged survival; FR167653 treatment also reduced expression of MCP-1 and IgG production in MRL-*Fas^lpr^* mice.[80]

Diabetic Nephropathy

In addition to the metabolic and hemodynamic abnormalities, infiltration of inflammatory cells, including macrophages, into diseased kidneys is an important histological feature that is associated with the progression of diabetic nephropathy.[32] Angiotensin II-dependent up-regulation of MCP-1 has been demonstrated to play a role in the genesis of glomerular and tubulointerstitial damage.[81] The glomerular recruitment of macrophages in streptozotocin-treated diabetic rats is regulated by angiotensin-stimulated MCP-1.[81] Therefore, activation of the renin-angiotensin system is an important determinant of the macrophage population in diabetic nephropathy, possibly by regulating certain chemokines. It is well accepted that in addition to their blood-pressurelowering effects, AT1 receptor antagonists are renoprotective in patients with type 2 diabetes mellitus with microalbuminuria.[82,83] More recently, combination treatment with an angiotensin-II receptor blocker and ACE inhibitor was found to be more effective in retarding the progression of nondiabetic renal diseases, in comparison with monotherapy.[84] Similarly, Utimura et al[85] reported the renoprotective effect of mycophenolate mofetil, which could have derived from its well-known anti-inflammatory properties that include restriction of lymphocyte and macrophage proliferation and modulation of the expression of adhesion molecules. These findings are consistent with the notion that inflammatory events are central to the pathogenesis of diabetic nephropathy.

Unilateral Ureteral Obstruction Model

A unilateral ureteral obstruction model is characterized by the interstitial infiltration of inflammatory cells, including macrophages, which gradually leads to the development of tubulointerstitial fibrosis, resulting in decreased renal function. Increased interstitial expression of

IL-1, TNF-α, MCP-1, TGF-β and type I collagen has been documented in this model.[86] Recently, the beneficial effects of Y-27632, a specific Rho-associated coiled-coil forming protein kinase (ROCK) inhibitor, were studied using the unilateral ureteral obstruction model.[87] In vivo studies have shown that Y-27632 treatment resulted in less alpha smooth muscle actin-positive cells, reduced numbers and expression of macrophages, MCP-1, TGF-β and α (I) collagen, and resulted in less interstitial fibrotic changes, in the unilateral ureteral obstruction model. It is therefore likely that the Rho-ROCK system may play an important role in fibrogenesis.[87] Similarly, blocking the chemokine receptor CCR1 using the nonpeptide antagonist BX471 resulted in reduced leukocyte infiltration, and subsequent improvement of renal fibrosis in unilateral ureteral obstruction.[88] Interestingly, BX471 was shown to be effective, even when it was used in the late stages of the disease, suggesting that CCR1 blockade may be useful in reducing early cellular infiltration and may modulate subsequent renal fibrosis, which is a major cause of end-stage renal failure.[88]

Effectiveness of Anti-Chemokine/Cytokine Therapy and Its Possible Therapeutic Implications in Renal Diseases

Agents that influence cAMP[89] or NF-κB, such as antioxidants, glucocorticoids and aspirin, can modulate the expression of chemokines/cytokines and improve renal pathology.[90] Clinically, during spontaneous or glucocorticoid therapy-induced convalescence in patients with inflammatory renal diseases, including acute glomerulonephritis, IgA nephropathy, lupus nephritis and crescentic glomerulonephritis, a reduced expression level of certain chemokines (IL-8, MCP-1, MIP-1α, fractalkine) and cytokines was detected.[2,91] In a separate study, Natori et al[92] assessed the pulse methylprednisolone (MP) dose required to exert these beneficial effects and its effect on the expression of cytokines/chemokines in an experimental model of crescentic glomerulonephritis. Numbers of glomerular and interstitial macrophages and T cells, as well as crescents, were reduced significantly by 5 mg/kg of MP, but a maximal effect was obtained by 30 mg/kg of MP. Urinary protein was reduced significantly in a 30 mg/kg group but not in other groups. The expression of chemokines was significantly inhibited by 5 mg/kg of MP. These results indicate that MP reduces the number of infiltrating mononuclear cells and formation of crescents in the rat model of crescentic nephritis in a dose-dependent fashion, despite the strong inhibition of chemokine expression at a lower dose. Therefore, the optimal dose of glucocorticoid remains to be determined clinically, although the expression of chemokines/chemokine receptors might be suppressed.

Furthermore, prostaglandin E1[93] and an AT1 receptor antagonist,[94] hydroxymethylglutaryl CoA reductase inhibitor[95] and ACE inhibitor[81] have been shown to inhibit the expression of certain chemokines/cytokines and to reduce the infiltration of inflammatory cells in experimental models of renal diseases.

Concluding Remarks

In this chapter, we have briefly summarized the current concept of renal inflammation that has resulted in a better conceptual understanding of the cellular and molecular basis of certain fibrotic renal diseases.[96-98] Based on in vitro and in vivo studies, it is likely that selective intervention of cytokines/chemokines, at the appropriate phase of a certain disease, may have the therapeutic potential for site- and phase-specific intervention into the progression of inflammatory renal diseases[99] (Tables 1, 2). Moreover, the development of humanized monoclonal antibodies, particular antagonists against cytokines/chemokines, or specific signal transduction pathways that can provide selective intrarenal blockage of bioactivities of involved

Table 2. Therapeutic strategy against cytokines/chemokines in renal diseases

1. Inhibition of gene expression of cytokines/chemokines
Interferon
Cyclosporin
Steroid
FK506
Mycophenolate mofetil
Vitamin D
Aspirin
HMG-CoA reductase inhibitors
Angiotensin converting enzyme inhibitors
Angiotensin receptor antagonists
All-trans-retinoic acid
Cannabinoid receptor agonists
Bindarit
2. Neutralization of cytokines/chemokines
Neutralizing antibodies
3. Inhibition of interaction between receptors and their ligands
Analogues (7ND)
Receptor antagonists
4. Inhibition of signal transduction
Kinase inhibitors

cytokines/chemokines, should have beneficial effects on modulation of renal inflammatory responses and subsequent progression of the disease process. It is apparent that various immunoinflammatory cytokines, chemokines, and adhesion molecules mediate the cell-cell and cell-matrix interactions to initiate and propagate various fibrotic renal diseases. Indeed, our understanding of the proinflammatory molecules involved in the pathogenesis of various renal diseases has provided new therapeutic choices, and led to the discovery of gene-based therapeutic options.

Acknowledgements

Figure 1 is adopted with modifications from the book entitled *Renal Fibrosis* (Contribution to Nephrology, Volume 139, p66-89), edited by M. S. Razzaque and T. Taguchi. Our apology goes to all the authors whose work could not be cited due to space limitation.

References

1. Jones DB. Glomerulonephritis. Am J Pathol 1953; 29:33-43.
2. Wada T, Yokoyama H, Matsushima K et al. Chemokines in renal diseases. Int Immunopharmacol 2001; 1:637-645.
3. Wolpe SD, Davatelis G, Sherry B et al. Macrophages secrete a novel heparin-binding protein with inflammatory and neutrophil chemotactic properties. J Exp Med 1988;167:570-580.
4. van Rooijen N, Sanders A. Elimination, blocking, and activation of macrophages: Three of a kind? J Leukoc Biol 1997; 62:702-709.
5. Wang JM, Griffin JD, Rambaldi A et al. Induction of monocyte migration by recombinant macrophage colony stimulating factor. J Immunol 1988; 141:575–579.
6. Lan HY, Nikolic-Paterson DJ, Mu W et al. Local macrophage proliferation in the progression of glomerular and tubulointerstitial injury in rat anti-GBM glomerulonephritis. Kidney Int 1995; 48:753-760.

7. Matsuda M, Shikata K, Makino H et al. Glomerular expression of macrophage colony-stimulating factor and granulocyte-macrophage colony-stimulating factor in patients with various forms of glomerulonephritis. Lab Invest 1996; 75:403-412.
8. Razzaque MS, Foster CS, Ahmed AR. Role of enhanced expression of m-CSF in conjunctiva affected by cicatricial pemphigoid. Invest Ophthalmol Vis Sci 2002; 43:2977-2983.
9. Lan HY, Yang N, Nikolic-Paterson DJ et al. Expression of macrophage migration inhibitory factor in human glomerulonephritis. Kidney Int 2000; 57:499-509.
10. Yang N, Nikolic-Paterson DJ, Ng YY et al. Reversal of established rat crescentic glomerulonephritis by blockade of macrophage migration inhibitory factor (MIF): Potential role of MIF in regulating glucocorticoid production. Mol Med 1998; 4:413-24.
11. Wada T, Schwarting A, Chesnutt MS et al. Nephritogenic cytokines and disease in MRL-Fas^{lpr} kidneys are dependent on multiple T cell subsets. Kidney Int 2001; 59:565-578.
12. Ohashi R, Shimizu A, Masuda Y et al. Peritubular capillary regression during the progression of experimental obstructive nephropathy. J Am Soc Nephrol 2002; 13:1795-805.
13. Imasawa T, Utsunomiya Y, Kawamura T et al. The potential of bone marrow-derived cells to differentiate to glomerular mesangial cells. J Am Soc Nephrol 2001; 12:1401-1409.
14. Ito T, Suzuki A, Imai E et al. Bone marrow is a reservoir of repopulating mesangial cells during glomerular remodeling. J Am Soc Nephrol 2001; 12:2625-2635.
15. Zlotnik A, Yoshie O. Chemokines: A new classification system and their role in immunity. Immunity 2000; 12:121-127.
16. Murphy PM, Baggiolini M, Charo IF et al. International union of pharmacology. XXII. Nomenclature for chemokine receptors. Pharmacol Rev 2000; 52:145-176.
17. Romagnani P, Lazzeri E, Lasagni L et al. IP-10 and Mig production by glomerular cells in human proliferative glomerulonephritis and regulation by nitric oxide. J Am Soc Nephrol 2002; 13:53-64.
18. Banas B, Wornle M, Berger T et al. Roles of SLC/CCL21 and CCR7 in human kidney for mesangial proliferation, migration, apoptosis, and tissue homeostasis. J Immunol 2002; 168:4301-4307.
19. Wada T, Tomosugi N, Naito T et al. Prevention of proteinuria by the administration of anti-interleukin 8 antibody in experimental acute immune complex-induced glomerulonephritis. J Exp Med 1994; 180:1135-1140.
20. Wada T, Yokoyama H, Furuichi K et al. Intervention of crescentic glomerulonephritis by antibodies to monocyte chemotactic and activating factor (MCAF/MCP-1). FASEB J 1996; 10:1418-1425.
21. Garin EH, Blanchard DK, Matsushima K et al. IL-8 production by peripheral blood mononuclear cells in nephrotic patients. Kidney Int 1994; 45:1311-1317.
22. Huber TB, Reinhardt HC, Exner M et al. Expression of functional CCR and CXCR chemokine receptors in podocytes. J Immunol 2002; 168:6244-6252.
23. Furuichi K, Wada T, Sakai N et al. Distinct expression of CCR1 and CCR5 in glomerular and interstitial lesions of human glomerular diseases. Am J Nephrol 2000; 20:291-299.
24. Moore KJ, Wada T, Barbee SD et al. Gene transfer of RANTES elicits autoimmune renal injury in MRL-Fas(lpr) mice. Kidney Int 1998; 53:1631-1641.
25. Parving HH, Osterby R, Anderson PW et al. Biology of renal cells in culture in The Kidney 6th edition. In: Brenner BM, ed. W.B.Saunders Company, 2000:93-191.
26. Holdsworth SR, Kitching AR, Tipping PG. Th1 and Th2 T helper cell subsets affect patterns of injury and outcomes in glomerulonephritis. Kidney Int 1999; 55:1198-1216.
27. Thornhill MH, Kyan-Aung U, Haskard DO. IL-4 increases human endothelial cell adhesiveness for T cells but not for neutrophils. J Immunol 1990; 144:3060-3065.
28. Kaplanski G, Marin V, Montero-Julian F et al. IL-6: A regulator of the transition from neutrophil to monocyte recruitment during inflammation. Trends Immunol 2003; 24:25-29.
29. Brady HR, McGinty A, Adler S. Cell-cell and cell-matrix interactions in The Kidney 6th edition. In: Brenner BM, ed. W.B. Saunders Company, 2000:192-214.
30. Luo Y, Lloyd C, Gutierrez-Ramos JC et al. Chemokine amplification in mesangial cells. J Immunol 1999; 163:3985-3992.
31. Schneider A, Panzer U, Zahner G et al. Monocyte chemoattractant protein-1 mediates collagen deposition in experimental glomerulonephritis by transforming growth factor-beta. Kidney Int 1999; 56:135-144.

32. Wada T, Furuichi K, Sakai N et al. Up-regulation of monocyte chemoattractant protein-1 in tubulointerstitial lesions of human diabetic nephropathy. Kidney Int 2000; 58:1492-1498.
33. Mezzano SA, Droguett MA, Burgos ME et al. Overexertion of chemokines, fibrogenic cytokines, and myofibroblasts in human membranous nephropathy. Kidney Int 2000; 57:147-158.
34. Wolf G, Jocks T, Zahner G et al. Existence of a regulatory loop between MCP-1 and TGF-beta in glomerular immune injury. Am J Physiol Renal Physiol 2002; 283:F1075-1084.
35. Luu NT, Rainger GE, Nash GB. Differential ability of exogenous chemotactic agents to disrupt transendothelial migration of flowing neutrophils. J Immunol 2000; 164:5961-5969.
36. Zernecke A, Weber KS, Erwing LP et al. Combinational model of chemokine involvement in glomerular monocyte recruitment: Role of CXC chemokine receptor 2 in infiltration dyring nephrotoxic nephritis. J Immunol 2001; 166:5755-5762.
37. Furuichi K, Wada T, Iwata Y et al. Administration of FR167653, a new anti-inflammatory compound, prevents renal ischemia-reperfusion injury in mice. Nephrol Dial Transplant 2002; 17:399-407.
38. Poom M, Megyesi J, Green RS et al. In vivo and in vitro inhibition of JE gene expression by glucocorticoids. J Biol Chem 1991; 266:22375-22379.
39. Xie Y, Sakatusume M, Nishi S et al. Expression, role, receptors, and regulation of osteopontin in the kidney. Kidney Int 2001; 60:1645-1657.
40. Miura M, Fu X, Zhang QW et al. Neutralization of GRO alpha and macrophage inflammatory protein-2 attenuates renal ischemia/reperfusion injury. Am J Pathol 2001; 159:2137-2145.
41. Furuichi K, Wada T, Iwata Y et al. Gene therapy expressing amino-terminal truncated monocyte chemoattractant protein-1 prevents renal ischemia-reperfusion injury. J Am Soc Nephrol 2003; 14:1066-1071.
42. Pascual M, Theruvath T, Kawai T et al. Strategies to improve long-term outcomes after renal transplantation. N Engl J Med 2002; 346:580-590.
43. Grone HJ, Weber C, Weber KS et al. Met-RANTES reduces vascular and tubular damage during acute renal transplant rejection: Blocking monocyte arrest and recruitment. FASEB J 1999; 13:1371-1383.
44. Strehlau J, Pavlakis M, Lipman M et al. Quantitative detection of immune activation transcripts as a diagnostic tool in kidney transplantation. Proc Natl Acad Sci USA 1997; 94:695-700.
45. Nagano H, Nadeau KC, Takada M et al. Sequential cellular and molecular kinetics in acutely rejecting renal allografts in rats. Transplantation 1997; 63:1101–1108.
46. Robertson H, Wheeler J, Morley AR et al. Beta-chemokine expression and distribution in paraffin-embedded transplant renal biopsy sections: Analysis by scanning laser confocal microscopy. Histochem Cell Biol 1998; 110:207-213.
47. Azuma H, Takahara S, Matsumoto K et al. Hepatocyte growth factor prevents the development of chronic allograft nephropathy in rats. J Am Soc Nephrol 2001; 12:1280-1292.
48. Song E, Zou H, Yao Y et al. Early application of Met-RANTES ameliorates chronic allograft nephropathy. Kidney Int 2002; 61:676-685.
49. Horuk R, Shurey S, Ng HP et al. CCR1-specific nonpeptide antagonist: Efficacy in a rabbit allograft rejection model. Immunol Lett 2001; 76:193-201.
50. Schuurman HJ, Menninger K, Audet M et al. Oral efficacy of the new immunomodulator FTY720 in cynomolgus monkey kidney allotransplantation, given alone or in combination with cyclosporine or RAD. Transplantation 2002; 74:951-960.
51. Grandaliano G, Gesualdo L, Ranieri E et al. Monocyte chemotactic peptide-1 expression and monocyte infiltration in acute renal transplant rejection. Transplantation 1997; 63:414-420.
52. Wada T, Furuichi K, Segawa C et al. MIP-1α and MCP-1 contribute crescents and interstitial lesions in human crescentic glomerulonephritis. Kidney Int 1999; 56:995-1003.
53. Sallusto F, Lanzavecchia A, Mackay CR. Chemokines and chemokine receptors in T-cell priming and Th1/Th2-mediated responses. Immunol Today 1998; 19:568-574.
54. Lloyd CM, Minto AW, Dorf ME et al. RANTES and monocyte chemoattractant protein-1 (MCP-1) play an important role in the inflammatory phase of crescentic nephritis, but only MCP-1 is involved in crescent formation and interstitial fibrosis. J Exp Med 1997; 185:1371-1380.

55. Fujinaka H, Yamamoto T, Takeya M et al. Suppression of anti-glomerular basement membrane nephritis by administration of anti-monocyte chemoattractant protein-1 antibody in WKY rats. J Am Soc Nephrol 1997; 8:1174-1178.
56. Tang WW, Qi M, Warren JS. Monocyte chemoattractant protein 1 mediates glomerular macrophage infiltration in anti-GBM Ab GN. Kidney Int 1996; 50:665-671.
57. Wu X, Dolecki GJ, Sherry B et al. Chemokines are expressed in a myeloid cell-dependent fashion and mediate distinct functions in immune complex glomerulonephritis in rat. J Immunol 1997; 158:3917-3924.
58. Tesch GH, Schwarting A, Kinoshita K et al. Monocyte chemoattractant protein-1 promotes macrophage-mediated tubular injury, but not glomerular injury, in nephrotoxic serum nephritis. J Clin Invest 1999; 103:73-80.
59. Topham PS, Csizmadia V, Soler D et al. Lack of chemokine receptor CCR1 enhances Th1 responses and glomerular injury during nephrotoxic nephritis. J Clin Invest 1999; 104:1549-1557.
60. Feng L, Xia Y, Yoshimura T et al. Modulation of neutrophil influx in glomerulonephritis in the rat with anti-macrophage inflammatory protein-2 (MIP-2) antibody. J Clin Invest 1995; 95:1009-1017.
61. Chen S, Bacon KB, Li L et al. In vivo inhibition of CC and CX3C chemokine-induced leukocyte infiltration and attenuation of glomerulonephritis in Wistar-Kyoto (WKY) rats by vMIP-II. J Exp Med 1998; 188:193-198.
62. Herlaar E, Brown Z. p38 MAPK signaling cascades in inflammatory disease. Mol Med Today 1999; 5:439-447.
63. Wada T, Furuichi K, Sakai N et al. Involvement of p38 mitogen-activated protein kinase followed by chemokine expression in crescentic glomerulonephritis. Am J Kidney Dis 2001; 38:1169-1177.
64. Wada T, Furuichi K, Sakai N et al. A new anti-inflammatory compound, FR167653, ameliorates crescentic glomerulonephritis in Wistar-Kyoto rats. J Am Soc Nephrol 2000; 11:1534-1541.
65. Banas B, Luckow B, Moller M et al. Chemokine and chemokine receptor expression in a novel human mesangial cell line. J Am Soc Nephrol 1999; 10:2314-2322.
66. Wenzel U, Schneider A, Valente AJ et al. Monocyte chemoattractant protein-1 mediates monocyte/macrophage influx in anti-thymocyte antibody-induced glomerulonephritis. Kidney Int 1997; 51:770-776.
67. Panzer U, Schneider A, Wilken J et al. The chemokine receptor antagonist AOP-RANTES reduces monocyte infiltration in experimental glomerulonephritis. Kidney Int 1999; 56:2107-2115.
68. Chen YM, Chien CT, Hu-Tsai MI et al. Pentoxifylline attenuates experimental mesangial proliferative glomerulonephritis. Kidney Int 1999; 56:932-943.
69. Schneider A, Harendza S, Zahner G et al. Cyclooxygenase metabolites mediate glomerular monocyte chemoattractant protein-1 formation and monocyte recruitment in experimental glomerulonephritis. Kidney Int 1999; 55:430-441.
70. Baldwin Jr S. The NF-kappa B and I kappa B proteins: New discoveries and insights. Annu Rev Immunol 1996; 14:649-83.
71. Ruiz-Ortega M, Bustos C, Hernandez-Presa MA et al. Angiotensin II participates in mononuclear cell recruitment in experimental immune complex nephritis through nuclear factor-kappa B activation and monocyte chemoattractant protein-1 synthesis. J Immunol 1998; 161:430-439.
72. Inan MS, Razzaque MS, Taguchi T. Pathological significance of renal expression of NFκB. Contrib Nephrol 2003; 139:90-101.
73. Guijarro C, Egido J. Transcription factor-kappa B (NF-kappa B) and renal disease. Kidney Int 2001; 59:415-24.
74. Zoja C, Liu XH, Donadelli R et al. Renal expression of monocyte chemoattractant protein-1 in lupus autoimmune mice. J Am Soc Nephrol 1997; 8:720-729.
75. Tesch GH, Maifert S, Schwarting A et al. Monocyte chemoattractant protein 1-dependent leukocytic infiltrates are responsible for autoimmune disease in MRL-Fas^{lpr} mice. J Exp Med 1999; 190:1813-1824.
76. Perez de Lema G, Maier H, Nieto E. Chemokine expression precedes inflammatory cell infiltration and chemokine receptor and cytokine expression during the initiation of murine lupus nephritis. J Am Soc Nephrol 2001; 12:1369-1382.
77. Yamada M, Yagita H, Inoue H et al. Selective accumulation of CCR4+ T lymphocytes into renal tissue of patients with lupus nephritis. Arthritis Rheum 2002; 46:735-740.

78. Ishikawa S, Sato T, Abe M et al. Aberrant high expression of B lymphocyte chemokine (BLC/CXCL13) by C11b+CD11c+ dendritic cells in murine lupus and preferential chemotaxis of B1 cells towards BLC. J Exp Med 2001; 193:1393-1402.
79. Zoja C, Corna D, Benedetti G et al. Bindarit retards renal disease and prolongs survival in murine lupus autoimmune disease. Kidney Int 1998; 53:726-734.
80. Iwata Y, Wada T, Furuichi K et al. p38 mitogen - activated protein kinase contributes to autoimmune renal injury in MRL-Fas^{lpr} mice. J Am Soc Nephrol 2003; 14:57-67.
81. Kato S, Luyckx VA, Ots M et al. Renin-angiotensin blockade lowers MCP-1 expression in diabetic rats. Kidney Int 1999; 56:1037-1048.
82. Parving HH, Lehnert H, Brochner-Mortensen J et al. The effect of irbesartan on the development of diabetic nephropathy in patients with type 2 diabetes. N Engl J Med 2001; 345:870-878.
83. Brenner BM, Cooper ME, de Zeeuw D et al. Effects of losartan on renal and cardiovascular outcomes in patients with type 2 diabetes and nephropathy. N Engl J Med 2001; 345:861-869.
84. Nakao N, Yoshimura A, Morita H et al. Combination treatment of angiotensin-II receptor blocker and angiotensin-converting-enzyme inhibitor in nondiabetic renal disease (COOPERATE): A randomised controlled trial. Lancet 2003; 361:117-124.
85. Utimura R, Fujihara CK, Mattar AL et al. Mycophenolate mofetil prevents the development of glomerular injury in experimental diabetes. Kidney Int 2003; 63:209-216.
86. Morrissey JJ, Klahr S. Differential effects of ACE and AT1 receptor inhibition on chemoattractant and adhesion molecule synthesis. Am J Physiol 1998; 274:F580-586.
87. Nagatoya K, Moriyama T, Kawada N et al. Y-27632 prevents tubulointerstitial fibrosis in mouse kidneys with unilateral ureteral obstruction. Kidney Int 2002; 61:1684-1695.
88. Anders HJ, Vielhauer V, Frink M et al. Chemokine receptor CCR-1 antagonist reduces renal fibrosis after unilateral ureter ligation. J Clin Invest 2002; 109:251-259.
89. Satriano JA, Hora K, Shan Z et al. Regulation of monocyte chemoattractant protein-1 and macrophage colony-stimulating factor-1 by IFN-γ, tumor necrosis factor-α, IgG aggregates, and cAMP in mouse mesangial cells. J Immunol 1993; 150:1971-1978.
90. Rangan GK, Wang Y, Tay YC et al. Inhibition of nuclear factor-kappa B activation reduces cortical tubulointerstitial injury in proteinuric rats. Kidney Int 1999; 56:118-134.
91. Furuichi K, Wada T, Iwata Y et al. Upregulation of fractalkine in human crescentic glomerulonephritis. Nephron 2001; 87:314-320.
92. Ou ZL, Nakayama K, Natori Y et al. Effective methylprednisolone dose in experimental crescentic glomerulonephritis. Am J Kidney Dis 2001; 37:411-417.
93. Jocks T, Zahner G, Freudenberg J et al. Prostaglandin E1 reduces the glomerular mRNA expression of monocyte-chemoattractant protein 1 in anti-thymocyte antibody-induced glomerular injury. J Am Soc Nephrol 1996; 7:897-905.
94. Wolf G, Schneider A, Helmchen U et al. AT1-receptor antagonists abolish glomerular MCP-1 expression in a model of mesangial glomerulonephritis. Exp Nephrol 1998; 6:112-120.
95. Park YS, Guijarro C, Kim Y et al. Lovastatin reduces glomerular macrophage influx and expression of monocyte chemoattractant protein-1 mRNA in nephrotic rats. Am J Kidney Dis 1998; 31:190-194.
96. Razzaque MS, Taguchi T. The possible role of colligin/HSP47, a collagen-binding protein, in the pathogenesis of human and experimental fibrotic diseases. Histol Histopathol 1999; 14:1199-1212.
97. Razzaque MS, Taguchi T. Cellular and molecular events leading to renal tubulointerstitial fibrosis. Med Electron Microsc 2002; 35:68-80.
98. Razzaque MS, Taguchi T. Factors that influence and contribute to the regulation of fibrosis. Contrib Nephrol 2003; 139:1-11.
99. Wada T, Matsushima K, Yokoyama H. Chemokines as therapeutic targets for renal diseases. Curr Med Chem Anti-inflammatory & Anti-Allergy Agents 2003; 2:175-190.

Chapter 3

Oxidative Stress, Lipoproteins and Angiotensin II:
The Unholy Triad in the Pathogenesis of Renal Fibrosis

Jan Galle, Thomas Quaschning and Stefan Seibold

Abstract

Renal fibrosis usually indicates irreversible tissue damage, irrespective of the initial cause. Thus, it is most relevant to understand mechanisms leading to renal fibrosis. Oxidative stress has emerged as an important factor contributing to tissue damage, and oxidative stress is enhanced in a variety of inflammatory disease states relevant for the kidney. It is therefore the purpose of this chapter to discuss the role of oxidative stress in the development of renal fibrosis. Inflammation is generally associated with enhanced oxidative stress, and since multiple factors contribute to inflammation [such as cytokines (e.g., interleukin-6, tumor necrosis factor α), infection, ischemia reperfusion injury, homocysteine, advanced glycation end products, atherogenic lipoproteins, or angiotensin II], multiple factors can cause enhanced oxidative stress. Here we will focus on the role of atherogenic lipoproteins, particularly oxidized low density lipoproteins, and the activated renin angiotensin system, for several reasons: firstly, these factors are well characterized as proinflammatory and as stimulators of superoxide-generating enzymes; secondly, the contribution of these factors to tubulointerstitial fibrosis has frequently been described; and thirdly, we already possess pharmacological tools to efficiently lower their activity. Thus, this chapter highlights the interplay of oxidative stress, atherogenic lipoproteins, and the renin angiotensin system in the pathophysiology of renal fibrosis and discusses potential treatment options.

Introduction

Reactive oxygen species (ROS) and antioxidative defense mechanisms are antagonists in oxidative homeostasis. An imbalance between formation of ROS and antioxidative defense mechanisms defines oxidative stress. In view of the profound biological effects of ROS, in recent years numerous clinical and experimental studies focused on detection of signs of oxidative stress in renal patients. There is good evidence indicating that uremia in general is associated with enhanced oxidative stress,[1,2] and treatment of uremic patients with hemodialysis or peritoneal dialysis has been suggested to particularly contribute to oxidative stress and reduced antioxidant levels in these patients.[3,4] The latter may result from hemodialysis membrane-induced activation of macrophages on the surface of dialysis membranes during the dialysis session. Loss or deficiency of antioxidant activity, e.g., vitamin E deficiency, could also contribute to

Fibrogenesis: Cellular and Molecular Basis, edited by Mohammed S. Razzaque.

enhanced oxidative stress in uremia. However, specific pathological processes involve enhanced formation of oxygen radicals, long before end stage renal disease is reached. Thus, many of the renal patients have to deal with enhanced oxidative stress early from the onset of their disease until renal replacement therapy. The development of renal fibrosis is one such example.

What Generates Oxidative Stress?

It is generally accepted that ROS such as hydrogen peroxide (H_2O_2) or hypochloride (HOCl), and free radicals such as superoxide (O_2^-), hydroxyl radical (OH·), and nitric oxide (NO·), are continuously formed in vivo.[5] Thus, detection of ROS per se does not yet define oxidative stress; however, in a situation where antioxidative defense mechanisms are attenuated, it is the imbalance between formation of ROS and defense mechanisms that creates oxidative stress. The balance between formation of ROS and antioxidative defense mechanisms depends on the activity of enzymes such as superoxide dismutases (SODs), catalase, NO-synthase, and glutathione peroxidase. This balance, however, is rather fragile, difficult to predict, and strongly depending on environmental conditions,[5] as illustrated by the example shown in Figure 1: once e.g., O_2^- is formed, the activity of SOD will transform it to H_2O_2. H_2O_2, in case of sufficient catalase activity, will react to harmless H_2O and O_2. However, too much SOD, in relation to H_2O_2-removing catalase, can be deleterious, giving rise to the formation of the highly reactive hydroxyl radical in the presence of metal ions such as Fe^{2+} or Cu^{2+} (Fenton reaction).[5] On the other hand, when there is too little SOD activity, OH. also can be produced from O_2^- via the Haber Weiss reaction. Different cellular enzymes, including mitochondrial oxidases, lipoxygenase, cyclooxygenase, myeloperoxidase, NADPH oxidase, xanthin oxidase, and, in case of L-arginine or tetrahydrobiopterin depletion, NO-synthase have been identified as cellular sources of ROS formation.[6-10]

Biological Effects of Oxidative Stress

Formation of ROS is part of the unspecific defense system of an organism against e.g., bacteria and other microbes. However, ROS may also affect cells of the host organism, in particular at sites of inflammation. The latter plays a role in a variety of human diseases and the following paragraph shall shed some light especially on the impact of ROS formation on development of progressive renal disease.

Biological Effects of Oxidative Stress on Renal Fibrosis

Renal interstitial fibrosis is a common histopathological end point in progressive renal disease of different etiology, characterized by an increased deposition of extracellular matrix and fibroblasts in the renal interstitium.[11] The accumulation of ECM thereby results from increased matrix synthesis as well as reduced matrix degradation. However, these late changes in renal disease are preceded by changes in cell number due to an imbalance of cellular proliferation and apoptosis and by changes in cell size due to cellular hypertrophy.[12,13] Although proliferation and hypertrophy of renal cells can be initially considered as a process to compensate for chronic destruction of surrounding nephrons, there is some evidence that they may also contribute to further injury of renal tissue leading to interstitial fibrosis.[14] Oxidative stress contributes substantially to the diverse processes leading to progressive renal fibrosis.[1,15-17] ROS can be formed in a variety of renal cell types including vascular cells, juxtaglomerular cells, tubular cells, podocytes, mesangial cells, and also invading PMNs.[1,18,19] It may be surmised that an altered oxidant/antioxidant balance probably precedes severe tissue damage and that a more or less constant level of oxidant stress can maintain ongoing destruction.[16]

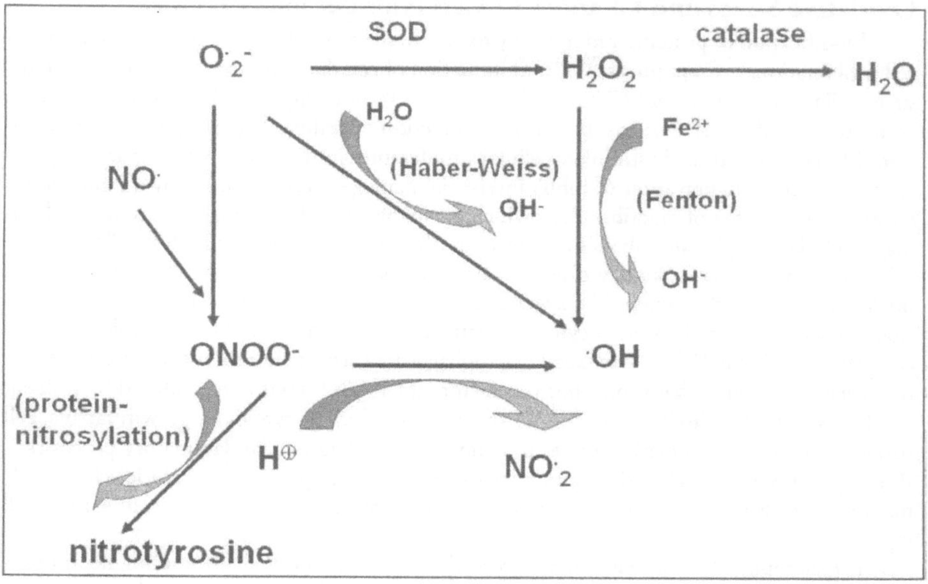

Figure 1. In the vascular system, main sources for superoxide anion (O_2^-) formation are xanthine oxidase, NAD(P)H oxidase, or NO-synthase.[57] O_2^- may react with NO to yield peroxynitrite ($ONOO^-$) which is rather stable but can rearrange to form nitrate and the highly reactive OH·. OH· can also result from the Haber Weiss and the Fenton reaction,[5] and may cause cellular damage and contribute to inflammation.[5]

Oxidative Stress and Accumulation of Extracellular Matrix

Transforming growth factor-beta (TGF-β) is a potent pro-fibrogenic factor in progressive renal disease, activated by oxidative stress.[20,21] TGF-β is not only a potent stimulus for the synthesis of ECM proteins such as fibronectin, laminin, collagens and proteoglycans. TGF-β also decreases ECM degradation by increasing the activity of tissue inhibitors of metalloproteinases (TIMPs) and by decreasing the activity of metalloproteinase.[11] TGF-β further stimulates the synthesis of matrix protein receptors such as integrins and osteopontin.[11] TGF-β also stimulates endothelin production, which in turn is also a potent stimulus for glomerulosclerosis and fibrinogenesis.[11] Furthermore TGF-β is a chemoattractant for fibroblasts and stimulates fibroblast proliferation.[11] A vicious circle results from stimulation of ROS production induced by TGF-β.[22] Since pro-fibrogenic factors involved in the development of progressive renal fibrosis such as basic fibroblast growth factor (bFGF), platelet-derived growth factor (PDGF) and connective tissue growth factor (CTGF) can either directly or indirectly be activated by ROS,[11,23] these molecules are closely linked to fibrogenesis in multiple ways.

ROS may also activate transcription factors such as NF-κB and AP-1.[15] NF-κB is located in the cytoplasm in an inactive form complexed to an inhibitor. The activated form of NF-κB is translocated to the nucleus and binds to DNA motifs present in the promoter of various genes regions that are directly involved in the pathogenesis of renal fibrosis.[11,15] NF-κB for example activates TGF-β1 via activation of transglutaminase,[11] and many of the chemoattractants for leucocytes and adhesion molecules are also regulated by NF-κB.[11]

Oxidative Stress and Changes in Cell Number and Cell Size

Modification of proteins and lipids by oxidative stress is believed to play a central role in cell proliferation,[1,2,24] apoptosis,[2-4,24] and induction of cellular hypertrophy.[14] Growth factors such as TGF-β, PDGF and CTGF activated by ROS not only stimulate tissue fibrosis via accumulation of ECM proteins, but also by induction of cellular hypertrophy and proliferation. ROS can also directly stimulate cellular proliferation in renal cells, without activation of growth factors.[15] Aggravation of tubulointerstitial damage is accompanied by a dramatic increase in the number of myofibroblasts which are involved in the pathogenetic sequence leading to tubular atrophy and fibrosis e.g., by a significant upregulation of TGF-β1.[16,25]

The cell number in kidney disease is also increased by invasion of cells such as mononuclear cells due to chemotaxis.[16] Oxygen radicals were identified as messengers for the expression of chemokines such as monocyte chemoattractant protein-1 (MCP-1) and colony-stimulating factor-1 (CSF-1).[16] The invaded macrophages are then a major source of TNF-α release, but it has to be noted that tumor necrosis factor-alpha (TNF-α) is also produced by resident renal cells.[11] TNF-α itself again activates transcription factors, cytokines, growth factors, cell surface receptors, cell adhesion molecules and other mediators of inflammatory processes.[11] Binding of TNF-α to cell surface receptors also results in apoptotic and necrotic cell death,[11] but ROS are also known to induce cellular apoptosis through other pathways in various cell types.[26]

Hypertrophy of renal cells emerges from chronic destruction of surrounding nephrons is a compensatory effect. ROS can also stimulate cellular hypertrophy directly or via activation of various growth factors.[11]

Oxidative Stress and Proteinuria

Proteinuria can result from inflammatory renal disease, which is additionally enforced by ROS, due to increased susceptibility to proteolytic damage and by inactivation of proteinase inhibitors.[15] Studies in vitro and in vivo have also shown that ROS can induce proteinuria without inflammatory renal disease either by degradation of glomerular basement membranes or even without apparent ultrastructural abnormalities.[15] Proteinuria has been linked to progressive tubulointerstitial disease in human and experimental renal diseases.[16] A high protein/albumin load additionally induces growth factor/cytokine expression such as CSF-1, MCP-1, TGFβ1 and PDGF and also stimulates the release of endothelin and fibronectin in tubular epithelia.[16]

Angiotensin II (Ang II) and Oxidative Stress

Mechanism of Ang II-Induced O_2^- Formation

In experiments with rat smooth muscle cell membranes, Ang II-induced stimulation of O_2^- formation could be inhibited by diphenylene iodinium, suggesting that O_2^- was produced by membrane bound NADH/NADPH oxidases.[27] The involvement of the NADH/NADPH oxidases was also detected in rabbit aortic adventitia preparations,[28] in cultured mesangial cells,[29] and in mouse tubular cells.[14] The Ang II stimulation was transmitted via the AT_1 receptor,[14,29,30] and involved the NADPH subunits p22(phox) (the alpha subunit of cytochrome b588) and part of the electron transfer component of the phagocytic NADPH oxidase) and p67(phox).

Functional Consequence of Ang II-Induced Oxidative Stress in Renal Tissue

A number of kidney diseases and their progression to end stage renal failure are enforced by the intercrine, autocrine, paracrine, and endocrine effects of Ang II.[11] All the compounds of the renin-angiotensin-aldosterone system, including substrate (angiotensinogen), enzymes involved in synthesis and degradation, as well as the receptors are present in the kidney.[11] The intrarenal concentration of Ang II is about one thousand fold greater than the circulating levels of Ang II.[11] Signaling of Ang II through the AT_1 receptor results in vasoconstriction, stimulation of growth and activation of fibroblasts and myocytes.[11] In contrast, activation of the AT_2 receptor can result in vasodilation, antiproliferative responses and apoptosis.[11] It appears that most of the damaging effects of Ang II are mediated by the AT_1 receptor.[11] In renal tubular cells (LLC-PK1 and mouse proximal tubule cells), Ang II-induced O_2^- formation was demonstrated. This intracellular accumulation of reactive oxygen species subsequently leads to phosphorylation and activation of p44/42 MAP kinases finally resulting in an increase in $p27^{kip1}$ expression, G1 phase arrest, and cell hypertrophy.[13] Increasing levels of Ang II can also upregulate the expression of other factors such as PDGF, bFGF, TGF-β1, TNF-α, osteopontin, vascular cell adhesion molecule-1 (VCAM-1), and NF-κB.[11] NF-κB is activated by Ang II in the kidney through the AT_1, as well as the AT_2 receptor.[11] Conversely, NF-κB stimulates the angiotensinogen gene expression.[11] This provides an autocrine reinforcing loop that upregulates Ang II production.[11] Inhibition of Ang II formation by angiotensin converting enzyme (ACE) inhibitors markedly decreases NF-κB activation in the kidney in the setting of renal disease.[11] Administration of an ACE inhibitor in an animal model of urethral obstruction also decreased the expression of TGF-β and collagen and ameliorated the degree of interstitial fibrosis, by downregulation of TGF-β synthesis.[11] Ang II also induces expression of the lectin-like OxLDL receptor-1 (LOX-1) receptor in vitro and in vivo, thereby stimulating cellular uptake of oxidized low density lipoproteins (OxLDL).[31]

Table 1 summarizes some nonhemodynamic, pleiotropic effects of Ang II which include increased expression of various growth factors, protooncogenes, and vasopeptides, aberrant growth responses and elaboration of inflammatory and fibrogenic cytokines. In consequence, these alterations contribute to changes in renal structure and function.

Lipoproteins and Oxidative Stress

OxLDL share many features with Ang II. Animal studies with cholesterol fed rabbits provided first indirect evidence for a role of LDL in the induction of oxidative stress. Aortas from hypercholesterolemic rabbits produced significantly more O_2^- than control aortas.[32,33] Later, our group could show directly that incubation of cultured HUVEC and of isolated arteries with oxidized LDL or Lp(a) stimulated O_2^- formation (Fig. 2).[34-37] Induction of free radical formation has also been demonstrated directly in macrophages after stimulation with Lp(a),[38] and indirectly in mesangial cells after stimulation with OxLDL.[39]

Mechanism of OxLDL-Induced O_2^- Formation

It is a general observation that oxidation of LDL enhances or is even a prerequisite for its capacity to stimulate O_2^- formation. This observation hints to metabolites of the lipid peroxidation process itself, and/or to the specific receptors for OxLDL which are distinct from the ApoB100 receptor for native LDL. During the lipoprotein oxidation, various more or less

Table 1. Renal effects of angiotensin II (Ang II)

Hemodynamic Effects	Pleiotropic Effects	
	Modulation of Renal Function	**Modulation of Renal Structure**
Renal vasoconstriction[58]	Increase in glomerular capillary permeability[61]	Induction of renal hypertrophy[71,72]
Aldosterone release[59]	Mesangial cell contraction[62]	Induction of cell proliferation[72]
Increase of glomerular capillary pressure[60]	Modification of tubular transport[63] Stimulation of mesangial uptake and processing of macromolecules[64] Modulation of nitric oxide release[29,65] Stimulation of endothelin production[66] Stimulation of cytokine production (VEGF, TGF-β)[61,67] Stimulation of superoxide production[13,27,68] Immunomodulatory effects[69,70]	Stimulation of extracellular matrix synthesis[73] Inhibition of extracellular matrix degradation[74]

stable products are formed, including lysophosphatidylcholine, aldehydic lipid peroxidation products, and fatty acids produced by phospholipase A_2.[40-42] Lysophosphatidylcholine, a by-product of cholesterol esterification, increases O_2^- formation in human endothelial cells[43] and in vascular smooth muscle cells via stimulation of protein kinase C.[44] A similar mechanism may take place also in renal cells. Recent data indicate that OxLDL stimulates O_2^- formation through its lectin-like receptor LOX-1.[45] Finally, OxLDL as well as its constituent LPC are strong stimulators of the NADPH-oxidase subunits p22phox and gp91phox.[46,47]

Functional Consequence of OxLDL-Induced Oxidative Stress in Renal Tissue

A large number of kidney diseases and their progression to end stage renal failure are aggravated by OxLDL.[15,16,48] Lipid peroxidation products have been found in kidney tissue within tubular and interstitial cells, and levels of renal lipid peroxidation products are significantly increased in renal disease.[16,49,50] In vitro the initial steps of glomerular and interstitial fibrosis are associated with increases in lipid peroxidation products and oxidative stress in the absence of inflammatory cell infiltration, hypertension, or hyperglycemia.[50] Renal tubular cells are able to internalize oxidized LDL, although the special receptor pathways are not yet known.[49] Oxidized LDL could get internalized in renal cells via scavenger receptors as seen e.g., in macrophages, or via uptake by the lectin-like LOX-1 as seen in endothelial cells. Oxidation of LDL can also take place in the renal cells themselves.[16,49] Atherogenic lipoproteins like Ox LDL and Lp(a) are capable of enhancing generation of ROS via stimulation of NADH/NADPH-dependent oxidases in endothelial cells, smooth muscle cells, juxtaglomerular cells and mesangial cells.[19] In contrast, native lipoproteins had no effect on O_2^- formation in endothelial cells or mesangial cells.[48] The molecular pathways that lead to interstitial fibrosis once LDL is oxidized in the kidney are not clear. Nevertheless, the elevation of procollagens is suggestive for an increased matrix synthesis in the kidney.[15,49] Thus, matrix synthesis could be a

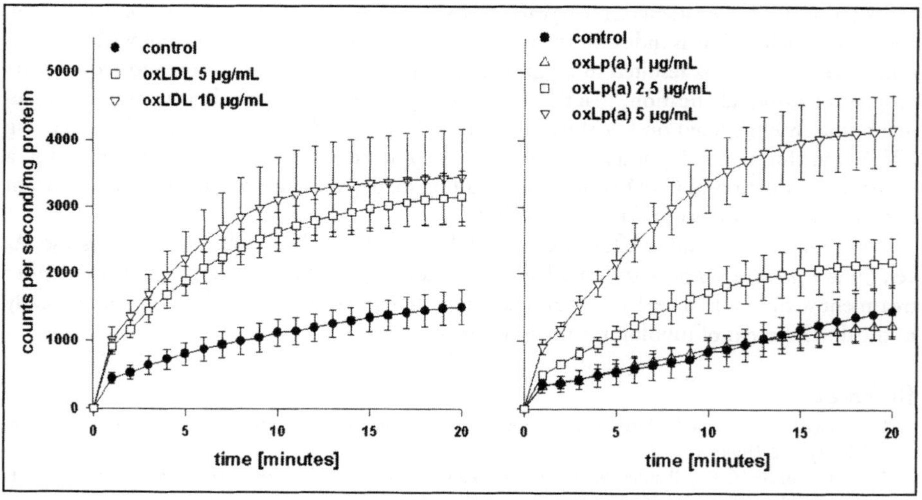

Figure 2. OxLDL and OxLp(a) stimulate O_2^- formation in cultured endothelial cells. Time course of the O_2^- formation in terms of chemiluminescence of lucigenin (counts per second/mg protein) in untreated cultured HUVEC (control), and in HUVEC treated with 5 and 10 µg/mL oxLDL (left panel), or treated with 1, 2.5, and 5 µg/mL oxLp(a) (right panel). The lipoproteins dose-dependently significantly stimulated O_2^- formation of the endothelial cells. Data are means ± SE of 8-10 independent experiments, $p < 0.05$ for oxLDL or oxLp(a) vs. control. Adapted from Galle et al: Lp(a) and LDL induce apoptosis in human endothelial cells and in rabbit aorta - role of oxidative stress. Kidney Int 1999; 55:1450-1461, with permission from the publisher.

direct consequence of LDL oxidation or an indirect one, regulated by the production of fibrogenic cytokines such as TGF-β, since OxLDL is known to stimulate the expression of TGF-β.[48,49] Impaired matrix turnover could also contribute to interstitial fibrosis, since LDL upregulates TIMP-1 expression, thereby inhibiting matrix degradation in rats.[49] Hypercholesterolemia in rats also downregulates urokinase type plasminogen activator (uPA) resulting in decreased matrix degradation and metalloproteinase activity due to decreased plasmin levels.[49]

A dual effect of both OxLDL and lysophosphatidylcholine (LPC) was described in endothelial cells and mesangial cells, including induction of proliferation at low concentrations and triggering apoptosis at higher concentration.[26,48] Both effects seem to be mediated by oxidative stress, since they could be overcome by antioxidants.[26] The induction of apoptosis by OxLDL can counteract proinflammatory stimulation for NF-κB-dependant genes, thereby overcoming protection from cell death.[26] Both OxLDL and LPC can also directly regulate NF-κB activity by a dose dependant regulation of Ik-Bα, the inhibitor of NF-κB.[26] In addition, OxLDL acts chemotactic for mononuclear cells in vitro and increases secretion of interleukin-8 in these cells.[16]

Antioxidant Studies in Uremia

Studies using antioxidant treatment to prevent vascular and other diseases revealed equivocal effects in nonrenal patients.[4,51-54] Since renal patients live under particularly pro-oxidative conditions,[3] the question raises whether there is particular benefit for this subset of patients. Unfortunately, only few antioxidant intervention studies with clinical endpoints have been published referring to renal patients. However, two studies recently reported beneficial effects of vitamin E treatment on lipid metabolism, atherosclerosis, and cardiovascular disease (CVD) in hemodialysis patients. Mune et al used vitamin E coated cellulose membrane dialyzers in end-stage renal disease patients for two years and measured oxidized LDL, LDL-malondialdehyde, and aortic calcification as index for the progression of atherosclerosis. Treatment with the vitamin E coated dialyzers resulted in a significant reduction of LDL oxidation and reduced the progress of aortic calcification.[55] In the SPACE study, the effect of high dose vitamin E supplementation (800 IU/day) on CVD was investigated in hemodialysis patients with preexisting CVD.[56] In this high risk population, vitamin E supplementation over two years resulted in a significant reduction of CVD endpoints including myocardial infarction, without significant effects on total mortality and mortality of CVD.

However, these studies refer to patients who already reached end stage renal disease. As yet, there is no intervention study available which could give an answer to the question whether treatment with a (still to be defined) antioxidative strategy could be beneficial for patients early in the development of tubulointerstitial disease.

References

1. Ichikawa I, Kiyama S, Yoshioka T. Renal antioxidant enzymes: Their regulation and function. Kidney Int 1994; 45:1-9.
2. Witko Sarsat V, Friedlander M, Capeillere Blandin CJ et al. Advanced oxidation protein products as a novel marker of oxidative stress in uremia. Kidney Int 1996; 49:1304-1313.
3. Toborek M, Wasik T, Drózdz M et al. Effect of hemodialysis on lipid peroxidation and antioxidant system in patients with chronic renal failure. Metabolism 1992; 41:1229-1232.
4. Epperlein MM, Nourooz-Zadeh J, Jayasena SD et al. Nature and biological significance of free radicals generated during bicarbonate hemodialysis. J Am Soc Nephrol 1998; 9:457-463.
5. Halliwell B. The role of oxygen radicals in human disease, with particular reference to the vascular system. Haemostasis 1993; 23(Suppl 1):118-126.
6. Carr AC, McCall MR, Frei B. Oxidation of LDL by myeloperoxidase and reactive nitrogen species—Reaction pathways and antioxidant protection. Arterioscler Thromb Vasc Biol 2000; 20:1716-1723.
7. Griendling KK, Sorescu D, Ushio-Fukai M. NAD(P)H oxidase—Role in cardiovascular biology and disease. Circ Res 2000; 86:494-501.
8. Vásquez-Vivar J, Kalyanaraman B. Generation of superoxide from nitric oxide synthase. FEBS Lett 2000; 481:305-306.
9. Böger RH, Böde-Boger SM, Phivthong-ngam L et al. Dietary L-arginine and α-tocopherol reduce vascular oxidative stress and preserve endothelial function in hypercholesterolemic rabbits via different mechanisms. Atherosclerosis 1998; 141:31-43.
10. Heitzer T, Brockhoff C, Mayer B et al. Tetrahydrobiopterin improves endothelium-dependent vasodilation in chronic smokers—Evidence for a dysfunctional nitric oxide synthase. Circ Res 2000; 86:E36-E41.
11. Klahr S, Morrissey JJ. The role of vasoactive compounds, growth factors and cytokines in the progression of renal disease. Kidney Int Suppl 2000; 75:S7-14.
12. Modi KS, Morrissey J, Shah SV et al. Effects of probucol on renal function in rats with bilateral ureteral obstruction. Kidney Int 1990; 38:843-850.
13. Hannken T, Schroeder R, Zahner G et al. Reactive oxygen species stimulate p44/42 mitogen-activated protein kinase and induce $p27^{Kip1}$: Role in angiotensin II - mediated hypertrophy of proximal tubular cells. J Am Soc Nephrol 2000; 11:1387-1397.
14. Hannken T, Schroeder R, Stahl RAK et al. Angiotensin II-mediated expression of $p27^{Kip1}$ induction of cellular hypertrophy in renal tubular cells depend on the generation of oxygen radicals. Kidney Int 1998; 54:1923-1933.
15. Klahr S. Oxygen radicals and renal diseases. Miner Electrolyte Metab 1997; 23:140-143.
16. Scheuer H, Gwinner W, Hohbach J et al. Oxidant stress in hyperlipidemia-induced renal damage. Am J Physiol Renal Physiol 2000; 278:F63-F74.
17. Kawada N, Moriyama T, Ando A et al. Increased oxidative stress in mouse kidneys with unilateral ureteral obstruction. Kidney Int 1999; 56:1004-1013.
18. Halliwell B. Antioxidant defence mechanisms: From the beginning to the end (of the beginning). Free Radic Res 1999; 31:261-272.

19. Galle J. Oxidative stress in chronic renal failure. Nephrol Dial Transplant 2001; 16:2135-2137.
20 Iglesias-De La Cruz MC, Ruiz-Torres P, Alcami J et al. Hydrogen peroxide increases extracellular matrix mRNA through TGF-beta in human mesangial cells. Kidney Int 2001; 59:87-95.
21. Lal MA, Brismar H, Eklof AC et al. Role of oxidative stress in advanced glycation end product-induced mesangial cell activation. Kidney Int 2002; 61:2006-2014.
22. Thannickal VJ, Day RM, Klinz SG et al. Ras-dependent and-independent regulation of reactive oxygen species by mitogenic growth factors and TGF-beta1. FASEB J 2000; 14:1741-1748.
23. Park SK, Kim J, Seomun Y et al. Hydrogen peroxide is a novel inducer of connective tissue growth factor. Biochem Biophys Res Commun 2001; 284:966-971.
24. Moriyama T, Kawada N, Nagatoya K et al. Oxidative stress in tubulointerstitial injury: Therapeutic potential of antioxidants towards interstitial fibrosis. Nephrol Dial Transplant 2000; 15 Suppl 6:47-49.
25. Kliem V, Johnson RJ, Alpers CE et al. Mechanisms involved in the pathogenesis of tubulointerstitial fibrosis in 5/6-nephrectomized rats. Kidney Int 1996; 49:666-678.
26. Heermeier K, Heinloth A, Galle J. OxLDL modulates the cell cycle via oxidative stress. In: Yoshikawa T, Toyokuni S, Yamamoto Y, Naito Y, eds. Free Radicals in Chemistry, Biology and Medicine. London: OICA International, 2001:299-304.
27 Griendling KK, Minieri CA, Ollerenshaw JD et al. Angiotensin II stimulates NADH and NADPH oxidase activity in cultured vascular smooth muscle cells. Circ Res 1994; 74:1141-1148.
28. Pagano PJ, Clark JK, Cifuentes Pagano ME et al. Localization of a constitutively active, phagocyte-like NADPH oxidase in rabbit aortic adventitia: Enhancement by angiotensin II. Proc Natl Acad Sci USA 1997; 94:14483-14488.
29. Jaimes EA, Galceran JM, Raij L. Angiotensin II induces superoxide anion production by mesangial cells. Kidney Int 1998; 54:775-784.
30. Yanagitani Y, Rakugi H, Okamura A et al. Angiotensin II type 1 receptor-mediated peroxide production in human macrophages. Hypertension 1999; 33:335-339.
31. Morawietz H, Rueckschloss U, Niemann B et al. Angiotensin II induces LOX-1, the human endothelial receptor for oxidized low density lipoprotein. Circulation 1999; 100:899-902.
32. Ohara Y, Peterson TE, Harrison DG. Hypercholesterolemia increases endothelial superoxide anion production. J Clin Invest 1993; 91:2546-2551.
33. Mügge A, Brandes RP, Böger RH et al. Vascular release of superoxide radicals is enhanced in hypercholesterolemic rabbits. J Cardivasc Pharmacol 1994; 24:994-998.
34. Galle J, Heinloth A, Schwedler S et al. Effect of HDL and atherogenic lipoproteins on formation of O_2^- and renin release in juxtaglomerular cells. Kidney Int 1997; 51:253-260.
35. Galle J, Schneider R, Heinloth A et al. Lp(a) and LDL induce apoptosis in human endothelial cells and in rabbit aorta: Role of oxidative stress. Kidney Int 1999; 55:1450-1461.
36. Galle J, Schneider R, Winner B et al. Glyc-oxidized LDL impair endothelial function more potently than oxidized LDL: Role of enhanced oxidative stress. Atherosclerosis 1998; 138:65-77.
37. Galle J, Bengen J, Schollmeyer P et al. Impairment of endothelium-dependent dilation by oxidized lipoprotein(a): Role of oxygen-derived radicals. J Am Soc Nephrol 1994; 5:578.
38. Hansen PR, Kharazmi AK, Jauhiainen M et al. Induction of free radical generation in human monocytes by lipoprotein(a). Eur J Clin Invest 1994; 24:497-499.
39 Sharma P, Reddy K, Franki N et al. Native and oxidized low density lipoproteins modulate mesangial cell apoptosis. Kidney Int 1996; 50:1604-1611.
40. Parthasarathy S, Steinbrecher U, Barnett J et al. Essential role of phospholipase A2 activity in endothelial cell-induced modification of low density lipoprotein. Proc Natl Acad Sci USA 1985; 82:3000-3004.
41. Yokoyama M, Hirata K, Miyake R et al. Lysophosphatidylcholine: Essential role in the inhibition of endothelium-dependent vasorelaxation by oxidized low density lipoprotein. Biochem Biophys Res Commun 1990; 168:301-308.
42. Esterbauer H, Jürgens G, Quehenberger O et al. Autoxidation of human low density lipoprotein: Loss of polyunsaturated fatty acids and vitamin E and generation of aldehydes. J Lipid Res 1987; 28:495-509.
43. Kugiyama K, Sugiyama S, Ogata N et al. Burst production of superoxide anion in human endothelial cells by lysophosphatidylcholine. Atherosclerosis 1999; 143:201-204.

44. Ohara Y, Peterson TE, Zheng B et al. Lysophosphatidylcholine increases vascular superoxide anion production via protein kinase C activation. Arterioscler Thromb 1994; 14:1007-1013.
45. Cominacini L, Rigoni A, Pasini AF et al. The binding of oxidized low density lipoprotein (ox-LDL) to ox-LDL receptor-1 reduces the intracellular concentration of nitric oxide in endothelial cells through an increased production of superoxide. J Biol Chem 2001; 276:13750-13755.
46. Heinloth A, Heermeier K, Raff U et al. Stimulation of NADPH Oxidase by oxidized LDL induces proliferation of human vascular endothelial cells. J Am Soc Nephrol 2000; 11:1819-1825.
47. Rueckschloss U, Galle J, Holtz J et al. Induction of NAD(P)H Oxidase by Oxidized Low-Density Lipoprotein in Human Endothelial Cells: Antiatherosclerotic Potential of HMG-CoA Reductase Inhibitor Therapy. Circulation 2001; 104:1767-1772.
48. Ding GH, Van Goor H, Ricardo SD et al. Oxidized LDL stimulates the expression of TGF-β and fibronectin in human glomerular epithelial cells. Kidney Int 1997; 51(1):147-154.
49. Ong AC, Moorhead JF. Tubular lipidosis: Epiphenomenon or pathogenetic lesion in human renal disease? Kidney Int 1994; 45:753-762.
50. Poirier B, Lannaud-Bournoville M, Conti M et al. Oxidative stress occurs in absence of hyperglycaemia and inflammation in the onset of kidney lesions in normotensive obese rats. Nephrol Dial Transplant 2000; 15:467-476.
51. Yusuf S, Dagenais G, Pogue J et al. Vitamin E supplementation and cardiovascular events in high-risk patients. The Heart Outcomes Prevention Evaluation Study Investigators. N Engl J Med 2000; 342:154-160.
52. Stephens NG, Parsons A, Schofield PM et al. Randomised controlled trial of vitamin E in patients with coronary disease: Cambridge Heart Antioxidant Study (CHAOS) [see comments]. Lancet 1996; 347:781-786.
53. GISSI. Dietary supplementation with n-3 polyunsaturated fatty acids and vitamin E after myocardial infarction: Results of the GISSI-Prevenzione trial. Lancet 1999; 354:447-455.
54. Kushi LH, Folsom AR, Prineas RJ et al. Dietary antioxidant vitamins and death from coronary heart disease in postmenopausal women. N Engl J Med 1996; 334:1156-1162.
55. Mune M, Yukawa S, Kishino M et al. Effect of vitamin E on lipid metabolism and atherosclerosis in ESRD patients. Kidney Int Suppl 1999; 71:S126-S129.
56. Boaz M, Smetana S, Weinstein T et al. Secondary prevention with antioxidants of cardiovascular disease in endstage renal disease (SPACE): Randomised placebo-controlled trial. Lancet 2000; 356:1213-1218.
57. Irani K. Oxidant signaling in vascular cell growth, death, and survival-A review of the roles of reactive oxygen species in smooth muscle and endothelial cell mitogenic and apoptotic signaling. Circ Res 2000; 87:179-183.
58. Harris RC, Cheng HF. The intrarenal renin-angiotensin system: A paracrine system for the local control of renal function separate from the systemic axis. Exp Nephrol 1996; 4(Suppl 1):2-7.
59. Morgan BJ, Lyson T, Scherrer U et al. Cyclosporine causes sympathetically mediated elevations in arterial pressure in rats. Hypertension 1991; 18:458-466.
60. Ziai F, Ots M, Provoost AP et al. The angiotensin receptor antagonist, irbesartan, reduces renal injury in experimental chronic renal failure. Kidney Int Suppl 1996; 57:S132-S136.
61. Pupilli C, Lasagni L, Romagnani P et al. Angiotensin II stimulates the synthesis and secretion of vascular permeability factor/vascular endothelial growth factor in human mesangial cells. J Am Soc Nephrol 1999; 10:245-255.
62. Wolf G, Ziyadeh FN. The role of angiotensin II in diabetic nephropathy: Emphasis on nonhemodynamic mechanisms. Am J Kidney Dis 1997; 29:153-163.
63. Cui XL, Douglas JG. Arachidonic acid activates c-jun N-terminal kinase through NADPH oxidase in rabbit proximal tubular epithelial cells. Proc Natl Acad Sci USA 1997; 94:3771-3776.
64. Wolf G. Angiotensin II: A pivotal factor in the progression of renal diseases. Nephrol Dial Transplant 1999; 14(Suppl 1):42-44.
65. Kihara M, Yabana M, Toya Y et al. Angiotensin II inhibits interleukin-1β-induced nitric oxide production in cultured rat mesangial cells. Kidney Int 1999; 55:1277-1283.
66. Rabelink TJ, Bakris GL. The renin-angiotensin system in diabetic nephropathy: The endothelial connection. Miner Electrolyte Metab 1998; 24:381-388.

67. Wolf G, Mueller E, Stahl RA et al. Angiotensin II-induced hypertrophy of cultured murine proximal tubular cells is mediated by endogenous transforming growth factor-beta. J Clin Invest 1993; 92:1366-1372.
68. Rajagopalan S, Kurz S, Munzel T et al. Angiotensin II-mediated hypertension in the rat increases vascular superoxide production via membrane NADH/NADPH oxidase activation. Contribution to alterations of vasomotor tone. J Clin Invest 1996; 97:1916-1923.
69. Radeke HH, Resch K. The inflammatory function of renal glomerular mesangial cells and their interaction with the cellular immune system. Clin Invest 1992; 70:825-842.
70. Wolf G, Ziyadeh FN, Thaiss F et al. Angiotensin II stimulates expression of the chemokine RANTES in rat glomerular endothelial cells. Role of the angiotensin type 2 receptor. J Clin Invest 1997; 100:1047-1058.
71. Wolf G, Neilson EG. Angiotensin II induces cellular hypertrophy in cultured murine proximal tubular cells. Am J Physiol 1990; 259(5 Pt 2):F768-F777.
72. Wolf G, Ziyadeh FN, Zahner G et al. Angiotensin II is mitogenic for cultured rat glomerular endothelial cells. Hypertension 1996; 27:897-905.
73. Kagami S, Border WA, Miller DE et al. Angiotensin II stimulates extracellular matrix protein synthesis through induction of transforming growth factor-beta expression in rat glomerular mesangial cells. J Clin Invest 1994; 93:2431-2437.
74. Singh R, Alavi N, Singh AK et al. Role of angiotensin II in glucose-induced inhibition of mesangial matrix degradation. Diabetes 1999; 48:2066-2073.

Chapter 4

Involvement of NF-κB in Renal Inflammation and Sclerosis

Laurent Baud, Bruno Fouqueray, Agnes Bellocq, Jean-Philippe Haymannn and Julie Peltier

Abstract

Nuclear factor-κB (NF-κB) comprises a family of transcription factors. They are thought to have a central role in the expression of genes involved in cell mobilization, cell proliferation and cell differentiation, and, hence, in inflammation, repair and fibrosis processes. In particular, NF-κB activation appears to drive a number of inflammatory diseases of the kidney and their progression to end-stage renal failure. Thus, targeting NF-κB activation would lead to the development of new pharmaceutical compounds that would provide novel treatment for these diseases.

Introduction

Inflammatory reaction comprises two successive stages, lesion and repair. The latter stage may be excessive, hence leading to cicatricial fibrosis. Cellular and molecular mechanisms responsible for the development of either stage are numerous, but involve in general a transcriptional control of gene expression. Among other transcription factors, NF-κB is likely to play a major role in this control. Thus, the present chapter will focus on the role of NF-κB in the development of both lesion and repair/fibrosis stages of inflammation within the kidney.

Structure and Activation Pathways of NF-κB

In mammals, the NF-κB family includes five members, that can homodimerize as well as form heterodimers with each other: p65/RelA, cRel, RelB, p100/p52, and p105/p50.[1-4] All of these proteins share a highly conserved domain (Rel homology region, RHR) comprising two Ig-like regions and responsible for interaction with other members of the family (carboxyterminal domain), DNA (aminoterminal domain), and inhibitory proteins. Studies in knockout mice have demonstrated the importance of each member of the NF-κB family. Only the lack of p65/RelA leads to embryonic lethality, suggesting that functional redundancy exists among other members.

NF-κB dimers are associated in the cytoplasm with inhibitory proteins (IκBα), including IκBα, IκBβ, and IκBε.[3,4] This form of NF-κB is inactive because IκB covers the nuclear localization sequence (NLS) contained in the RHR domain. Thus, NF-κB activation requires the degradation of IκBα, a process involving four steps:

Fibrogenesis: Cellular and Molecular Basis, edited by Mohammed S. Razzaque.

1. IκBα phosphorylation at serine residues 32 and 36 by a specific kinase, IκBα kinase (IKK), which is activated through phosphorylation by a kinase belonging to the mitogen activated protein kinase kinase kinase (MAPKKK) family,
2. recognition of phosphorylated IκBα by the type 3 ubiquitin-protein ligase complex,
3. polyubiquitinylation of IκBα at lysine residues 21 and 22, and
4. degradation of IκBα by the proteasome.

In addition to this conventional pathway, others have been described, in particular an alternative pathway involving calpains. These cysteine proteases bind the proline, glutamine, serine, and threonine (PEST) sequence in the carboxyterminal domain of IκBα, and promote its degradation.[5] Thus, calpain inhibitors prevent NF-κB activation in divers models of inflammation.

Once IκBα has been degraded, NLS is uncovered allowing the nuclear translocation of NF-κB. Within the nucleus, NF-κB controls the transcription of genes involved in inflammatory reaction through two different mechanisms. First, by its binding to specific response elements in DNA, NF-κB recruits coactivators such as the cAMP response element binding protein (CREB)-binding protein (CBP) that in turn control transcription process. Second, by its interaction with other transcription factors such as activator protein-1 (AP-1), NF-κB alters the efficiency of these factors.

Origin of NF-κB Activation in Inflammatory Reaction

Stimuli responsible for NF-κB activation in the two successive stages of inflammatory reaction are numerous. They include for instance vasoactive peptides (e.g., angiotensin II and endothelin-1), reactive oxygen species, cytokines [e.g., interleukin (IL)-1 and tumor necrosis factor (TNF)-α], ligands for adhesion molecules, immune complexes, virus, and bacterial products.[3] The consensus pathway for NF-κB activation in response to bacterial products has been characterized recently. These products act through binding to members of the Toll-like receptor (TLR) family: components of Gram-positive bacteria bind TLR2 while lipopolysaccharide (LPS) of Gram-negative bacteria bind TLR4. Both receptors trigger NF-κB activation via a MyD88-dependent intracellular signaling pathway. Within the kidney, these TLRs are localized mainly in epithelial cells of the distal and proximal tubules and in Bowman's capsule.[6]

The microenvironment of inflammatory lesions is characterized by a low pH. In a recent study,[7] we have demonstrated that this acidification is also responsible for NF-κB activation in macrophages. The response involves an amplification loop in which TNF-α is produced under the control of NF-κB and, in turn, activates NF-κB. Another report has described identical mechanisms in a different cell model.[8]

Monitoring of NF-κB Activation

While detection of NF-κB activity is easy in vitro, it is much more difficult in vivo. Initially, this evidence has been based on the expression of genes which requires NF-κB activation.[3] Afterwards, three effective techniques have been developed:

1. electrophoretic mobility shift assay (EMSA) in which nuclear proteins such as NF-κB are identified in tissues by their capacity to bind specific DNA probes, hence limiting DNA probe migration in gel electrophoresis,
2. immunohistochemical analysis of tissue sections with specific antibodies which recognize only the active form of NF-κB, and
3. Southwestern analysis in which the presence of active NF-κB in tissue sections is demonstrated by the hybridization of a specific labeled DNA probe.

Very recently, an interesting technique has been developed to monitor NF-κB activation in vivo and in real time.[9] In transgenic mice that express a luciferase reporter gene driven by NF-κB response elements, NF-κB-induced luminescence is sufficiently intense to be detected externally with an ultrasensitive camera. By using this technique, the authors have demonstrated the activation of NF-κB in kidneys of mice given inflammatory factors such as LPS, Il-1α, and TNF-α.

Role of NF-κB in the Development of Renal Inflammation

NF-κB activation has been evidenced in the kidney in various inflammatory diseases of the glomerulus and/or the interstitium, as for instance in human IgA nephropathy, in experimental glomerulonephritis (e.g.induced by the injection to rat of antibodies directed against glomerular basement membrane or thymocytes), and in experimental tubulointerstitial disorders (e.g., induced by unilateral ureteral obstruction, UUO).[3]

NF-κB activation plays a pivotal role in the development of these inflammatory diseases, mainly through the transcription of genes encoding pro-inflammatory cytokines, chemokines, adhesion molecules, growth factors, and enzymes (e.g., type 2 cyclooxygenase, COX2, inducible nitric oxide synthase, iNOS).[3] Interestingly, the expression of such genes occurs in two temporally distinct phases.[10] Genes whose promoter is constitutively acetylated and, hence, immediately accessible to NF-κB are transcribed first, immediately after NF-κB activation. These genes include genes encoding macrophage inflammatory protein-2 (MIP-2), manganese superoxide dismutase (MnSOD), and IκBα. The second wave of transcription requires the formation of transcrition factor complexes that direct the hyperacetylation of gene promoters. These genes with regulated and late accessibility include genes encoding macrophage chemoattractant protein-1 (MCP-1), IL-6, and RANTES.

Thus, the response to NF-κB appears to change with time. This possibility has been confirmed nicely in two recent studies. A first report has shown that macrophage production of TNF-α diminishes when LPS challenges are repeated sequentially.[11] This LPS tolerance is explained by a change in NF-κB composition rather than in NF-κB activation: repeated treatment with LPS results in increased expression of p50 instead of other κB family members. As a consequence, p50 homodimers form, translocate to nucleus, and interact with TNF-α gene promoter, hence inhibiting LPS-induced TNF-α mRNA expression. Lawrence et al[12] have completed this study by analyzing an established model of inflammation. In this model, the onset of inflammation is associated with the formation of cRel-p50 heterodimers that are involved in the expression of genes encoding pro-inflammatory cytokines (e.g., TNF-α) and anti-apoptotic proteins (e.g., Bcl-2). During the resolution of inflammation, cRel-p50 heterodimers are replaced by p50-p50 homodimers that are responsible for the expression of genes encoding anti-inflammatory cytokines (e.g., transforming growth factor-β, TGF-β) and pro-apoptotic proteins (e.g., Bax and p53). Thus, NF-κB participates successively to the onset and to the resolution of inflammatory reaction.

Role of NF-κB in the Development of Renal Fibrosis

Fibrosis of the renal interstitium occurs in response to inflammatory processes either in the tubulointerstitium (e.g., during ureteral obstruction) or in the glomerulus. In experimental UUO, interstitial volume increases as a consequence of fibroblast proliferation, inflammatory cell infiltration, and extracellular matrix deposition. These changes are associated with activation of the renin angiotensin system. In turn, angiotensin II formation is responsible for NF-κB activation in tubular and interstitial cells. Indeed, NF-κB activation is not observed in animals given angiotensin converting enzyme inhibitors[13] and in angiotensin type 1a receptor (AT1a) deficient mice.[14] Activation of NF-κB in response to angiotensin II would be increased by an amplification loop in which TNF-α is produced under the control of NF-κB and, in turn, activates NF-κB. Accordingly, UUO does not result in NF-κB activation in renal interstitium of TNF-α receptor (TNFR1 and TNFR2) knockout mice.[15]

Glomerular lesions responsible for severe and persistent proteinuria precede in general tubulointerstitial inflammation and fibrosis, and these latter events participate in the progression of renal damage. The underlying mechanisms involve also NF-κB activation. Urinary proteins (mainly albumin) bind to megalin and cubilin, two multiligand receptors at the apical pole of epithelial cells in proximal tubules. Subsequent endocytosis process leads to protein kinase C activation and, hence, reactive oxygen species production through mechanisms requiring both membrane NAD(P)H oxidase and mitochondrial respiratory chain.[17] In turn, reactive oxygen species (e.g., hydrogen peroxide) promote NF-κB activation. This response is amplified by the expression of vasoactive substances such as angiotensin II and endothelin-1.[18]

NF-κB activation plays a key role in fibrosis process by inducing both growth and differentiation of fibroblasts (Fig. 1). Fibroblast growth results from proliferation and/or resistance to apoptosis. NF-κB is involved in the progression of cell cycle in fibroblasts, as suggested by ealier observations that NF-κB activity is elevated during the G_0 to G_1 cell cycle transition.[19] The evidence for anti-apoptotic effects of NF-κB has been provided initially by gene knockout studies in which p65-deficient mice die during embryonic development, through apoptosis of hepatocytes.[19] Subsequent studies have demonstrated that NF-κB activates genes encoding anti-apoptotic factors in various cells including fibroblasts.[20]

Myofibroblasts derive from both differentiation of interstitial cells and transdifferentiation of epithelial cells from proximal tubules. NF-κB is involved in this latter event since its activation leads to expression of vimentin, a marker of epithelial cell dedifferentiation, and fibroblast specific protein-1 (FSP-1), a marker of fibroblast differentiation.[21] In addition, NF-κB activation increases the availability of TGF-β1 which is involved in both differentiation of interstitial cells and transdifferentiation of epithelial cells. Two mechanisms would explain the importance of NF-κB. First, NF-κB activation results in expression of genes encoding chemokines (e.g., MCP-1 and RANTES), adhesion molecules (e.g., intercellular adhesion molecule-1 and vascular cell adhesion molecule-1), and fractalkine that combines properties of both chemoattractant and adhesion molecules. As a consequence, macrophages and lymphocytes invade the interstitium and release growth factors such as TGF-β. Second, NF-κB activation leads to expression of transglutaminase gene. In turn, transglutaminase activates latent TGF-β1.[21] This enzyme exacerbates fibrosis by crosslinking extracellular matrix proteins as well.

NF-κB Pathway As a Target for New Anti-Inflammatory and Anti-Fibrotic Treatments

As NF-κB plays a key role in inflammatory and fibrotic processes, it would be a target for new therapeutic strategies. NF-κB activation and action are potentially prevented:

1. by limiting the expression or the activity of hormones and mediators responsible for NF-κB activation. For instance, administration of angiotensin converting enzyme inhibitors, by decreasing angiotensin II availability, reduces NF-κB activation in UUO model;[13]
2. by increasing the expression of IκBα. This result can be obtained by using anti-inflammatory cytokines such as IL-10, IL-11, and IL-13;[21]
3. by preventing IKK-dependent phosphorylation of IκBα. To this aim, it is possible to transfer the gene encoding a truncated form of IκBα that lacks phosphorylation sites.[23] The phosphorylation of IκBα can also be prevented by cell exposure to 15-deoxy-$\Delta^{12\text{-}14}$-prostaglandin J_2 (15d PGJ_2). This prostaglandin D_2 metabolite contains a cyclopentenone ring that reacts with critical cysteine residues in IKKβ, thereby blocking its kinase activity.[24] By this mechanism, 15d PGJ_2 prevents IL-1β-induced activation of NF-κB in mesangial cells in vitro and, hence, limits MCP-1 expression. Similarly, aspirin and sodium salicylate reduce the activity of IKKβ by preventing the binding of ATP[22] while antioxidants reduce this activity by preventing the oxidation of critical cysteine residues in kinases and phosphatases;[1]

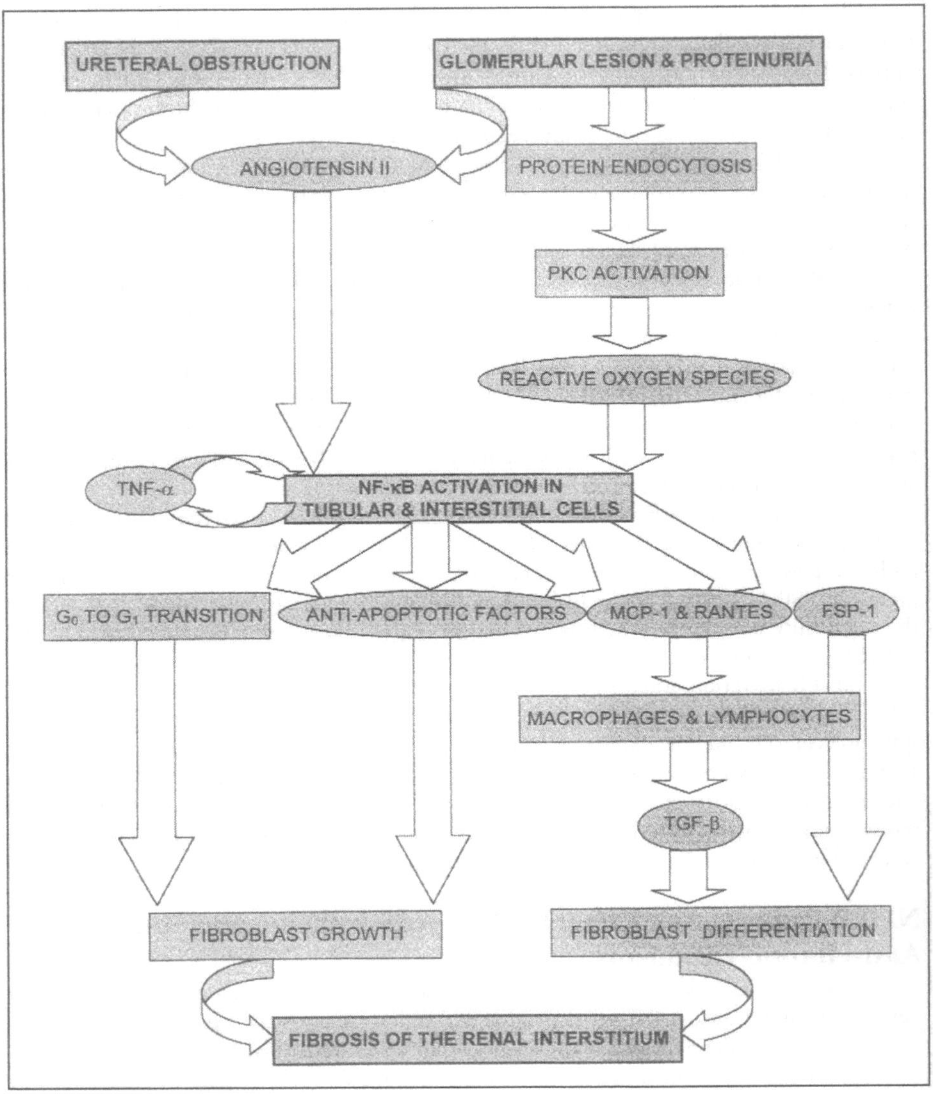

Figure 1. Role of NF-κB activation in the renal fibrogenesis response.

4. by limiting IκBα degradation by proteasome enzymes or calpains. For instance, cyclosporin A, an immunosuppressive agent, serves as a non-competitive inhibitor of proteasome enzymes and thus prevents IκBα degradation without modifying its phosphorylation and polyubiquitinylation;[22]
5. by preventing NF-κB binding to specific response elements in the promoter region of target genes. To this aim, it is possible to transfer NF-κB decoy oligodeoxynucleotides. This strategy has been shown efficient in vitro, and in vivo to limit inflammatory processes induced in the kidney by TNF-α administration,[25] antiglomerular basement membrane serum injection,[26] or renal transplantation.[27]

Interestingly, glucocorticoids, that are widely used for their anti-inflammatory properties, inhibit NF-κB pathway by interfering with at least two steps: in inducing expression of IκBα and in limiting the transactivation properties of NF-κB.[22] Indeed, glucocorticoid receptor can directly bind and inactivate NF-κB; alternatively, a competition can occur between NF-κB and glucocorticoid receptor for limited amounts of the coactivator CBP.

In conclusion, NF-kB pathway is clearly involved in inflammation and fibrosis processes within the kidney. Thus, this pathway represents a potential target for new therapeutic strategies.

References

1. Chen F, Castranova V, Shi X, Demers LM, New insights into the role of nuclear factor-κB, a ubiquitous transcription factor in the initiation of diseases. Clin Chem 1999; 45:7-17.
2. Perkins ND, The Rel/ NF-κB family: friend and foe. Trends Cell Biol 2000; 25:434-440.
3. Guijarro C, Egido J, Transcription factor-κB (NF-κB) and renal disease. Kidney Int 2001; 59:415-424.
4. Tak PP, Firestein GS, NF-κB: a key role in inflammatory diseases. J Clin Invest 2001; 107:7-11.
5. Shumway SD, Maki M, Miyamoto S. The PEST domain of IκBα is necessary and sufficient for in vitro degradation by μ-calpain. J Biol Chem 1999; 274:30874-30881.
6. Wolfs TGAM, Buurman WA, van Schadewijk A et al. In vivo expression of Toll-like receptor 2 and 4 by renal epithelial cells: IFN-γ and TNF-α mediated up-regulation during inflammation. J Immunol 2002; 168:1286-1293.
7. Bellocq A, Suberville S, Philippe C et al. Low environmental pH is responsible for the induction of nitric-oxide synthase in macrophages. Evidence for involvement of nuclear factor-kB activation. J Biol Chem 1998; 273:5086-5092.
8. Shi Q, Le X, Wang B et al. Regulation of interleukin-8 expression by cellular pH in human pancreatic adenocarcinoma cells. J Interferon Res 2000; 20:1023-1028.
9. Carlsen H, Moskaug JO, Fromm SH et al. In vivo imaging of NF-κB activity. J Immunol 2002; 168:1441-1446.
10. Saccani S, Pantano S, Natoli G. Two waves of nuclear factor κB recruitment to target promoters. J Exp Med, 2001; 193:1351-1359.
11. Bohuslav J, Kravchenko VV, Parry GCN, et al. Regulation of an essential innate immune response by the p50 subunit of NF-κB. J Clin Invest 1998; 102:1645-1652.
12. Lawrence T, Gilroy DW, Colville-Nash PR, et al. Possible new role for NF-κB in the resolution of inflammation. Nature Med 2001; 7:1291-1297.
13. Morrissey J, Klahr S. Transcription factor NF-κB regulation of renal fibrosis during ureteral obstruction. Semin Nephrol 1998; 18:603-611.
14. Satoh M, Kashihara N, Yamasaki Y, et al. Renal interstitial fibrosis is reduced in angiotensin II type 1a receptor-deficient mice. J Am Soc Nephrol 2001; 12:317-325.
15. Klahr S, Urinary tract obstruction. Semin Nephrol 2001; 21:133-145.
16. Remuzzi G, Bertani T, Pathophysiology of progressive nephropathies. N Engl J Med 1998; 339:1448-1456.
17. Morigi M, Macconi D, Zoja C et al. Protein overload-induced NF-κB activation in proximal tubular cells requires H_2O_2 through a PKC-dependent pathway. J Am Soc Nephrol 2002; 13:1179-1189.
18. Gomez-Garre D, Largo R, Tejera N, et al. Activation of NF-κB in tubular epithelial cells of rats with intense proteinuria. Role of angiotensin II and endothelin-1. Hypertension 2001; 37:1171-1178.
19. Chen F, Castranova V, Shi X, New insights into the role of nuclear factor-κB in cell growth regulation. Am J Pathol 2001; 159:387-397.
20. Karin M, Lin A, NF-κB at the crossroads of life and death. Nature Immunol 2002; 3:221-227.
21. Klahr S, Morrissey JJ. The role of vasoactive compounds, growth factors and cytokines in the progression of renal disease. Kidney Int 2000; 57:S7-S14.
22. Yamamoto Y, Gaynor RB, Therapeutic potential of inhibition of the NF-κB pathway in the treatment of inflammation and cancer. J Clin Invest 2001; 107:135-142.
23. Hirahashi J, Takayanagi A, Hishikawa K, et al. Overexpression of truncated I kappa B alpha potentiates TNF-alpha-induced apoptosis in mesangial cells. Kidney Int 2000; 57:959-968.

24. Rovin BH, Lu L, Cosio A. Cyclopentenone prostaglandins inhibit cytokine-induced NF-κB activation and chemokine production by human mesangial cells. J Am Soc Nephrol 2001; 12:1659-1667.
25. Tomita N, Morishita R, Tomita S, et al. Inhibition of TNF-α, induced cytokine and adhesion molecule. Exp Nephrol 2001; 9:181-190.
26. Tomita N, Morishita R, Lan HY, et al. In vivo administration of a nuclear transcription factor-κB decoy suppresses experimental crescentic glomerulonephritis. J Am Soc Nephrol 2000; 11:1244-1252.
27. Vos IHC, Govers R, Gröne HJ, et al. NF-κB decoy oligodeoxynucleotides reduce monocyte infiltration in renal allografts. FASEB J 2000; 14:815-822.

Chapter 5

Low-Density Lipoprotein and Glomerulosclerosis

Hyun Soon Lee

Abstract

Hypercholesterolemia is a common feature of nephrosis or uremia. Dietary hypercholesterolemia aggravates the renal injury in experimental focal segmental glomerulosclerosis (FSGS). Hypercholesterolemia is mainly due to increased level of low-density lipoprotein (LDL). Accumulation of apolipoprotein B–containing lipoproteins or LDL is frequently shown in the diseased human glomeruli. Furthermore, oxidized LDL (Ox-LDL) has been demonstrated in the lesions of FSGS or mesangial areas. Ox-LDL in the diseased glomeruli could recruit the circulating monocytes leading to the accumulation of macrophages and subsequent foam cell formation. Increased release of macrophage-derived products could produce an altered mesangial cell matrix biosynthesis. In cultured mesangial cells, LDL is susceptible to oxidative modification, and stimulates mRNA and/or protein expression of α1(I), α1(III) and α1(IV) collagen, laminin, fibronectin and transforming growth factor-β1 (TGF-β1). LDL also upregulates plasminogen activator inhibitor-1 (PAI-1), urokinase-type plasaminogen activator (uPA) and tissue-type plasminogen activator (tPA) expression after prolonged incubation times in mesangial cells. LDL-induced plasminogen activator inhibitory activity was greater than plasminogen activator activity. LDL increased protein kinase C (PKC) and mitogen-activated protein kinase (MAPK) activity as well as bioactive TGF-β secretion in cultured mesangial cells. LDL-induced α1(IV) collagen, PAI-1 or fibronectin overexpression in mesangial cells was abrogated by inhibition or downregulation of PKC or by administration of anti-TGF-β suggesting that LDL stimulates mesangial matrix protein expression through induction of PKC or TGF-β. Altogether, these effects of LDL on altered regulation of the mesangial matrix synthesis/degradation might have a pathophysiological function in the pathogenesis of glomerulosclerosis.

Introduction

Cholesterol is a molecule of great biomedical significance. Cholesterol is used for cell growth and membrane synthesis, but cholesterol accumulation in blood vessels leads to the development of atherosclerotic plaque-like lesions. Cholesterol is carried in the blood as part of complex lipoprotein particles, such as low-density lipoprotein (LDL) (reviewed in refs. 1,2). LDL is an emulsion of cholesteryl ester stabilized by surface phospholipid, unesterified cholesterol and apolipoprotein B (apo B). Disruption of these components by certain proteases or

Fibrogenesis: Cellular and Molecular Basis, edited by Mohammed S. Razzaque.

oxidation could be involved in the formation of larger esterified choelsterol-rich lipid particles that deposit in the extracellular space, a key process for the development of atherosclerosis.[3,4]

Hypercholesterolemia and lipoprotein abnormalities are common features of the nephrotic syndrome and uremia.[5-8] Secondary hypercholesterolemia associated with nephrotic syndrome appears to play a role in the progression of original glomerular injury to focal segmental glomerulosclerosis (FSGS). FSGS represents a pathological hallmark of progressive glomerular injury. The evidence suggesting that hypercholesterolemia modulates original glomerular injury was primarily obtained from experimental animals, in which either cholesterol-feeding or endogenous hyperlipidemia aggravates proteinuria and glomerulosclerosis.[9-14]

Atherosclerotic lesions contain necrotic lipid-rich cores surrounded by proliferating vascular smooth muscle cells and connective tissue as well as lipid-containing foam cells. In a similar fashion, the lesions of FSGS show macrophage-derived foam cells and lipid deposits as well as accumulation of mesangial matrix[15-17] (Fig. 1) suggesting that the pathogenesis of glomerulosclerosis is analogous to that of atherosclerosis.[16,18-22] Glomerular mesangial cells and vascular smooth muscle cells are closely related in terms of origin, histochemistry, and contractility.[19] Mesangial cells are not only a major target of lipid-induced glomerular damage but also a major site of matrix protein synthesis in chronic glomerular disease. In this regard, studies to examine the role of LDL on subsequent glomerulosclerosis have mainly focused on the changes of mesangial cells exposed to LDL or hypercholesterolemia.

Effects of Lipoprotein Abnormalities on the Glomerulus in Experimental Animals

Most of the animal studies to determine the sequelae of hypercholesterolemia and dyslipoproteinemia on the kidney have been performed in the rats. Yet, normal rats have very low plasma levels of LDL,[23] and therefore lipid studies in the rats may not reflect the consequences of dyslipoproteinemia in humans. Long-term cholesterol feeding to normal rats was associated with glomerular lipid deposition, increased mesangial expansion and a modest degree of glomerulosclerosis.[12] Even short-term cholesterol feeding to normal rats induced an enhanced glomerular mRNA expression for α1 chain of type IV [α1(IV)] collagen, tissue inhibitor of metalloproteinase and transforming growth factor-β1 (TGF-β1) as well as increased mesangial immunostaining for type IV collagen, fibronectin and laminin.[24] In guinea pigs, glomerular enlargement, mesangial expansion, and proteinuria were evident after only 1 month of increased circulating cholesterol.[25]

In addition, a high cholesterol diet aggravates proteinuria and glomerulosclerosis in rats with another underlying cause of renal injury, such as chronic puromycin aminonucleoside nephrosis[9,13,14] hypertension,[26] or reduced renal mass.[10,12,27] These animal studies support the hypothesis that secondary hypercholesterolemia in human nephrotics may play a role in the progression of original glomerular injury to FSGS.

Effects of LDL on Mesangial Cells

Receptor-Mediated Uptake of LDL or Modified LDL by Mesangial Cells

In vivo clearance of LDL is mediated by cell surface receptors that recognize apo B.[28] When intracellular cholesterol stores are adequate, LDL receptors are downregulated. This prevents intracellular overaccumulation of cholesteryl esters.[29] Indeed, the existence of the high-affinity LDL receptor in all human cells allows the body to maintain plasma LDL levels below the threshold range for atherosclerosis and at the same time to supply its cells with adequate amounts of cholesterol.[1,29] Cultured human or rat mesangial cells possess specific receptors for LDL[30-34] on the clathrine-coated pits.[33,34]

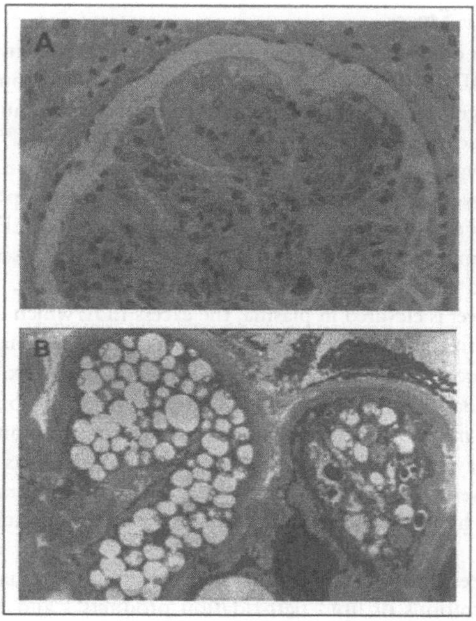

Figure 1. A) Light micrograph of a glomerulus showing lipid containing foam cells and mesangial matrix expansion in the lesions of focal segmental glomerulosclerosis (FSGS). (PAS stain). B) Electron micrograph of a glomerulus showing lipid-containing foam cells in the sclerotic lesion.

In contrast to LDL receptor, scavenger receptor is not downregulated by an increase in intracellular cholesterol and can lead to intracellular lipid loading.[35] Rat mesangial cells are made up of contractile smooth muscle-like cells and bone marrow-derived phagocytic cells.[34,37] Of these, only phagocytic cells express large number of scavenger receptors for oxidized LDL (Ox-LDL), which can result in foam cell formation.[34] Some researchers reported that specific receptors for modified LDL are almost negligible in cultured human mesangial cells,[33,34] whereas others suggested their existence.[38] The origin of glomerular foam cells in human glomerular disease is controversial. Despite the prevailing view that they are of macrophage origin, Moorhead and colleagues[39,40] conjectured that even human mesangial cells could transform to glomerular foam cells. Recently, Ruan et al[41] suggested that cultured human mesangial cells can permit unregulated intracellular accumulation of LDL with foam cell formation influenced by inflammatory cytokines, and this needs further confirmation.

Growth-Stimulating Effect of LDL on Mesangial Cells

LDL stimulates rapid and transient expression of platelet-derived growth factor (PDGF) B mRNA in cultured human mesangial cells.[33,42] LDL also induces a slight but significant increase in [^{3}H]-thymidine incorporation into mesangial DNA.[33,42] PDGF B-chain is the most potent mitogen for mesangial cells in culture[43-46] and seems to regulate mesangial matrix production in vitro[47,48] or in vivo[49-51] Thus, LDL-induced upregulation of PDGF in diseased glomeruli may be related to mesangial expansion affecting both cells and matrix eventually leading to glomerulosclerosis.

LDL and Oxidative Stress

Superoxide (O_2^-) or other reactive oxygen species are known to be involved in the mediation of renal injury.[52,53] NADPH-oxidase complex is the most familiar source of O_2^-,[54,55] and is present not only in phagocytic mononuclear cells but also in cultured human mesangial cells.[56] LDL is susceptible to oxidative modification in cultured mesangial cells,[42,57-59] which are known to produce reactive oxygen species.[52,60] During lipoprotein oxidation, various products are formed including aldehydic lipid peroxidation products,[61,62] lysophosphatidylcholine[63,64] and fatty acids produced by phospholipase A_2.[65] Generally, oxidation of LDL is a prerequisite for the ability of LDL to stimulate O_2^- production.[66] The biological responses triggered by Ox-LDL are associated with lipid peroxidation derivatives (reviewed in ref. 67).

When the LDL level is elevated in plasma, the excess LDL which is not taken up by cells might be accumulated in the extracellular space. In a number of human glomerular diseases, apo B-containing lipoproteins accumulate within the mesangial matrix.[68] Increased collagen[69] or other extracellular matrix[70] in the diseased glomeruli may contribute to the retention of LDL in the mesangial matrix. When the LDL is trapped for a prolonged period of time in mesangial matrix, it may be oxidized due to depletion of antioxidants found in plasma,[71] or due to free radicals derived from glomerular mesangial cells[52,60] or infiltrating macrophages.[72,73] In experimental FSGS, hypercholesterolemia appears to make lipoproteins more susceptible to oxidation.[14]

Oxidative stress seems to be increased in uremic patients. Plasma from uremic patients contains high concentrations of lipid peroxidation products[74] and autoantibodies against Ox-LDL.[75] Major modifications could occur in the chemical properties of LDL from patients with uremia and renal patients, and these modifications would increase susceptibility of LDL to oxidation.[76,77] Ox-LDL has been demonstrated by immunostaining in the glomeruli of both patients[78] and experimental animals,[14,79] indicating that oxidative modification of LDL occurs in vivo (Fig. 2). Patients with heavy Ox-LDL accumulation in the sclerotic segments of glomeruli have more advanced renal disease than those with mesangial Ox-LDL.[78] Dietary supplementation of lipid-soluble antioxidant, vitamin E or probucol, reduced the intensity of Ox-LDL staining[14] and attenuated the renal injury in rats with experimental FSGS.[14,80] These results suggest that lipid peroxidation of LDL plays a role in the progression of glomerulosclerosis.

Ox-LDL in the diseased glomeruli can recruit the circulating monocytes[81] leading to the accumulation of macrophages in the glomeruli, whereas antioxidants inhibit the process.[14] Activated macrophages secrete cytokines, growth factors, vasoactive substances, coagulation factors, reactive oxygen substances and proteolytic enzymes.[72] Furthermore, macrophages have a large number of scavenger receptors and accumulate Ox-LDL within cells, resulting in foam cell formation. When the foam cells die, the release of cytotoxic components could cause a loss of glomerular cells and stimulate extracellular matrix secretion eventually leading to glomerulosclerosis as originally proposed in atherogenesis.[71,82]

LDL Stimulates Extracellular Matrix Synthesis in Cultured Mesangial Cells

Glomerular mesangial matrix mainly consists of type IV collagen, laminin B1 and fibronectin.[83,84] Besides these extracellular matrix proteins, type I and type III collagens are present within the cell and in the medium in cultured mesangial cells.[85,86] LDL stimulated mRNA and/or protein expression of $\alpha 1$(I), $\alpha 1$(III) and $\alpha 1$(IV) collagen, fibronectin or laminin in cultured mesangial cells in a dose-dependent manner.[42,58,87-90] LDL during its incubation with cultured mesangial cells is oxidatively modified,[42,57-59] suggesting that oxidation is involved in LDL-induced mesangial matrix gene regulation. Cu^{++}-catalyzed Ox-LDL also stimulated the collagen or other extracellular matrix synthesis in mesangial cells,[42,90] whereas vitamin E and anti-Ox-LDL caused a marked reduction in collagen mRNA stimulated by LDL[42]

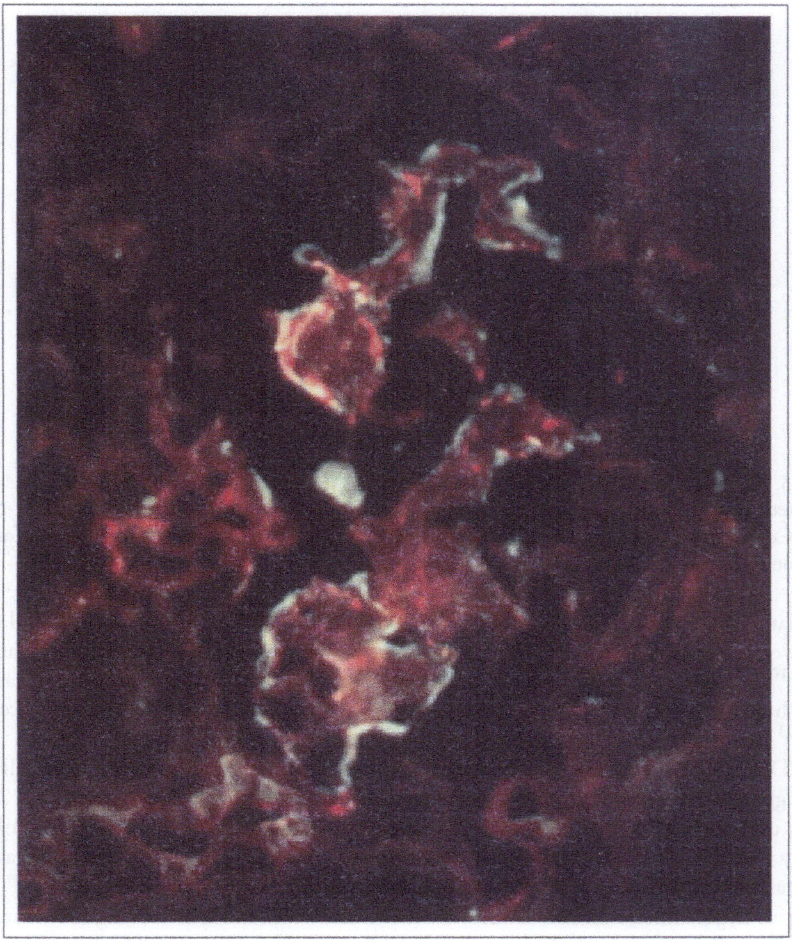

Figure 2. Dual immunofluorescence for oxidized LDL (Ox-LDL) (FITC-labeled and shown in green) and apolipoprotein B (apo B) (rhodamine-labeled and shown in red) in a glomerulus with FSGS. Apo B-containing lipoprotein and Ox-LDL are located in the sclerotic segments.

(Fig. 3). These results suggest that LDL, particularly lipid peroxidation products of LDL, may be implicated in the development of glomerulosclerosis by facilitating excessive mesangial matrix generation.

LDL Regulates Plasminogen Activator/Inhibitor in Cultured Mesangial Cells

The extracellular matrix can be degraded, and a distortion of the balance between extracellular matrix synthesis and turnover may result in an abnormal extracellular matrix accumulation in renal disease. Matrix metalloproteinases and plasminogen activator/plasmin system are known to play a key role in matrix degradation.[91-94] Cultured human mesangial cells produce urokinase-type plasminogen activator (uPA),[91,95] tissue-type plasminogen activator (tPA) and plasminogen activator inhibitor-1 (PAI-1).[96-98] Plasminogen is activated to plasmin by the

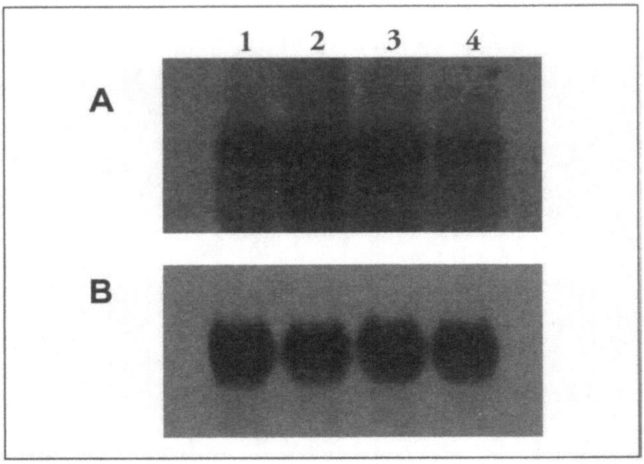

Figure 3. Northern blot analysis of α1(IV) collagen mRNA in human mesangial cells (HMC). Cells were incubated with serum-free DMEM alone (lane 1) or with the addition of 200 μg/ml LDL (lane 2), 100 μg/ml Cu^{++}-catalyzed Ox-LDL (lane 3), or both 200 μg/ml LDL and 50 μM vitamin E (lane 4) for 48 h. The blots were hybridized with [^{32}P]-labeled cDNAs for: α1(IV) collagen (A), and β-actin (B).

enzymatic activity of uPA and tPA, and plasmin mediates the degradation of extracellular matrix by cultured mesangial cells.[91,99] The rate of plasmin production is primarily regulated by PAI-1,[100] which also inhibits the plasmin-dependent activation of matrix metalloproteinases.[101] PAI-1 has been implicated in renal disease as being a mediator of extracellular matrix accumulation[93] and as a feedback mechanism to limit vascular fibrinolysis.[102,103]

LDL upregulated PAI-1, uPA and tPA expression in cultured human mesangial cells after prolonged incubation times, but short incubation times with LDL rather decreased 2.4 kb PAI-1, uPA and tPA mRNA levels[104] (Fig. 4). LDL-induced plasminogen activator inhibitory activity was greater than plasminogen activator activity in mesangial cells, suggesting that LDL causes a plasminogen activator/inhibitor imbalance favoring accumulation of matrix.[104] Thus, persistent hypercholesterolemia in chronic renal disease could lead to mesangial matrix accumulation and eventual renal fibrosis partly linked to the impaired matrix degradation.

Intracellular Signaling Process in Association with LDL-Induced Mesangial Matrix Synthesis

The stimulation by LDL of quiescent mesangial cells, which in turn oversecretes matrix proteins, might require a number of coordinated cellular events. Protein kinase C (PKC) is activated by physiological second messenger diacylglycerol, a product of phosphatidylinositol.[105] Upon stimulation of cell surface receptors, there is an immediate and transient increase in diacylglycerol levels. Tumor promoting phorbol esters, such as phorbol myristate acetate, possess a molecular structure that is similar to that of diacylglycerol, and cause a prolonged activation of PKC and downregulate the enzyme.

LDL induced two peaks of PKC activation in cultured human mesangial cells[104] (Fig. 5). In contrast to the transient first PKC activation,[88,89,104] the second one persisted for over 9 hours and gradually decreased to basal level at hour 18 after the start of LDL administration.[104] LDL increased membrane-associated PKC-α, -β1 and -δ, with a decreased cytosol content in mesangial cells.[104]

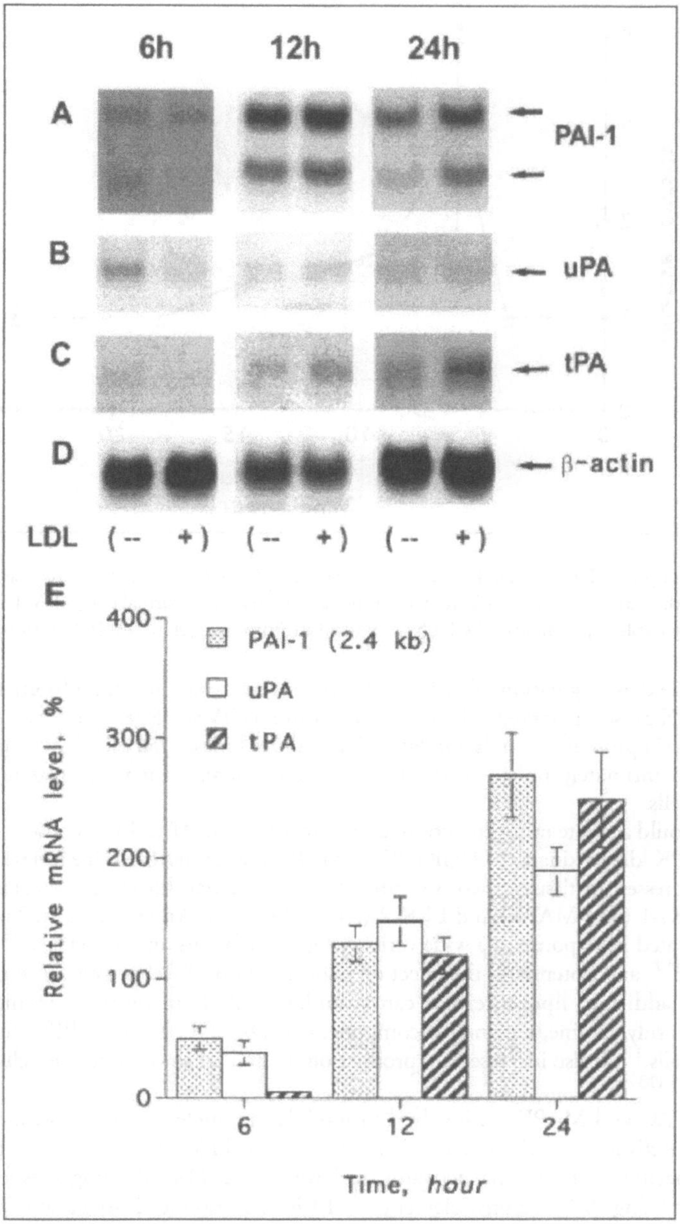

Figure 4. Northern blot analysis of plasminogen activator (PA) inhibitor-1 (PAI-1), urokinase-type PA (uPA) and tissue-type PA (tPA) mRNA in HMC. Cells were incubated with serum-free DMEM alone (-) or with the addition of 200 µg/ml LDL (+) for 6, 12, or 24 h. The blots were hybridized with [^{32}P]-labeled cDNAs for: PAI-1 (A), uPA (B), tPA (C), and β-actin (D). E) quantitative expression of PAI-1, uPA and tPA mRNA abundance after correcting for the β-actin signal. The PAI-1, uPA, and tPA mRNA levels of treated HMC are expressed as percentage increases above the mRNA levels of untreated controls. Values are means ± SD of 3 separate experiments. (From Song et al[104] with permission.)

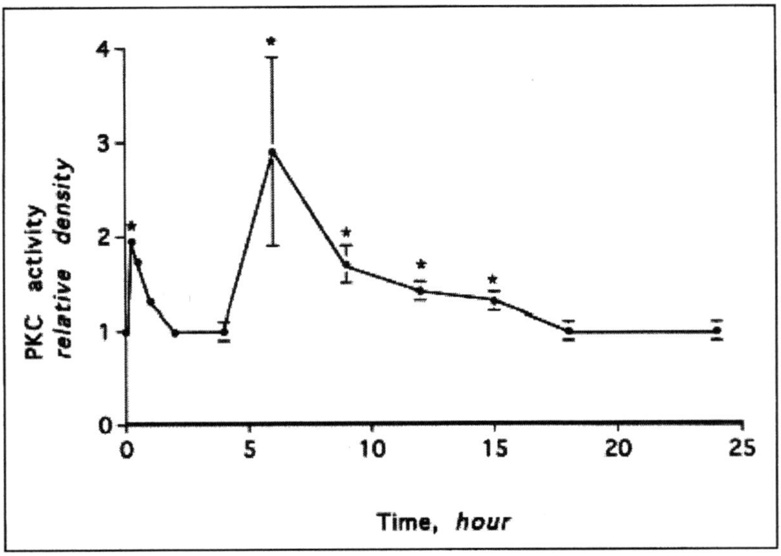

Figure 5. Effects of LDL on protein kinase C (PKC) activity of HMC. Data are expressed as density units, calculated as the ratio of PKC-specific activity of the sample to that of normal control. Values are means ± SD of 3 comparable experiments. *$P < 0.05$ vs. controls. (From Song et al ref. 104, with permission.)

Both the downregulation of PKC by phorbol myristate acetate and inhibition of PKC by GF-109203X caused a marked reduction in α1(I) and α1(IV) collagen and PAI-1 mRNA[88,104] and fibronectin protein[89] stimulation by LDL in cultured mesangial cells. These results suggest that PKC is intimately linked to the LDL-induced enhanced matrix protein synthesis in mesangial cells.

PKC could activate mitogen-activated protein kinase (MAPK) by a pathway that involves Raf-1 (MAPK kinase kinase, MAP3K).[106-108] MAPK is a serine/threonine kinase found ubiquitously expressed in tissues. Two isoforms of MAPK, extracellular signal-regulated protein kinase (ERK)-1 (p44 MAPK) and ERK-2 (p42 MAPK), are known to exist. These isoforms can be activated in response to a wide variety of growth factors and mitogens.[108] LDL stimulates ERK-2[109] and potentiates the effect of vasopression on ERK activation in rat mesangial cells.[110] In addition, lipoprotein(a) can stimulate MAPK in human mesangial cells.[111] Lysophosphatidylcholine, a principal component of Ox-LDL, activates PKC and MAPK in mesangial cells.[112] It also increases O_2^- production in vascular smooth muscle cells via stimulation of PKC.[113]

Both PKC and MAPK are involved in modulating nuclear events associated with gene expression. Activated protein-1 (AP-1) family, which includes the Jun and Fos proteins, is a sequence-specific transcriptional activator (reviewed in ref. 114). Binding sites for AP-1 were described as being PKC response elements.[115] ERK activation leads to elevated AP-1 activity via c-fos induction.[116] The ability of TGF-β to induce the expression of several genes such as PAI-1, type I collagen and TGF-β1 itself depends on specific AP-1 DNA-binding sites in the promoter regions of these genes.[117,118] Ox-LDL stimulates the DNA binding activity of AP-1 in cultured vascular smooth muscle cells,[119] suggesting that LDL oxidation may also be involved in activation of AP-1 in mesangial cells.

LDL Stimulates Mesangial Matrix Synthesis Through Induction of TGF-β Expression

TGF-β is a key factor to stimulate extracellular matrix deposition in the kidney,[120,121] and its overproduction has been implicated in the pathogenesis of glomerulosclerosis.[122] Intrarenal infusion of the TGF-β1 gene causes glomerulosclerosis[123] and antisense to TGF-β ameliorates experimental nephropathy.[124] In mouse mesangial cells, TGF-β stimulates collagen and fibronectin synthesis.[125]

TGF-β1 is secreted as a latent complex, which could be activated by reactive oxygen species.[126] Most so-called growth factors signal through a receptor tyrosine kinase. By contrast, the TGF-β family of intercellular mediators signals through an initial serine-threonine kinase.[127] Activated TGF-β binds to three different types of serine/threonine kinase receptor.[128] The type II receptor kinases are constitutively active: upon ligand binding, the type II receptors activate the type I receptor kinases,[129] which then activate intracellular substrates, such as Smad2 and Smad3. Activated or phosphorylated Smad2 and Smad3 form a complex with Smad4.[130] The complexes translocate to the nucleus and activate transcription of specific TGF-β target genes including the extracellular matrix component PAI-1, type I collagen, fibronectin and TGF-β1 itself.[131,132] TGF-β and Smad signaling pathway is present in human mesangial cells.[133] Smads are stimulated by TGF-β1 to generate mesangial collagen production.[133,134] In addition, the activated TGF-β type I receptor kinases can activate MAPK families.[134,135]

Cholesterol feeding to normal or nephrotic rats significantly increased glomerular TGF-β1, α1(IV) collagen, or fibronectin mRNA levels.[24,136] In cultured human mesangial cells, LDL stimulated TGF-β mRNA expression and bioactive TGF-β secretion.[88] The LDL-induced upregulation of α1(I) and α1(IV) collagen gene expression was inhibited by the application of anti-TGF-β neutralizing antibody, suggesting that TGF-β is intimately linked to the onset of enhanced collagen mRNA expression.[88] In murine mesangial cells exposed to LDL, anti-TGF-β also prevented the expected increase in fibronectin synthesis.[89] These results suggest that TGF-β functions as a key signaling mediator in the pathway by which LDL upregulates mesangial matrix synthesis. As yet it is not clear whether Smad signaling is activated by LDL in mesangial cells.

Effects of LDL on Glomerular Epithelial Cells

Glomerular visceral epithelial cells or podocytes are highly specialized terminally differentiated cells interlinked by slit diaphragms. They are known to play a key role in maintaining normal glomerular permselectivity. In various human glomerular diseases, apo B-rich lipoproteins accumulate not only in the mesangium but also within the glomerular basement membrane.[68] Takemura et al[137] claimed that glomerular epithelial cells express both LDL receptors and scavenger receptors in human nephritic kidneys. Previous animal studies suggested that the dietary hypercholesterolemia mainly activates glomerular mesangial cells.[12,24,25] Joles et al,[138] however, recently suggested that lipid abnormalities by cholesterol feeding in uninephrectomized rats aggravate renal injury primarily via podocyte rather than via mesangial cell damage. In rats with high cholesterol diet, increased expression of heat shock protein 47, a collagen-specific stress protein, was demonstrated mainly in glomerular epithelial cells together with increased collagen expression, suggesting that phenotypically altered glomerular epithelial cells by hypercholesterolemia might play a role in renal fibrosis.[139]

Cultured human glomerular epithelial cells take up LDL via a receptor-mediated pathway.[140,141] Ox-LDL stimulates fibronectin expression by a mechanism involving expression of

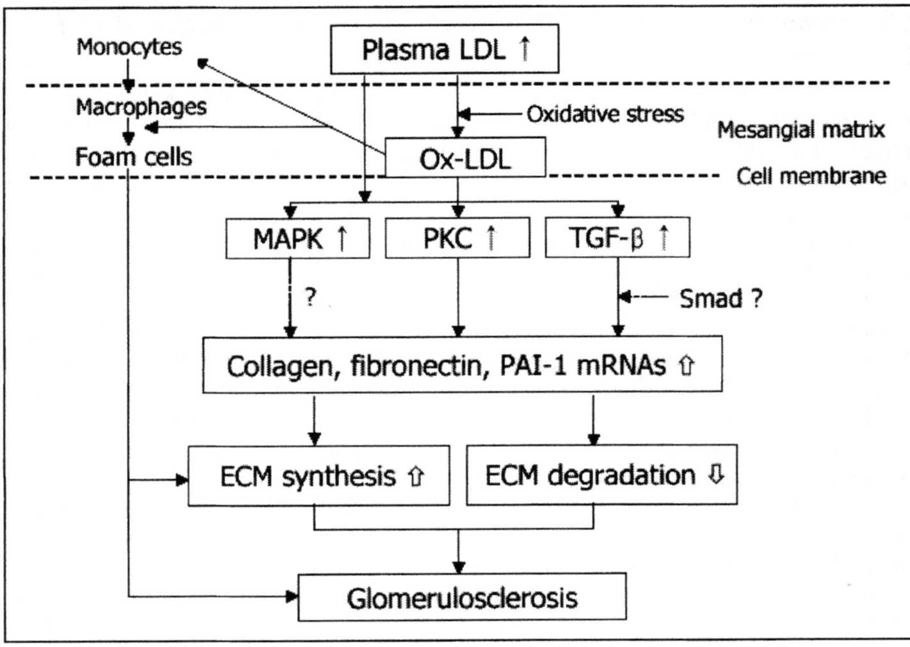

Figure 6. Schematic diaphragm depicting how LDL could lead to mesangial matrix accumulation in relation to the development of glomerulosclerosis. LDL, particularly Ox-LDL, increases PKC and mitogen-activated protein kinase (MAPK) activity and bioactive transforming growth factor-β (TGF-β) secretion in mesangial cells leading to the enhanced expression of collagen, fibronectin and PAI-1. Also, Ox-LDL in the diseased glomeruli could recruit the circulating monocytes leading to the accumulation of macrophages and subsequent foam cell formation resulting in an altered mesangial cell matrix biosynthesis. ECM, extracellular matrix.

TGF-β in cultured human glomerular epithelial cells.[142] Nonetheless, the selective analysis of podocytes in vitro was difficult, because only undifferentiated podocytes of questionable cellular origin were available in culture.[143-145] Recently, Saleem et al[146] have developed a conditionally immortalized human podocyte cell line demonstrating markers of differentiated in vivo podocytes, such as nephrin, podocin, CD2AP and synaptopodin. There is growing evidence suggesting that the damage to glomerular visceral epithelial cells is related to the development of FSGS.[147-151] Thus, these cells in culture expressing podocytic markers would be valuable in the future to examine whether LDL-induced podocyte injury is related to subsequent renal fibrosis.

Concluding Remarks

As shown in Figure 6, LDL is susceptible to oxidative modification by activated mesangial cells or macrophages. LDL, particularly Ox-LDL, increases PKC and MAPK activity as well as bioactive TGF-β secretion in mesangial cells leading to the enhanced expression of collagen, fibronectin and PAI-1. Also, Ox-LDL in the diseased glomeruli could recruit the circulating monocytes leading to the accumulation of macrophages and subsequent foam cell formation. Increased release of macrophage-derived products could produce an altered mesangial cell matrix biosynthesis. In conclusion, these effects of LDL on altered regulation of the mesangial matrix synthesis/degradation might have a pathophysiological function in the pathogenesis of glomerulosclerosis.

References

1. Goldstein JL, Brown MS. The low-density lipoprotein pathway and its relation to atherosclerosis. Ann Rev Biochem 1977; 46:897-930.
2. Gotto AM Jr, Pownall HJ, Havel RJ. Introduction to the plasma membrane. Methods Enzymol 1986; 128:3-41.
3. Piha M, Lindstedt L, Kovanen PT. Fusion of proteolyzed low-density lipoprotein in the fluid phase. A novel mechanism generating atherogenic lipoprotein particles. Biochemistry 1995; 34:10120-10129.
4. Kruth HS. Cholesterol deposition in atherosclerotic lesions. In: Robert Bittman, ed. Subcellular Biochemistry Cholesterol: Its functions and metabolism in biology and medicine. New York: Plenum Press; 1997:319-362.
5. Appel GB, Blum CB, Chien S et al. The hyperlipidemia of the nephrotic syndrome. Relation to plasma albumin concentration, oncotic pressure, and viscosity. N Engl J Med 1985; 312:1544-1548.
6. Joven J, Villabona C, Vilella E et al. Abnormalities of lipoprotein metabolism in patients with the nephrotic syndrome. N Engl J Med 1990; 323:579-584.
7. Warwick GL, Caslake MJ, Boulton-Jones JM et al. Low-density lipoprotein metabolism in the nephrotic syndrome. Metabolism 1990; 39:187-192.
8. Wheeler DC, Bernard DB. Lipid abnormalities in the nephrotic syndrome: causes, consequences and treatment. Am J Kidney Dis 1994; 23:331-346.
9. Diamond JR, Karnovsky MJ. Exacerbation of chronic aminonucleoside nephrosis by dietary cholesterol supplementation. Kidney Int 1987; 32:671-677.
10. Gröne H-J, Walli A, Gröne E et al. Induction of glomerulosclerosis by dietary lipids. A functional and morphologic study in the rat. Lab Invest 1989; 60:433-446.
11. Kasiske BL, O'Donnell MP, Cleary MP et al. Effects of reduced renal mass on tissue lipids and renal injury in hyperlipidemic rats. Kidney Int 1989; 35:40-47.
12. Kasiske BL, O'Donnell MP, Schmitz PG et al. Renal injury of diet-induced hypercholesterolemia in rats. Kidney Int 1990; 37:880-891.
13. Pesek-Diamond I, Ding G, Frye J et al. Macrophages mediate adverse effects of cholesterol feeding in experimental nephrosis. Am J Physiol (Renal Fluid Electrolyte Physiol. 32) 1992; 263:F776-F783.
14. Lee HS, Jeong JY, Kim BC et al. Dietary antioxidant inhibits lipoprotein oxidation and renal injury in experimental focal segmental glomerulosclerosis. Kidney Int 1997; 51:1151-1159.
15. Lee HS, Spargo BH. Significance of renal hyaline arteriolosclerosis in focal and segmental glomerulosclerosis. Nephron 1985; 41:86-93.
16. Grond J, van Goor H, Elema JD. Histochemical analysis of focal segmental hyalinosis and sclerosis lesions in various rat models of experimental nephrotic syndrome. Kidney Int 1986; 29:945-946.
17. Magil AB, Cohen AH. Monocytes and focal glomerulosclerosis. Lab Invest 1989; 61:404-409.
18. Moorhead JF, Chan MK, El-Nahas M et al. Lipid nephrotoxicity in chronic progressive glomerular and tubulointerstitial disease. Lancet 1982; 2:1309-1311.
19. Diamond JR, Karnovsky MJ. Focal and segmental glomerulosclerosis: Analogies to atherosclerosis. Kidney Int 1988; 33:917-924.
20. Keane WF, Kasiske BL, O'Donnell MP, Kim Y. The role of altered lipid metabolism in the progression of renal disease: experimental evidence. Am J Kidney Dis 1991; 17(Suppl 1):38-42.
21. Keane WF. Lipids and the kidney. Kidney Int 1994; 46:910- 920.
22. Gröne EF, Walli AK, Gröne HJ et al. The role of lipids in nephrosclerosis and glomerulosclerosis. Atherosclerosis 1994; 107:1-13.
23. Chapman MJ. Animal lipoproteins: Chemistry, structure and comparative aspects. J Lipid Res 1980; 21:789-853.
24. Guijarro C, Kasiske BL, Kim Y et al. Early glomerular changes in rats with dietary-induced hypercholesterolemia. Am J Kidney Dis 1995; 26:152-161.
25. Al-Shebeb T, Frohlich J, Magil AB. Glomerular disease in hypercholesterolemic guinea pigs: A pathogenetic study. Kidney Int 1988; 33:498-507.
26. Tolins JP, Stone BF, Raij L. Interactions of hypercholesterolemia and hypertension in initiation of glomerular injury. Kidney Int 1992; 41:1254-1261.

27. Rayner HC, Ward L, Walls J. Cholesterol feeding following unilateral nephrectomy in the rat leads to glomerular hypertrophy. Nephron 1991; 57:453-459.
28. Carew TE, Pittman RC, Steinberg D. Tissue sites of degradation of native and reductively methylated [^{14}C] sucrose-labeled low density lipoprotein in rats. Contribution of receptor-dependent and receptor-independent pathways. J Biol Chem 1982; 257:8001-8008.
29. Goldstein JL, Brown MS, Anderson RGW et al. Receptor-mediated endocytosis: concepts emerging from the LDL receptor system. Ann Rev Cell Biol 1985; 1:1-39.
30. Wasserman J, Santiago A, Rifici V et al. Interactions of low density lipoprotein with rat mesangial cells. Kidney Int 1989; 35:1168-1174.
31. Wheeler DC, Fernando RI, Gillett MPT et al. Characterization of the binding of low-density lipoproteins to cultured rat mesangial cells. Nephrol Dial Transplant 1991; 6:701-708.
32. Gupta S, Rifici V, Crowley S et al. Interactions of LDL and modified LDL with mesangial cells and matrix. Kidney Int 1992; 41:1161-1169.
33. Gröne EF, Abboud HE, Höhne M et al. Actions of lipoproteins in cultured human mesangial cells: modulation by mitogenic vasoconstrictors. Am J Physiol (Renal Fluid Electrolyte Physiol) 1992; 263:F686-F696.
34. Lee HS, Koh HI. Visualization of binding and uptake of oxidized low density lipoproteins by cultured mesangial cells. Lab Invest 1994; 71:200-208.
35. Brown MS, Goldstein JL. Lipoprotein metabolism in the macrophage: implications for cholesterol deposition in atherosclerosis. Annu Rev Biochem 1983; 52:223-261.
36. Schreiner GF, Unanue ER. Origin of the rat mesangial phagocyte and its expression of the leukocyte common antigen. Lab Invest 1984; 51:515-523.
37. Striker GE, Striker LJ. Glomerular cell culture. Lab Invest 1985; 53:122-131.
38. Schlondorff D. Cellular mechanisms of lipid injury in the glomerulus. Am J Kidney Dis 1993; 22:72-82.
39. Moorhead JF. Lipids and the pathogenesis of kidney disease. Am J Kidney Dis 1991; 17(Suppl1):65-70.
40. Moorhead JF, Brunton C, Varghese Z. Glomerular atherosclerosis. Miner Electrolyte Metab 1997; 23:287-290.
41. Ruan XZ, Varghese Z, Powis SH et al. Dysregulation of LDL receptor under the influence of inflammatory cytokines: A new pathway for foam cell formation. Kidney Int 2001; 60:1716-1725.
42. Lee HS, Kim BC, Kim YS et al. Involvement of oxidation in LDL-induced collagen gene regulation in mesangial cells. Kidney Int 1996; 50:1582-1590.
43. Schultz PJ, DiCorleto PE, Silver BJ et al. Mesangial cells express PDGF mRNAs and proliferate in response to PDGF. Am J Physiol 1988; 255:F674-F684.
44. Silver BJ, Jaffer FE, Abboud HE. Platelet-derived growth factor synthesis in mesangial cells: induction by multiple peptide mitogens. Proc Natl Acad Sci USA 1989; 86:1056-1058.
45. Abboud HE. Platelet-derived growth factor and mesangial cells. Kidney Int 1992; 41:581-583.
46. Abboud HE, Grandaliano G, Pinzani M et al. Actions of platelet-derived growth factor isoforms in mesangial cells. J Cell Physiol 1994; 158:140-150.
47. Doi T, Vlassara H, Kirstein M et al. Receptor-specific increase in extracellular matrix production in mouse mesangial cells by advanced glycosylation end products is mediated via platelet-derived growth factor. Proc Natl Acad Sci USA 1992; 89:2873-2877.
48. Yamabe H, Osawa H, Kaizuka M et al. Platelet-derived growth factor, basic fibroblast growth factor, and interferon γ increase type IV collagen production in human fetal mesangial cells via a transforming growth-factor-β-dependent mechanism. Nephrol Dial Transplant 2000; 15:872-876.
49. Johnson RJ, Raines E, Floege J et al. Inhibition of mesangial cell proliferation and matrix expansion in glomerulonephritis in the rat by antibody to platelet-derived growth factor. J Exp Med 1992; 175:1413-1416.
50. Floege J, Eng E, Young BA et al. Infusion of platelet-derived growth factor or basic fibroblast growth factor induces elective glomerular mesangial cell proliferation and matrix accumulation in rats. J Clin Invest 1993; 92:2952-2962.
51. Floege J, Ostendorf T, Janssen U et al. Novel approach to specific growth factor inhibition in vivo: antagonism of platelet-derived growth factor in glomerulonephritis by aptamers. Am J Pathol 1999; 154:169-179.

52. Baud L, Ardaillou R. Reactive oxygen species: production and role in the kidney. Am J Physiol 1986; 251:F765-F776.
53. Diamond JR. The role of reactive oxygen species in animal models of glomerular disease. Am J Kidney Dis 1992; 19:292-300.
54. Babior BM, Kipnes RS, Curnutte JT. Biological defense mechanisms: the production of leukocytes of superoxide, a potential bactericidal agent. J Clin Invest 1973; 52:741-744.
55. Rossi F. The O_2^- forming NADPH oxidase of the phagocytes: nature, mechanisms of activation and function. Biochim Biophys Acta 1986; 853:65-89.
56. Radeke HH, Cross AR, Hancock JT et al. Functional expression of NADPH oxidase components (α- and β-subunits of cytochrome b558 and 45-kDa flavoprotein) by intrinsic human glomerular mesangial cells. J Biol Chem 1991; 266:21025-21029.
57. Keane WF, O'Donnell MP, Kasiske BL et al. Oxidative modification of low-density lipoproteins by mesangial cells. J Am Soc Nephrol 1993; 4:187-194.
58. Rovin BH, Tan LC. LDL stimulates mesangial fibronectin production and chemoattractant expression. Kidney Int 1993; 43:218-225.
59. Wheeler DC, Chana RS, Topley N et al. Oxidation of low density lipoprotein by mesangial cells may promote glomerular injury. Kidney Int 1994; 45:1628-1636.
60. Radeke HH, Meier B, Topley N et al. Interleukin I-α and tumor necrosis factor-α induce oxygen radical production in mesangial cells. Kidney Int 1990; 37:767-775.
61. Esterbauer H, Jürgens G, Quehenberger O et al. Autooxidation of human low density lipoprotein: loss of polyunsaturated fatty acids and vitamin E and generation of aldehydes. J Lipid Res 1987; 28:495-509.
62. Steinbrecher UP. Oxidation of human low density lipoprotein results in derivatization of lysine residues of apolipoprotein B by lipid peroxide decomposition products. J Biol Chem 1987; 262:3603-3608.
63. Yokoyama M, Hirata K, Miyake R et al. Lysophosphatidylcholine: essential role in the inhibition of endothelium-dependent vasorelaxation by oxidized low density lipoprotein. Biochim Biophys Res Commun 1990; 168:301-308.
64. Liu S-Y, Lu X, Choy S et al. Alteration of lysophosphatidylcholine content in low density lipoprotein after oxidative modification: relationship to endothelium dependent relaxation. Cardivasc Res 1994; 28:1476-1481.
65. Parthasarathy S, Steinbrecher U, Barnett J et al. Essential role of phospholipase A_2 activity in endothelial cell-induced modification of low density lipoprotein. Proc Natl Acad Sci USA 1985; 82:3000-3004.
66. Galle J, Heermeier K. Angiotensin II and oxidized LDL: an unholy alliance creating oxidative stress. Nephrol Dial Transplant 1999; 14:2585-2589.
67. Hajjar DP, Haberland ME. Lipoprotein trafficking in vascular cells. J Biol Chem 1997; 272:22975-22978.
68. Lee HS, Lee JS, Koh HI et al. Intraglomerular lipid deposition in routine biopsies. Clin Nephrol 1991; 36:67-75.
69. Hoover GA, McCormick S, Kalant N. Interaction of native and cell-modified low density lipoprotein with collagen gel. Arteriolosclerosis 1988; 8:525-534.
70. Srinivasan SR, Vijayagopal P, Eberle K et al. Low-density lipoprotein binding affinity of arterial wall proteoglycans: characteristics of a chondroitin sulfate proteoglycan subfraction. Biochim Biophys Acta 1989; 1006:159-166.
71. Steinberg D, Parthasarathy S, Carew TE et al. Beyond cholesterol. Modification of low-density lipoprotein that increases its atherogenicity. N Engl J Med 1989; 320:915-924.
72. Nathan CF. Secretory products of macrophages. J Clin Invest 1987; 79:319-326.
73. Rie J, Silbiger S, Neugarten J. Glomerular macrophages in nephrotoxic serum nephritis are activated to oxidize low-density lipoprotein. Am J Kidney Dis 1995; 26:362-367.
74. Richard MJ, Arnaud J, Jurkovitz C et al. Trace elements and lipid peroxidation abnormalities in patients with chronic renal failure. Nephron 1991; 57:10-15.
75. Maggi E, Bellazzi R, Gazo A et al. Autoantibodies against oxidatively-modified LDL in uremic patients undergoing dialysis. Kidney Int 1994; 46:869-876.

76. Maggi E, Beliazzi R, Falschi F et al. Enhanced LDL oxidation in uremic patients: An additional mechanism for accelerated atherosclerosis? Kidney Int 1994; 45:876-883.
77. Sutherland WHF, Walker RJ, Ball M et al. Oxidation of low density lipoproteins from patients with renal failure or renal transplants. Kidney Int 1995; 48:227- 236.
78. Lee HS, Kim YS. Identification of oxidized low density lipoprotein in human renal biopsies. Kidney Int 1998; 54:848-856.
79. Magil AB, Frohlich JJ, Innis SM et al. Oxidized low-density lipoprotein in experimental focal glomerulosclerosis. Kidney Int 1993; 43:1243-1250.
80. Trachtman H, Schwob N, Maesaka J et al. Dietary vitamin E supplementation ameliorates renal injury in chronic puromycin aminonucleoside nephropathy. J Am Soc Nephrol 1995; 5:1811-1819.
81. Quinn MT, Parthasarathy S, Fong LG et al. Oxidatively modified low density lipoproteins: a potential role in recruitment and retention of monocyte/macrophages during atherogenesis. Proc Natl Acad Sci USA 1987; 84:2995-2998.
82. Brown MS, Goldstein JL. Scavenging for receptors. Nature 1990; 343:508-509.
83. Scheinman JI, Foidart J-M, Michael AF. The immunohistology of glomerular antigens. V. The collagenous antigens of the glomerulus. Lab Invest 1980; 43:373-381.
84. Ebihara I, Suzuki S, Nakamura T et al. Extracellular matrix component mRNA expression in glomeruli in experimental focal glomerulosclerosis. J Am Soc Nephrol 1993; 3:1387-1397.
85. Ardaillou N, Bellon G, Nivez MP et al. Quantification of collagen synthesis by cultured human glomerular cells. Biochim Biophys Acta 1989; 991:445-452.
86. Scheinman JI, Tanaka H, Haralson M et al. Specialized collagen mRNA and secreted collagens in human glomerular epithelial, mesangial, and tubular cells. J Am Soc Nephrol 1992; 2:1475-1483.
87. Kim SB, Kang SA, Cho YJ et al. Effects of low density lipoprotein on type IV collagen production by cultured rat mesangial cells. Nephron 1994; 67:327-333.
88. Lee HS, Kim BC, Hong HK et al. LDL stimulates collagen mRNA synthesis in mesangial cells through induction of PKC and TGF-β expression. Am J Physiol (Renal Physiol 46) 1999; 277:F369-F376.
89. Studer RK, Craven PA, DeRubertis FR. Low-density lipoprotein stimulation of mesangial cell fibronectin synthesis: Role of protein kinase C and transforming growth factor-β. J Lab Clin Med 1995; 125:86-95.
90. Roh DD, Kamanna V, Kirschenbaum MA. Oxidative modification of low-density lipoprotein synthesis enhances mesangial cell protein synthesis and gene expression of extracellular matrix proteins. Am J Nephrol 1998; 18:344-350.
91. Baricos WH, Cortez SL, El-Dahr SS et al. ECM degradation by cultured human mesangial cells is mediated by a PA/plasmin/MMP-2 cascade. Kidney Int 1995; 47:1039-1047.
92. Stetler-Stevenson WG. Dynamics of matrix turnover during pathologic remodeling of the extracellular matrix. Am J Pathol 1996; 148:1345-1350.
93. Rerolle J-P, Hertig A, Nguyen G et al. Plasminogen activator inhibitor type I is a potential target in renal fibrogenesis. Kidney Int 2000; 58:1841-1850.
94. Martin J, Eynstone L, Davies M et al. Induction of metalloproteinases by glomerular mesangial cells stimulated by proteins of the extracellular matrix. J Am Soc Nephrol 2001; 12:88-96.
95. Glass II WF, Kreisberg JI, Troyer DA. Two-chain urokinase, receptor, and type I inhibitor in cultured human mesangial cells. Am J Physiol 1993; 264:F532-F539.
96. Lacave R, Rondeau E, Ochi S et al. Characterization of a plasminogen activator and its inhibitor in human mesangial cells. Kidney Int 1989; 35:806-811.
97. Hagege J, Peraldi MN, Rondeau E et al. Plasminogen activator inhibitor 1 deposition in the extracellular matrix of cultured mesangial cells. Am J Pathol 1992; 141:117-128.
98. Peraldi MN, Rondeau E, Medcalf RL et al. Cell-specific regulation of plasminogen activator inhibitor 1 and tissue type plasminogen activator release by human kidney mesangial cells. Biochim Biophys Acta 1992; 1134:189-196.
99. Wong AP, Cortez SL, Baricos WH. Role of plasmin and gelatinase in extracellular matrix degradation by cultured rat mesangial cells. Am J Physiol 1992; 263:F1112-F1118.
100. Sprengers ED, Kluft C. Plasminogen activator inhibitors. Blood 1987; 69:381-387.
101. Murphy G, Ward R, Gavrilovic J et al. Physiological mechanisms for metalloproteinase activation. Matrix Suppl 1992; 1:224-230.

102. Rondeau E, Mougenot B, Lacave R et al. Plaminogen activator inhibitor I in renal fibrin deposits of human nephropathies. Clin Nephrol 1990; 33:55-60.
103. Lee HS, Park SY, Moon KC et al. mRNA expression of urokinase and plasminogen activator inhibitor-1 in human crescentic glomerulonephritis. Histopathology 2001; 39:182-188.
104. Song CY, Kim BC, Hong HK et al. Biphasicregulation of plasminogen activator/inhibitor by LDL in mesangial cells. Am J Physiol Renal Physiol 2002; 283:H423-H430.
105. Nishizuka Y. Protein kinase C and lipid signaling for sustained cellular responses. FASEB J 1995; 9:484-496.
106. Howe CR, Leevers SJ, Gomez N et al. Activation of the MAP kinase pathway by the protein kinase raf. Cell 1992; 71:335-342.
107. Kyriakis JM, App H, Zhang X-F et al. Raf-1 activates MAP kinase kinase. Nature 1992; 358:417-421.
108. Seger R, Krebs EG. The MAPK signaling cascade. FASEB J 1995; 9:726-735.
109. Bassa BV, Roh DD, Kirschenbaum MA et al. Atherogenic lipoproteins stimulate mesangial cell p42 mitogen-activated protein kinase. J Am Soc Nephrol 1998; 9:488-496.
110. Ishikawa S, Kawasumi M, Okada K et al. Low density lipoprotein enhances the cellular action of arginine vasopressin in rat glomerular mesangial cells in culture. J Clin Invest 1994; 93:2710-2717.
111. Mondorf UF, Piiper A, Herrero M et al. Lipoprotein(a) stimulates mitogen activated protein kinase in human mesangial cells. FEBS Lett 1998; 441:205-208.
112. Bassa BV, Roh DD, Vaziri ND et al. Lysophosphatidylcholine activates mesangial cell PKC and MAP kinase by PLCγ-1 and tyrosine kinase-Ras pathways. Am J Physiol (Renal Physiol. 46) 1999; 277:F328-F337.
113. Ohara Y, Peterson TE, Zheng B et al. Lysophosphatidylcholine increases vascular superoxide anion production via protein kinase C activation. Arterioscler Thromb 1994; 14:1007-1013.
114. Angel P, Karin M. The role of Jun, Fos and the AP-1 complex in cell-proliferation and transformation. Biochim Biophys Acta 1991; 1072:129-157.
115. Angel P, Allegretto EA, Okino ST et al. Oncogene jun encodes a sequence-specific trans-activator similar to AP-1. Nature 1988; 332:166-171.
116. Karin M. The regulation of AP-1 activity by mitogen- activated protein kinases. J Biol Chem 1995; 270:16483-16486.
117. Keeton MR, Curriden SA, van Zonneveld AJ et al. Identification of regulatory sequences in the type I plasminogen activator inhibitor gene responsive totransforming growth factor beta. J Biol Chem 1991; 266:23048-23052.
118. Chang E, Goldberg H. Requirements for transforming growth factor-β regulation of the pro-α2(I) collagen and plasminogen activator inhibitor-1 promotors. J Biol Chem 1995; 270:4473-4477.
119. Ares MPS, Kallin B, Eriksson P et al. Oxidized LDL induces transcription factor activator protein-1 but inhibits activation of nuclear factor-kappa B in human vascular smooth muscle cells. Arterioscler Thromb Vasc Biol 1995; 15:1584-1590.
120. Roberts AB, Heine VI, Flanders KC et al. Transforming growth factor-β. Major role in regulation of extracellular matrix. Ann NY Acad Sci 1990; 580:225-232.
121. Border WA, Noble NA. Transforming growth factor β in tissue fibrosis. N Engl J Med 1994; 331:1286-1292.
122. Kopp JB, Factor VM, Mozes M et al. Transgenic mice with increased plasma levels of TGF-β1 develop progressive renal disease. Lab Invest 1996; 74:991-1003.
123. Isaka Y, Fujiwara Y, Ueda N et al. Glomerulosclerosis induced by in vivo transfection of transforming growth factor-β or platelet-derived growth factor gene into the rat kidney. J Clin Invest 1993; 92:2597-2601
124. Akagi Y, Isaka Y, Arai M et al. Inhibition of TGF-β1 expression by antisense oligonucleotides suppressed extracellular matrix accumulation in experimental glomerulonephritis. Kidney Int 1996; 50:148-155.
125. MacKay K, Striker LJ, Stauffer JW et al. Transforming growth factor-β. Murine glomerular receptors and responses of isolated glomerular cells. J Clin Invest 1989; 83:1160-1167.
126. Barcellos-Hoff MH, Dix TA. Redox-mediated activation of latent transforming growth factor-β1. Mol Endocrinol 1996; 10:1077-1083.

127. Piek E, Heldin C-H, ten Dijke P. Specificity, diversity and regulation of TGF-β superfamily signaling. FASEB J 1999; 13:2105-2124.
128. Blobe GC, Schiemann WP, Lodish HF. Role of transforming growth factor β in human disease. N Engl J Med 2000; 342:1350-1358.
129. Wrana JL, Attisano L, Wieser R et al. Mechanism of activation of the TGF-β receptor. Nature 1994; 370:341-347.
130. Nakao A, Imamura T, Souchelnytskyi S et al. TGF-β receptor-mediated signalling througn Smad2, Smad3 and Smad4. EMBO J 1997; 16:5353-5362.
131. Dennler S, Itoh S, Vivien D et al. Direct binding of Smad3 and Smad4 to critical TGFβ-inducible elements in the promotor of human plasminogen activator inhibitor-type I gene. EMBO J 1998; 17:3091-3100.
132. Massague J, Chen YG. Controlling TGF-β signalling. Genes Dev 2000; 14:627-644.
133. Poncelet A-C, de Caestecker MP, Schnaper HW. The TGF-β/SMAD signaling pathway is present and functional in human mesangial cells. Kidney Int 1999; 56:1354-1365.
134, Hayashida T, Poncelet A-C, Hubchak SC et al. TGF-β1 activates MAP kinase in human mesangial cells: A possible role in collagen expression. Kidney Int 1999; 56:1710-1720.
135. Sano Y, Harada J, Tashiro S et al. ATF-2 is a common nuclear target of Smad and TAK1 pathways in transforming growth factor-beta signaling. J Biol Chem 1999; 274:2761-2767.
136. Ding G, Pesek-Diamond I, Diamond JR. Cholesterol, macrophages, and gene expression of TGF-β1 and fibronectin during nephrosis. Am J Physiol (Renal Fluid Electrolyte Physiol. 33) 1993; 264:F577-F584.
137. Takemura T, Yoshioka K, Aya N et al. Apolipoproteins and lipoprotein receptors in glomeruli in human kidney diseases. Kidney Int 1993; 43:918-927.
138. Joles JA, Kunter U, Janssen U et al. Early mechanisms of renal injury in hypercholesterolemic or hypertriglyceridemic rats. J Am Soc Nephrol 2000; 11:669-683.
139. Razzaque MS, Taguchi T. Role of glomerular epithelial cell-derived heat shock protein 47 in experimental lipid nephropathy. Kidney Int 1999; 56(Suppl 71):S256-S259
140. Gröne H-J, Walli AK, Gröne E et al. Receptor mediated uptake of apo B and apo E rich lipoproteins by human glomerular epithelial cells. Kidney Int 1990; 37:1449-1459.
141. Rayner HC, Horsburgh T, Brown SL et al. Receptor-mediated endocytosis of low-density lipoprotein by cultured human glomerular cells. Nephron 1990; 55:292-299.
142. Ding G, van Goor H, Ricardo SD et al. Oxidized LDL stimulates the expression of TGF-β and fibronectin in human glomerular epithelial cells. Kidney Int 1997; 51:147-154.
143. Holthöfer H, Saino K, Miettinen A. Rat glomerular cells do not express podocytic markers when cultured in vitro. Lab Invest 1991; 65:548-557.
144. Weinstein T, Cameron R, Katz A et al. Rat glomerular epithelial cells in culture express characteristics of parietal, not visceral, epithelium. J Am Soc Nephrol 1992; 3:1279-1287.
145. Orikasa M, Iwagana T, Takahashi-Iwanaga H et al. Macrophagic cells outgrowth from normal rat glomerular culture: possible metaplastic change from podocytes. Lab Invest 1996; 75:719-733.
146. Saleem MA, O'Hare MJ, Reiser J et al. A conditionally immortalized human podocyte cell line demonstrating nephrin and podocin expression. J Am Soc Nephrol 2002; 13:630-638.
147. Fries JWU, Sandstrom DJ, Meyer TW, Rennke HG. Glomerular hypertrophy and epithelial cell injury modulate progressive glomerulosclerosis in the rat. Lab Invest 1989; 60:205-218.
148. Kriz W, Elger M, Nagata M et al. The role of podocytes in the development of glomerular sclerosis. Kidney Int. 1994; 45(Suppl 42):S64-S72.
149. Floege J, Kriz W, Schulze M et al. Basic fibroblast growth factor augments podocyte injury and induces glomerulosclerosis in rats with experimental membranous nephropathy. J Clin Invest 1995; 96:2809-2819.
150. Barisoni L, Kriz W, Mundel P et al. The dysregulated podocyte phenotype: A novel concept in the pathogenesis ofcollapsing idiopathic focal segmental glomerulosclerosis and HIV-associated nephropathy. J Am Soc Nephrol 1999; 10:51-61.
151. Kim HW, Moon KC, Park SY et al. Differential expression of platelet-derived growth factor and transforming growth factor-β in relation to progression of IgA nephropathy. Nephrology 2002; 7(Suppl):S131-S139.

Chapter 6

Molecular Developments in the Treatment of Renal Fibrosis

Gavin J. Becker and Tim D. Hewitson

Abstract

Progressive renal disease is associated with the development of fibrosing lesions not only in the glomerulus, but also in the interstitial and vascular compartments of the kidney, in a process that involves the mesenchymally derived, phenotypically similar, mesangial cell, myofibroblast and vascular smooth muscle cell. The similarities in the pathogenesis of all three processes means that the search for rational treatment strategies for any one may be of universal benefit to the others.

Potential therapeutic strategies target fibrosis both indirectly and directly. Indirect therapies alter the environment the kidney operates in such as by controlling blood pressure, hyperlipidemia and hyperglycaemia. As our understanding of the mechanisms of fibrosis increase, we are developing more direct treatment strategies that target the vasoactive mediators, growth factors and cell signaling pathways that regulate renal fibrogenesis. Finally attempts to increase collagen degradation and maintain blood supply are likely to reduce the damage resulting from aberrant collagen synthesis.

The continuing advances in cellular and molecular biology mean that we are becoming more aware of how cells interrelate with each other and their environment. Measures that specifically interfere with fibrosis can therefore be expected to improve prognosis not only in progressive renal failure but also in progressive fibrosing diseases in many other organs.

Progressive renal disease is associated with the concurrent development of fibrosing lesions not only in the glomerulus, but also in the interstitial and vascular compartments of the kidney (Fig. 1). Though attention is usually directed separately to the three processes—glomerulosclerosis, tubulointerstitial fibrosis and vascular sclerosis—all three eventually occur and the fundamental pathology is similar.[1]

Renal fibrosis or sclerosis refers to the replacement of renal parenchyma with connective tissue, in a process that resembles the generalised chronic inflammation that occurs elsewhere. Initiating injury, recruitment of inflammatory cells (neutrophils, macrophages, T-cells), generation and release of profibrotic growth factors, proliferation and matrix synthesis, and finally matrix remodelling are the sequential but overlapping events. Increases in both the number and activity of matrix producing cells is responsible for matrix deposition, with the balance between this and remodelling determining the extent of scarring.

Fibrogenesis: Cellular and Molecular Basis, edited by Mohammed S. Razzaque.

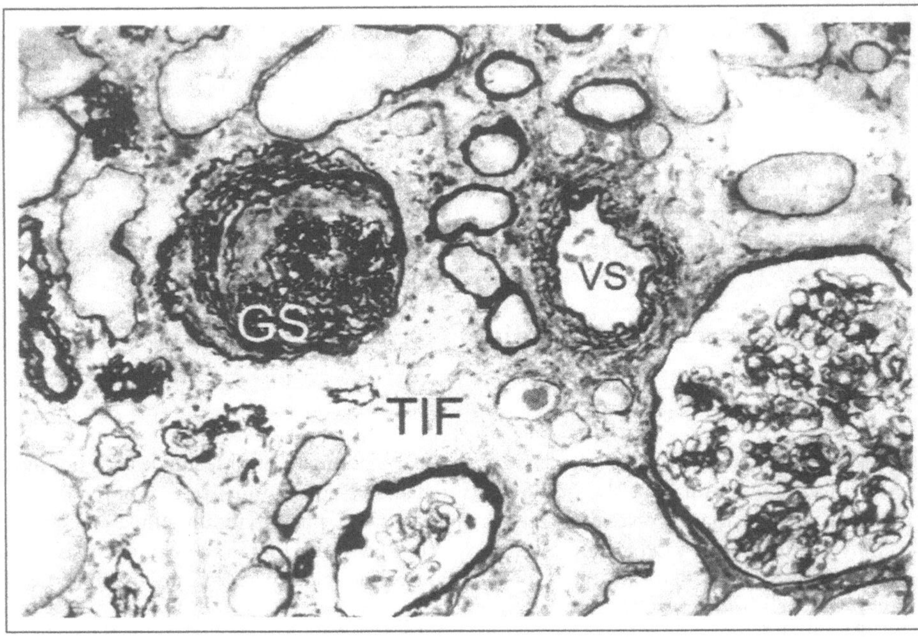

Figure 1. Histopathology of progressive renal disease indicates that renal scarring manifests itself in three forms: glomerulosclerosis (GS), tubulointerstitial fibrosis (TIF) and vascular sclerosis (VS).

Initiation

Glomerulosclerosis

Direct insult to the glomerulus, either injures or stimulates the intrinsic cells of the glomerulus to proliferate, secrete pro-inflammatory factors, undergo cell death or necrosis, and to lay down extracellular matrix.[2] The accumulation of extracellular matrix further acts as a reservoir for growth factors and may further exacerbate injury.

Important from the viewpoint of mechanisms of progressive glomerulosclerosis has been the observation that renal injury with loss of nephrons promotes progressive glomerulosclerosis in nephrons not directly injured by the primary process. Following the landmark work of Brenner and his colleagues,[3] a variety of processes have been demonstrated to produce ongoing injury and sclerosis in the glomeruli.[4] These range from haemodynamic injury to the glomerular cells via enhanced filtration (hyperfiltration) and stretch,[5] to the deleterious effects of mediator molecules such as angiotensin II (AII),[6] endothelin (ET),[7] reactive oxygen species (ROS),[8] platelet derived growth factor (PDGF)[9] and transforming growth factor β1 ($TGF\beta_1$).[10]

Interstitial Fibrosis

Recently attention has refocused on the old observation that in virtually all forms of progressive renal failure tubulointerstitial fibrosis not only occurs but also better parallels the decline in glomerular filtration rate than does the degree of glomerulosclerosis.[11,12] A great deal of work has shown that a major mechanism is via tubular epithelial cell stimulation,[13] which in turn lead to secretion into the interstitium of a similar range of profibrotic factors as are released by the glomerular cells.[14] Important examples are AII, ET, $TGF\beta_1$, ROS and a variety of other growth factors, cytokines and inflammatory mediators.[12,15] The major mechanisms for tubular cell stimulation relate to the increased workload and albumin resorption per tubule as the number of nephrons decrease and the consequent relative hypoxia.[16,17] Other profibrogenic activities of the "overworked" tubular cell include effects of complement activation[15] and ammoniagenesis.[18]

Vascular Sclerosis

There is no doubt that patients with chronic renal failure are at an increased risk of cardiovascular death, estimated at twenty times age and sex matched controls, leading to the concept of "accelerated atherosclerosis" in renal failure.[19]

Intimal proliferation of small renal arteries accompanies atherosclerosis in the aorta and coronary artery. Similarly, histopathology studies have shown that loss of renal function is accompanied by intimal proliferation of renal arterioles, even in the absence of hypertension.[20] This chronic vascular disease displays many of the hallmarks of an inflammatory process and is consequent upon release of chemokines, cytokines and other inflammatory mediators by vascular endothelial cells, which combined with the well recognised vascular haemodynamic factors, provide a stimulus for fibrosis.

Fibrogenesis

Significance of the Renal Mesenchymal Cell

In all three forms of renal fibrosis outlined above, the major matrix producing cell is a mesenchyme derived fibrogenic cell, juxtapositioned between resident endothelial and epithelial cells (Fig. 2).

In the glomerulus, the mesangial cell[21] is the mesenchymal element pivotal to the process of glomerulosclerosis.[22] Recruitment of inflammatory cells and paracrine release of prosclerotic factors causes mesangial cells to proliferate,[23] acquire the contractile protein alpha smooth muscle actin (αSMA),[24] and to produce matrix components[21,25] and an array of cytokines and growth factors.[26] Likewise, other intrinsic pathways such as formation of glycated end products (AGE)[27] and the serine protease thrombin[28] may act directly on mesangial cells to further increase autocrine generation of the above factors.

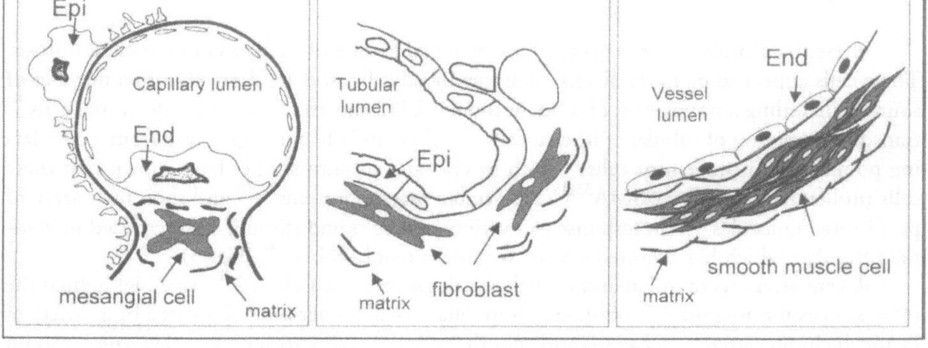

Figure 2. Juxtapositioning of mesenchymal cells and resident intrinsic epithelial (Epi) and endothelial (End) cells in the glomerulus, tubulointerstitium and vasculature. In each case injury to intrinsic endothelial and epithelial cells leads to paracrine stimulation of adjacent mesenchymal cells, represented by the mesangial cell, fibroblast and vascular smooth muscle cell respectively.

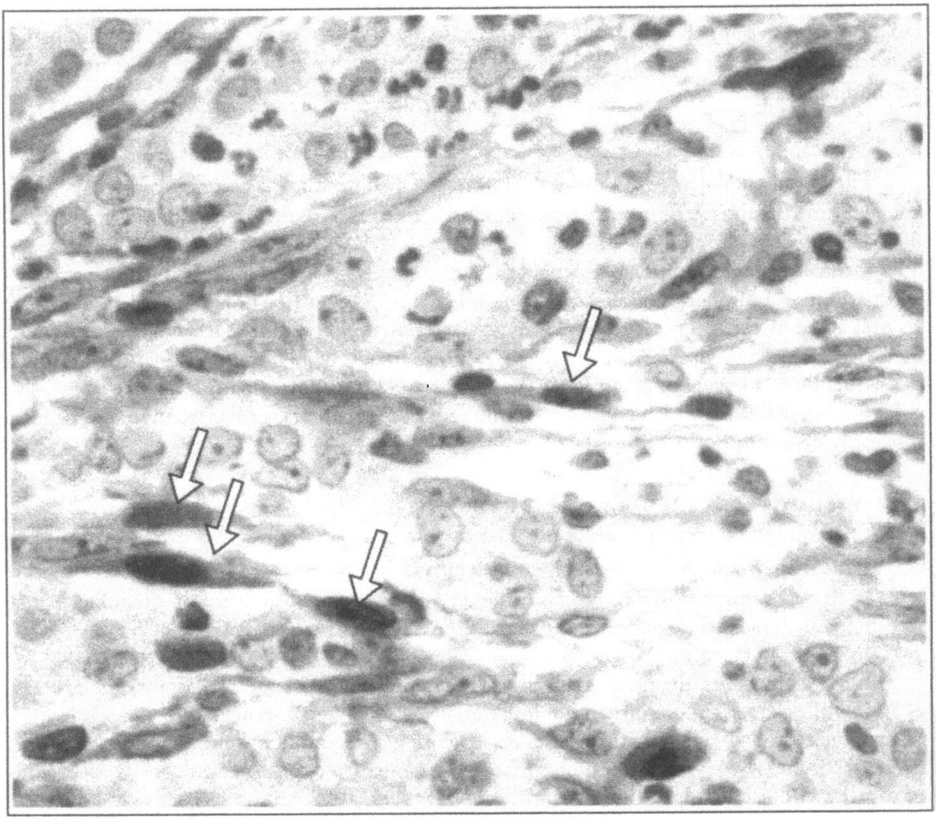

Figure 3. Local proliferation of myofibroblasts is a major source of matrix producing cells in tubulointerstitial fibrosis. Double labelling for the myofibroblast marker αSMA (red) and the thymidine analogue 5-bromo-2′ deoxyuridine (brown) localises DNA replication in interstitial myofibroblasts (arrows) in an experimental model of renal infection.[29] To view online color figure go to http://www.eurekah.com/eurekahlogin.php?chapid=1927&bookid=83&catid=28.

Interstitial fibroblasts are a central mesenchymal feature of all forms of interstitial fibrosis. These cells appear to be more diverse than mesangial cells, and are derived from a number of sources, including a population of resident interstitial fibroblasts,[29] vascular adventitial cells,[30] transdifferentiation of tubular epithelial cells[31] and possibly haematogenously from a circulating population of progenitor cells.[32] Both in vivo and in vitro studies have shown that these cells proliferate,[29] acquire αSMA[33,35] and secrete matrix proteins[29,36] and their own array of profibrotic molecules[37,38] in response to a variety of insults and stimuli. The activated interstitial fibroblast which has acquired αSMA is called a myofibroblast[34] (Fig. 3).

Recent attempts to isolate intra-renal vascular smooth muscle cells[39] have highlighted the interest in examining the role of these cells in renal vascular sclerosis. Stimulated by a variety of factors including stretch, ischaemia and endothelium derived factors, nonrenal vascular smooth muscle cells proliferate[40,41] and produce extracellular matrix,[41] in the process of which they also migrate to the intimal layer.[42] Again the release of AII,[6] ET,[43] ROS,[44] PDGF[45] and $TGF\beta_1$[46] in response to injury are recognised prosclerotic factors.[47]

Clearly therefore, although glomerular mesangial cells, interstitial myofibroblasts and vascular smooth muscle cells are located at separate sites and fulfil a variety of differing functions, it is increasingly obvious that similarities between the three cell types outweigh the differences.

Functional Significance of αSMA Expression

Perhaps one of the most intriguing aspects of renal fibrogenesis is the accumulation of cells which share the αSMA positive characteristic of smooth muscle cells. Expression of αSMA and its incorporation into stress fibres is a defining feature of activated mesangial cells, myofibroblasts and vascular smooth muscle cells. What therefore is the functional significance of αSMA expression?

Several models of wound healing emphasise the importance of the contractile myofibroblast in reducing wound area. Both renal fibroblasts[48] and mesangial cells[49] have been shown to contract collagen lattices in vitro, in a process mediated by β1 integrins.[48] Experimental in vivo models of renal scarring establish that renal scarring is a fibro-contractive disease, with morphometric studies showing that the increase in matrix density is partly due to contraction of the renal parenchyma,[50] endstage renal failure kidneys being smaller when fibrosed.

The selective incorporation of αSMA into stress fibers may also retard motility. Alpha-SMA expressing breast myofibroblasts are less motile than their αSMA negative counterparts, while this isoform is downregulated in smooth muscle cells translocating during atherosclerosis.[51] Finally, if we accept evidence from cardiac models, myofibroblasts produce collagen at a greater rate than their fibroblast progeny, an intrinsic property that cannot be accounted for by differences in response to $TGF\beta_1$.[52]

Molecular Regulation of Renal Fibrogenesis

Fibroblasts, mesangial cells and vascular smooth muscle cells are activated by a diverse range of stimuli. In addition to receptor signal transduction, nonreceptor signaling, for example by nitric oxide, offer opportunity for abrogation by blockade.

Receptor Binding

Receptors acting at the membrane level transduce extracellular signals into a cytosolic one, where subsequent communication of signals involves translocation of intracellular signaling components. These membrane receptors take various forms. Arguably the most important fibrogenic factors, $TGF\beta_1$ and PDGF, signal through a receptor serine/threonine kinase and tyrosine kinase respectively.

$TGF\beta_1$ belongs to a superfamily of structurally related regulatory proteins which include three mammalian TGF isoforms ($TGF\beta_1$ -β_2, -β_3), activin/inhibin and bone morphogenic proteins. The most abundant isoform, $TGF\beta_1$, both increases collagen synthesis and decreases collagenase expression in a variety of mesenchymal cells. Signaling through functional cooperation of the Smad proteins, the TGF-β receptor phosphorylates receptor-regulated Smads (Smad2, Smad3), a pathway that is strictly regulated at various levels, including inhibition by Smad7 which complexes with Smad2 and 3 to prevent phosphorylation.[53]

PDGF is secreted as a homodimer, composed of two A-,B-,C- or D- chains or a heterodimer composed of an A- and B-chain which complex variously with α or β receptor subunits. The PDGF B chain, as part of either PDGF-BB or PDGF-AB, is a potent mesenchymal cell mitogen,[54] while various PDGF-BB isoforms mediate chemotaxis, contraction and the activity of other factors.[9] Induction of PDGF-B type receptors on smooth muscle cells is seen in atherosclerosis and renal allograft rejection.[55]

Likewise the janus tyrosine kinase family is important for cytokine signaling, while G-protein transmembrane receptors signal through associated effector enzymes such as Ras GTPases, a convergent point in signal cascades for many mitogens implicated in proliferative renal disease.[56]

Finally, integrins are of particular importance in mesenchymal cells. Acting as a bridge between extracellular matrix and the cellular cytoskeleton, integrins generate tension within the cell and initiate signaling events, in many cases mediated by activation of focal adhesion kinase, a member of the tyrosine kinase family.[57]

Second Messenger Pathways

The activation and translocation of kinases from the cytoplasm to the nucleus provide the link between extracellular signals and gene expression. Again these pathways take many forms and include, amongst others, the mitogen activated protein kinase, phosphatidylinositol 3-kinase and protein kinase C cascades. In each case the end result is the phosphorylation and activation of nuclear transcription factors, with gene transcription being controlled by the interaction of transcription factors with specific DNA sequences known as promoter or enhancer motifs.[58]

Various transcription factor families have established roles in renal fibrosis and sclerosis. Nuclear factor-κB (NF-κB), perhaps the best known of the nuclear transcription families, promotes the upregulation of cytokines involved in inflammation.[59] Mesenchymal cell differentiation is under the control of NF-κB[60] and MyoD, a member of the basic helix-loop-helix family,[61] while various members of the Kruppel-like transcription factor family regulate collagen transcription.[58] Conversely, peroxisome proliferator activated receptor γ (PPARγ) is active in quiescent mesenchymal cells,[58] suggesting that it may block fibrogenesis.

Treatment Options Available and Developing

Historically, treatment strategies for preventing progressive renal failure have concentrated on limiting glomerulosclerosis. Similarities between the mechanisms of interstitial, glomerular and vascular sclerosis mean that attempts to develop treatment strategies for any one may be of universal benefit in the others.

Clearly the first step in any treatment strategy is the removal of the injury, however this area is not the domain of this Chapter. Instead we will concentrate mostly on those therapies directed at decreasing ongoing fibrosis independently of the initiating injury.

Anti-Inflammatory

Interference with the early pro-inflammatory pathways may reduce fibrogenesis by limiting the activity of chemotactic cytokines and therefore leucocyte migration into tissues.[62] Potentially, cytokine activity can be directly inhibited by a number of strategies. Antibody inhibition, soluble receptors and single stranded oligonucleotoide inhibitors (aptamers) have achieved some experimental success. Antisense oligonuleotides, decoy double strand oligonucleotides and cleavage of RNA with ribozymes may all suppress gene expression, if target cells can be transfected.

Promising targets in the renal context include monocyte chemoattractant protein-1,[62] regulated upon activation normal T-cell expressed and secreted (RANTES),[62] interleukin-1β[63] and interleukin-8,[62] upregulation of which has been consistently observed during inflammatory disease, with clinically relevant strategies seeking to inhibit chemokine production,[64] neutralise activation[65] or antagonise receptors.[66]

Indirect Therapies

The last 20 years has seen a burgeoning of studies directed at slowing the rate of progression of renal failure and fibrosis.[67] Usually therapies were initially developed in a rat model since the rat, like man, experiences progressive renal fibrosis once sufficient renal damage has occurred. These therapies have been moved (with some difficulty) to application in patients with renal disease.[68]

In many cases the initial hypothesis related to altering the environment in which the kidney functioned. Increased understanding of the molecular mechanisms involved has lead to reevaluation of many of these. This has resulted in a better appreciation of the intrarenal molecular mechanisms responsible for progressive renal fibrosis, and in some cases to recognition that the effect is at least partially due to direct interactions between drugs and cells in the kidney.

Dietary Protein Intake

Reducing dietary protein intake slows progressive renal damage.[69] Not surprisingly a wide variety of intrarenal changes are believed responsible.[70] A lower dietary protein intake reduces single-nephron glomerular filtration by physiological mechanisms largely mediated by angiotensin. Accordingly the multiple processes leading to progressive glomerulosclerosis and tubulointerstitial fibrosis consequent upon hyperfiltration, tubular overwork and albumin resorption as outlined previously in this chapter, are ameliorated. There is still much to be learnt regarding the molecular events responsible for dietary protein induced hyperfiltration.

Blood Pressure

Reduction in systemic blood pressure by pharmacological means appears to have an even more profound effect, reducing not only glomerular hyperfiltration.[71,72] and hence tubular work, but also because of direct effects of the antihypertensive molecules on intrinsic renal cells. Hence angiotensin receptor antagonists and calcium channel blockers are believed to be renoprotective through a variety of means, both directly and indirectly related to fibrogenesis, as outlined later in this chapter.

Hyperlipidemia

Control of hyperlipidemia, a common accompaniment to renal disease, slows renal failure progression in man.[73] The improvement is not only due to abrogation of direct deleterious effects of lipids on the kidney[74] but also because, at least in the case of hydroxymethyl glutaryl coenzyme A reductase (HMGCoA) inhibitors, the pharmacological agents used to reduce circulating lipid levels have direct antifibrotic effects with the kidney,[75] affecting profibrotic activities of mesangial cells,[76] myofibroblasts[77] and vascular smooth muscle cells.[78]

Hyperglycaemia

Reducing hyperglycaemia is proven to be effective in abrogating progressive diabetic renal disease in man.[79] Glucose, glycosylated protein (advanced glycosylation end products - AGEs) and glycosylated lipoprotein have been shown to directly stimulate profibrotic activities of mesenchymal cells.[80] When better understood, this may lead to development of molecules to block these profibrotic effects of glucose and glycosylated products.

Direct Antagonists of Mesenchymal Fibrogenesis

A variety of molecules have been shown to directly influence or interfere with the behaviour of renal profibrotic mesenchymal cells, particularly if we accept experiments involving nonrenal fibroblasts or vascular smooth muscle cells as being representative of their intrarenal counterparts. As sclerosis in all three renal settings is associated with the activities of a mesenchyme-derived cell stimulated by factors either derived from adjacent cells (inflammatory, endothelial or epithelial cells) or inherent to the uremic milieu, interference with this final common pathway has the potential to ameliorate all three.

Vasoactive Mediators

Angiotensin Antagonists

Angiotensin II (AII) receptors are found on mesangial cells, fibroblasts and vascular smooth muscle cells.[81,82] AII directly stimulates proliferation and matrix synthesis by all these cell types,[82,85] and administration of angiotensin converting enzyme inhibitors (ACEI), which inhibit AII production, or angiotensin type I receptor blockers, which inhibit AII action more directly, have been shown to reduce renal failure progression in experimental animals and patients with chronic renal disease.[86,87] Importantly, the effect is greater than would be expected from blood pressure alone, making it likely that intrarenal effects on glomerular haemodynamics and on renal cells are both responsible.

Endothelin Antagonists

Endothelin (ET) receptors are also found on all three mesenchymal cell types in the kidney. ET receptor antagonists have been shown to reduce vascular injury,[88] mesangial cell proliferation[89] and accumulation of new fibroblasts[90] in various experimental nephropathies. Differential effects[91] may however indicate important differences in the role of ET-A and ET-B receptor sub-types in mesenchymal injury.

Nitric Oxide Stimulation

Nonreceptor mediated pathways are likewise implicated. Increased nitric oxide generation through arginine supplementation improves tubulointerstitial fibrosis after ureteric obstruction,[92] presumably via a nitric oxide mediated reduction in oxidative stress.

Calcium-Channel Blockers

Various examples of this heterogenous group of drugs have been demonstrated to inhibit proliferation of mesangial and vascular smooth muscle cells.[93,94] Used as antihypertensive agents it is difficult to discern whether this direct effect has an impact in humans with renal disease.[95]

Growth Factors

Despite the function of all these cells being influenced by a myriad of factors and their interactions in a highly redundant system, perhaps surprisingly several findings now suggest that changes in even a single cytokine/growth factor can be effective in abrogating renal fibrogenesis.

A hierarchy exists amongst the profibrotic growth factors, with the most compelling evidence being for $TGF\beta_1$ and PDGF. A variety of strategies have been used experimentally to inhibit the activity of polypeptide growth factors.

$TGF\beta_1$ antisense oligonucleotides have been used successfully as an intervention in both glomerulonephritis[96] and interstitial fibrosis.[97] Anti-serum to $TGF\beta_1$ has ameliorated experimental glomerular sclerosis[98] and vascular intimal hyperplasia.[99] Pirfenidone[100] and decorin[101] have been shown to reduce experimental fibrosis, acting at the level of the latent and activated molecule respectively. Angiotensin converting enzyme inhibitors and AII receptor antagonists reduce but do not normalise aberrant $TGF\beta_1$ production,[102,103] at least in experimental renal disease. Treatments that target inactivation of Smad2 by overexpression of Smad7 may inhibit recruitment of myofibroblasts by inhibiting epithelial-myofibroblast transdifferentiation.[104]

Given that $TGF\beta_1$ affects a variety of cell types, long term suppression may be problematic. Such concerns point to connective tissue growth factor (CTGF) as a possible target. Widely recognised as a downstream mediator of $TGF\beta_1$ activity, expression of CTGF appears to be confined to aberrant fibrogenesis in mesangial cells, fibroblasts and atherosclerosis.[105]

An alternative approach is to take advantage of the naturally occurring interactions between various endogenously expressed growth factors. Hepatocyte growth factor (HGF) is inversely related to $TGF\beta_1$ expression, with neutralisation of endogenous HGF accelerating progression of renal fibrosis,[106] and exogenous administration linked to decreased $TGF\beta_1$ expression and progression.[106,107] In a similar manner, bone morphogenic protein-7, a member of the $TGF\beta$ superfamily, preserves renal function and prevents interstitial fibrosis after unilateral obstruction, acting as a counterbalance to $TGF\beta_1$ activity.[108]

Oligonucleotide aptamers have been developed to antagonise PDGF activity by binding with high affinity to the PDGF molecule.[109] Selective inhibitors of a PDGF specific receptor tyrosine kinase reduce mesangial cell proliferation and matrix synthesis in mesangial proliferative glomerulonephritis[110] and attenuate, albeit to a lesser degree, interstitial fibrosis after unilateral ureteric obstruction.[111]

In nonrenal fibroblasts at least, blocking of extracellular regulated kinase activity prevents growth factor induced proliferative response.[112]

Transcription Factors

Though the coordinated activity of several transcription factors seems necessary to control expression of any fibrogenic protein, the recognition that some individual transcription factors are able to regulate cell phenotype has provided potential therapeutic targets. A synthetic double stranded DNA decoy sequence has successfully ameliorated experimental crescentic glomerulonephritis by blocking promoter binding of NF-κB.[113] In mesangial cells cultured in high glucose, PPARγ agonists suppress $TGF\beta_1$ mediated expression of type I collagen mRNA and protein,[114] while the PPARγ agonist troglitazone abrogates glomerulosclerosis in both diabetic[115] and nondiabetic[116] rat models.

Some of the most immediate therapeutic strategies are provided by investigations that have focused on agents introduced for quite different indications.

So-called "statins" (HMGCoA inhibitors) introduced to lower blood cholesterol have now been shown to have variable effects on fibrogenesis. Although statins have been shown to reduce extrarenal vascular smooth muscle mitogenesis both in vivo and in vitro,[78] effects of individual statins differ, with proliferation being inhibited by lovastatin,[77] simvastatin[78] but not pravastatin,[78] an effect expected to confer differential benefit in vasculopathy. Pravastatin, simvastatin and lovastatin have variously been shown to reduce in vitro collagen secretion by mesangial cells[76] and cortical fibroblasts.[77,117] At the molecular level, statins block prenylation of Ki-Ras, an important intracellular protein required for fibroblast proliferation.[118] The more specific prenylation inhibitors being developed for cancer therapy,[119] are therefore likely to be valuable potential antagonists of the uncontrolled mesenchymal proliferation that accompanies fibrogenesis.

Phosphodiesterase inhibitors (the most well-known of which is theophylline used in the treatment of asthma), increase the intracellular levels of cyclic nucleotides which theoretically could interfere with mesenchymal cell function. Pentoxifylline has now been shown to inhibit mitogenesis and collagen synthesis in cultures of mesangial cells[120] and renal fibroblasts[121,122] and attenuate progressive fibrosis after sub-total renal ablation.[123] Dipyridamole, another drug which increases intracellular cyclic nucleotide levels, inhibits mesangial cell,[124] aortic smooth muscle cell[125] and interstitial fibroblast[126] function, though it was originally intended as an antiplatelet agent.

The long recognised benefits of heparin in experimental models of progressive renal disease[127] may be due not only to anti-thrombotic properties but also its ability to inhibit mitogenesis in mesangial cells,[128] fibroblasts[129] and smooth muscle cells.[130]

Halofuginone, commonly used as a coccidiostat in chickens, is a potent in vitro inhibitor of both mesangial cell proliferation and collagen I synthesis[131] and has been shown to inhibit smooth muscle cell proliferation after experimental intimal injury.[132]

Finally, immunosuppressive agents have been shown to reduce proliferation of not only T and B-lymphocytes but also vascular smooth muscle cells[133] and mesangial cells,[134] effects mediated at least in part by preventing degradation of I-κB, the inhibitory subunit of NF-κB. The in vivo abrogation of fibroblast accumulation after sub-total nephrectomy[135] suggests that mycophenolate mofetil is also a direct antagonist of fibroblast proliferation.

Collagen Assembly and Degradation

The fact that collagen synthesis involves several unique post-translational reactions provides a number of potential therapeutic targets. Inhibition of prolyl-4-hydroxylase and lysyl oxidase, enzymes involved in triple helix formation, has successfully reduced nonrenal scarring, albeit with marked side effects.[136]

The recognition that renal scarring is a consequence of both increased collagen synthesis and decreased breakdown has inevitably led to a search for agents which promote collagen degradation. Such strategies have been given further impetus by the clinical reality that extensive scarring is often already present at presentation.

Relaxin is a naturally occurring polypeptide hormone generated in increased quantities during the latter stages of gestation. Consistent with its role in softening the pubic symphysis and uterine cervix in preparation for parturition,[137] antifibrotic effects have been demonstrated in experimental models of rapidly fibrosing chemically induced renal papillary necrosis[138] and renal ablation,[139] data suggesting that this may relate at least in part to its in vitro ability to increase collagenase synthesis.[140,141] Likewise, HGF has the ability to increase collagenase activity in cultures of tubule epithelial cells.[106] Pirfenidone, a pyridone compound, reduces ongoing fibrosis in both chronic anti-thy1 glomerulonephritis[142] and unilateral ureteric obstruction models of renal sclerosis[143,144] ostensively through a variety of mechanisms indirectly increasing collagen degradation.

Angiogenesis

Deficiency of vascular endothelial growth factor (VEGF) is implicated in the chronic ischaemia and hypoxia which results from impaired angiogenesis. In several models of progressive renal failure, a decrease in VEGF expression correlates well with less microvascular density, with exogenous VEGF improving renal function and reducing fibrosis after sub-total renal ablation.[145,146] Preliminary findings suggest that exogenous HGF has similar effects in experimental glomerulonephritis.[147]

Future Perspectives

As we learn more about the mechanisms of fibrogenesis, studies in both the kidney and elsewhere will continue to identify potential therapeutic targets. Future efforts directed at in vivo studies with gene deletion or over expression will determine the pathological significance of each of these mechanisms.

While much of what we have learnt about sclerosis and fibrosis has been a consequence of recognising the similarity between healing and scarring, the challenge remains to target anti-fibrotic therapies and limit inappropriate scarring without compromising the biology of healing.

References

1. Diamond JR, Karnovsky MJ. Focal and segmental glomerulosclerosis: Analogies to atherosclerosis. Kidney Int 1988; 33:917-924.
2. Cybulsky AV. Growth factor pathways in proliferative glomerulonephritis. Curr Opin Nephrol Hypertension 2000; 9:217-223.
3. Brenner BM, Lawler EV, Mackenzie HS. The hyperfiltration theory: A paradigm shift in nephrology. Kidney Int 1996; 49:1774-1777.
4. Wardle N. Glomerulosclerosis: The final pathway is clarified, but can we deal with the triggers? Nephron 1996; 73:1-7.
5. Harris RC, Akai Y, Yasuda T et al. The role of physical forces in alterations of mesangial cell function. Kidney Int 1994; 45(s45):s17-s21
6. Johnson RJ, Alpers CE, Yoshimura A et al. Renal injury from angiotensin II-mediated hypertension. Hypertension 1992; 19:464-474.
7. Gomez-Garre D, Ruiz-Ortega M, Ortego M et al. Effects and interactions of endothelin-1 and angiotensin II on matrix protein expression and synthesis and mesangial cell growth. Hypertension 1996; 27:885-892.
8. Hahn S, Krieg RJ, Hisano S et al. Vitamin E suppresses oxidative stress and glomerulosclerosis in rat remnant kidney. Pediatr Nephrol 1999; 13:195-198.
9. Johnson RJ, Floege J, Couser WG et al. Role of platelet-derived growth factor in glomerular disease. J Am Soc Nephrol 1993; 4:119-128.
10. Isaka Y, Akagi Y, Ando Y et al. Cytokines and glomerulosclerosis. Nephrol Dial Transplant 1999; 14:30-32.
11. Risdon RA, Sloper JC, de Wardener HE. Relationship between renal function and histological changes found in renal biopsy specimens from patients with persistent glomerular nephritis. Lancet 1968; I:363-366.
12. Becker GJ, Hewitson TD. The role of tubulointerstitial injury in chronic renal failure. Curr Opin Nephrol Hypertension 2000; 9:133-148.
13. Healy E, Brady HR. Role of tubule epithelial cells in the pathogenesis of tubulointerstitial fibrosis induced by glomerular disease. Curr Opin Nephrol Hypertension 1998; 7:525-530.
14. Johnson DW, Saunders HJ, Baxter RC et al. Paracrine stimulation of human renal fibroblasts by proximal tubule cells. Kidney Int 1998; 54:747-757.
15. Ong ACM. Tubulointerstitial actions of endothelins in the kidney: Roles in health and disease. Nephrol Dial Transplant 1996; 11:251-257.
16. Sahai M, Mei C, Schrier RW et al. Mechanisms of chronic hypoxia-induced renal cell growth. Kidney Int 1999; 56:1277-1288.
17. Fine LG, Orphanides C, Norman JT. Progressive renal disease: The chronic hypoxia hypothesis. Kidney Int 1998; 53:S74-S78
18. Nath KA, Hostetter MK, Hostetter TH. Increased ammoniagenesis as a determinant of progressive renal injury. Am J Kidney Dis 1991; 17:654-657.
19. Lindner A, Charra B, Sherrard DJ et al. Accelerated Atherosclerosis in Prolonged Maintenance Hemodialysis. N Engl J Med 1974; 290 No.13:697-701.
20. Bos WJW, Demircan MM, Weening JJ et al. Renal vascular changes in renal disease independent of hypertension. Nephrol Dial Transplant 2001; 16:537-541.
21. Striker LM-M, Killen PD, Chi E et al. The composition of glomerulosclerosis. I. Studies in focal sclerosis, crescentic glomerulonephritis, and membranoproliferative glomerulonephritis. Lab Invest 1984; 51:181-191.
22. Johnson RJ, Floege J, Yoshimura A et al. The activated mesangial cell: A glomerular "myofibroblast"? J Am Soc Nephrol 1992; 2:s190-s197
23. Johnson RJ, Lida H, Alpers E et al. Expression of smooth muscle cell phenotype by rat mesangial cells in immune complex nephritis.Alpha-smooth muscle actin is a marker of mesangial cell proliferation. J Clin Invest 1991; 87:847-858.
24. Alpers CE, Hudkins KL, Gown AM et al. Enhanced expression of "muscle-specific" actin in glomerulonephritis. Kidney Int 1992; 41:1134-1142.
25. Stokes MB, Holler S, Cui Y et al. Expression of decorin, biglycan, and collagen type I in human renal fibrosing disease. Kidney Int 2000; 57:487-498.

26. Couser WG, Johnson RJ. Mechanisms of progressive renal disease in glomerulonephritis. Am J Kidney Dis 1994; 23:193-198.
27. Oldfield MD, Bach LA, Forbes JM et al. Advanced glycation end products cause epithelial-myofibroblast transdifferentiation via the receptor for advanced glycation end products (RAGE). J Clin Invest 2001; 108:1853-1863.
28. Cunningham MA, Rondeau E, Chen X et al. Protease-activated receptor 1 mediates thrombin-dependent, cell-mediated renal inflammation in crescentic glomerulonephritis. J Exp Med 2000; 191:455-461.
29. Hewitson T, Wu H, Becker GJ. Interstitial myofibroblasts in experimental renal infection and scarring. Am J Nephrol 1995; 15:411-417.
30. Wiggins R, Goyal M, Merritt S et al. Vascular adventitial cell expression of collagen I messenger ribonucleic acid in anti- glomerular basement membrane antibody induced crescentic nephritis in the rabbit. A cellular source for interstitial collagen synthesis in inflammatory renal disease. Lab Invest 1993; 68:557-565.
31. Jinde K, Nikolic-Paterson DJ, Huang XR et al. Tubular phenotypic change in progressive tubulointerstitial fibrosis in human glomerulonephritis. Am J Kidney Dis 2001; 38:761-769.
32. Grimm PC, Nickerson P, Jeffery J et al. Neointimal and tubulointerstitial infiltration by recipient mesenchymal cells in chronic renal-allograft rejection. N Engl J Med 2001; 345:93-97.
33. Alpers CE, Hudkins KL, Floege J et al. Human renal cortical interstitial cells with some features of smooth muscle cells participate in tubulointerstitial and crescentic glomerular injury. J Am Soc Nephrol 1994; 5:201-210.
34. Goumenos DS, Brown CB, Shortland J et al. Myofibroblasts, predictors of progression of mesangial IgA nephropathy? Nephrol Dial Transplant 1994; 9:1418-1425.
35. Hewitson TD, Becker GJ. Interstitial myofibroblasts in IgA glomerulonephritis. Am J Nephrol 1995; 15:111-117.
36. Tang WW, Van GY, Qi M. Myofibroblast and α_1(III) collagen expression in experimental tubulointerstitial nephritis. Kidney Int 1997; 51:926-931.
37. Han DC, Isono M, Hoffman BB et al. High glucose stimulates proliferation and collagen I synthesis in renal cortical fibroblasts: Mediation by autocrine activation of TGFbeta progression. J Am Soc Nephrol 1999; 10:1891-1899.
38. Strutz F, Zeisberg M, Hemmerlein B et al. Basic fibroblast growth factor expression is increased in human renal fibrogenesis and may mediate autocrine fibroblast proliferation. Kidney Int 2000; 57:1521-1538.
39. Endlich N, Endlich K, Taesch N et al. Culture of vascular smooth muscle cells from arteries of the rat kidney. Kidney Int 2000; 57:2468-2475.
40. Betz E, Faller-Becker P, Wolberg-Buchholz K et al. Proliferation of smooth muscle cells in the inner and outer layers of the tunica media of arteries: an in vitro study. J Cell Physiol 1991; 147:385-395.
41. Dilley RJ, McGeachie JK, Prendergast FJ. A review of the proliferative behaviour, morphology and phenotypes of vascular smooth muscle. Atherosclerosis 1987; 63:99-107.
42. Akyürek LM, Paul LC, Funa K et al. Smooth muscle cell migration into intima and adventitia during development of transplant vasculopathy. Transplantation 1996; 62:1526-1529.
43. Kanse SM, Wijelath E, Kanthou C et al. The proliferative responsiveness of human vascular smooth muscle cells to endothelin correlates with endothelin receptor density. Lab Invest 1995; 72:376-382.
44. Greene EL, Lu G, Zhang D et al. Signaling events mediating the additive effects of olsic acid and angiotensin II on vascular smooth muscle migration. Hypertension 2001; 37:308-312.
45. Nabel EG, Yang Z, Liptay S et al. Recombinant platelet-derived growth factor B gene expression in porcine arteries induces intimal hyperplasia in vivo. J Clin Invest 1993; 91:1822-1829.
46. Okada Y, Katsuda S, Watanabe H et al. Collagen synthesis of human arterial smooth muscle cells: Effects of platelet-derived growth factor, transforming growth factor-beta 1 and interleukin-1. Acta Pathol Jpn 2001; 43:160-167.
47. Campbell JH, Campbell GR. Cell biology of atherosclerosis. J Hypertens 1994; 12 (Suppl 10):S129-S132
48. Kelynack KJ, Hewitson TD, Nicholls KM et al. Human renal fibroblast contraction of collagen I lattices is an integrin mediated process. Nephrol Dial Transplant 2000; 15:1766-1772.

49. Kitamura M, Maruyama N, Mitarai T et al. Extracellular matrix contraction by cultured mesangial cells: Modulation by transforming growth factor-beta and matrix components. Exp Mol Pathol 1992; 56:132-143.
50. Hewitson TD, Darby IA, Bisucci T et al. Evolution of tubulointerstitial fibrosis in experimental renal infection and scarring. J Am Soc Nephrol 1998; 9:632-642.
51. Ronnov-Jessen L, Peterson OW. A function for filamentous alpha smooth muscle actin: Retardation of motility in fibroblasts. J Cell Biol 1996; 134:67-80.
52. Petrov VV, Fagard RH, Lijnen S. Stimulation of collagen production by transforming growth factor-β during differentiation of cardiac fibroblasts to myofibroblasts. Hypertension 2002; 39:258-263.
53. Schnaper HW, Hayashida T, Poncelet A-C. It's a Smad World: Regulation of TGF-β signaling in the kidney. J Am Soc Nephrol 2002; 13:1126-1128.
54. Floege J. Glomerular remodelling: Novel therapeutic approaches derived from the apparently chaotic growth factor network. Nephron 2002; 91:582-587.
55. Fellstrom B, Klareskog L, Heldin CH et al. Platelet-derived growth factor receptors in the kidney - upregulated expression in inflammation. Kidney Int 1989; 36:1099-1102.
56. Khwaja A, Connolly JO, Hendry BM. Prenylation inhibitors in renal disease. Lancet 2000; 355:741-744.
57. Yasuda T, Kondo S, Owada M et al. Integrins and the cytoskeleton: Focal adhesion kinase and paxillin. Nephrol Dial Transplant 2002; 14:58-60.
58. Mann DA, Smart DE. Transcriptional regulation of hepatic cell activation. Gut 2002; 50:891-896.
59. Guijarro C, Egido J. Transcription factor-$_k$B (NF-$_k$B) and renal disease. Kidney Int 2001; 59:415-424.
60. Klahr S, Morrissey JJ. The role of vasoactive compounds, growth factors and cytokines in the progression of renal disease. Kidney Int 2000; suppl 75:s7-s14
61. Mayer DC, Leinwand LA. Sarcomeric gene expression and contractility in myofibroblasts. J Cell Biol 1997; 139:1477-1484.
62. Rovin BH. Chemokine blockade as a therapy for renal disease. Curr Opin Nephrol Hypertension 2000; 13:225-232.
63. Vesey DA, Cheung CWY, Cuttle L et al. Interleukin-1β induces human proximal tubule cell injury, α-smooth muscle actin expression and fibronectin production. Kidney Int 2002; 62:31-40.
64. Zoja C, Corna D, Benedetti G et al. Bindarit retards renal disease and prolongs survival in murine lupus autoimmune disease. Kidney Int 1998; 53:726-734.
65. Panzer U, Thaiss F, Zahner G et al. Monocyte chemoattractant protein-1 and osteopontin differentially regulate monocytes recruitment iin experimental glomerulonephritis. Kidney Int 2002; 59:1762-1769.
66. Grone H-J, Weber C, Weber KSC et al. Met-RANTES reduces vascular and tubular damage during acute renal transplant rejection: Blocking monocyte arrest and recruitment. FASEB J 1999; 13:1371-1383.
67. Becker GJ, Whitworth JA, Ihle BU et al. Prevention of progression in nondiabetic chronic renal failure. Kidney Int 1994; 45 Suppl. 45:S167-S170
68. Burgess E. Conservative treatment to slow deterioration of renal function: Evidence-based recommendations. Kidney Int 1999; 55:S17-S25
69. Brenner BM, Meyer TW, Hostetter TH. Dietary protein intake and the progressive nature of kidney disease: The role of haemodynamically mediated glomerular injury in the pathogenesis of progressive glomerular sclerosis in ageing, renal ablation and intrinsic renal disease. N Engl J Med 1982; 307:652-659.
70. Heidland A, Sebekova K, Ling H. Effect of low-protein diets on renal disease: Are nonhaemodynamic factors involved? Nephrol Dial Transplant 1995; 1512-1514.
71. Herbert LA, Kusek JW, Greene T et al. Effect of blood pressure control on progressive renal disease in black and whites. Hypertension 1997; 30:428-435.
72. Klahr S. Role of dietary protein and blood pressure in the progression of renal disease. Kidney Int 1996; 49:1783-1786.
73. Fried LF, Orchard TJ, Kasiske BL. Effect of lipid reduction on the progression of renal disease: A meta-analysis. Kidney Int 2001; 59:260-269.

74. Schlöndorff D. Cellular mechanisms of lipid injury in the glomerulus. Am J Kidney Dis 1993; 22:72-82.
75. Buemi M, Senatore M, Corica F et al. Are there potential nonlipid-lowering uses of statins in the kidney? Nephron 2001; 89:363-368.
76. Nishimura M, Tanaka T, Yasuda T et al. Effect of pravastatin on type IV collagen secretion and mesangial cell proliferation. Kidney Int 1999; 56 suppl. 71:s97-s100.
77. Kelynack KJ, Hewitson TD, Martic M et al. Lovastatin downregulates renal myofibroblast function in vitro. Nephron 2002; 91:701-709.
78. Rosenson RS, Tangney CC. Antiatherothrombotic properties of statins. JAMA 1998; 279:1643-1650.
79. The diabetes control and complications trial research group: The effect of intensive treatment of diabetes on the development and progression of long-term complications in insulin-dependent diabetes mellitus. New Eng J Med 1993; 329:977-986.
80. Kim YS, Kim BC, Song CY et al. Advanced glycosylation end products stimulate collagen mRNA synthesis in mesangial cells mediated by protein kinase C and transformaing growth factor-beta. J Lab Clin Med 2002; 138:59-68.
81. Matsubara H, Sugaya T, Murasawa S et al. Tissue-specific expression of human angiotensin II AT1 and AT2 receptors and cellular localization of subtype mRNAs in adult human renal cortex using in situ hybridization. Nephron 1998; 80:25-34.
82. Ruiz-Ortega M, Egido J. Angiotensin II modulates cell growth-related events and synthesis of matrix proteins in renal interstitial fibroblasts. Kidney Int 1997; 52:1497-1510.
83. Nahman S, Rothe KL, Falkenhain ME et al. Angiotensin II induction of fibronectin biosynthesis in cultured human mesangial cells- association with CREB transcription factor activation. J Lab Clin Med 1996; 127:599-611.
84. Orth SR, Weinreich T, Bonisch S et al. Angiotensin II induces hypertrophy and hyperplasia in adult human mesangial cells. Exp Nephrol 1995; 3:23-33.
85. Mifune M, Sasamura H, Shimizu-Hirota R et al. Angiotensin II type 2 receptors stimulate collagen synthesis in cultures vascular smooth muscle cells. Hypertension 2000; 36:845-850.
86. Jafar TH, Schmid CH, Landa M et al. Angiotensin-converting enzyme inhibitors and progression of nondiabetic renal disease. Ann Intern Med 2002; 135:73-87.
87. Hilgers KF, Mann JFE. ACE inhibitors versus AT_1 receptor antagonists in patients with chronic renal disease. J Am Soc Nephrol 2002; 13:1100-1108.
88. Ferrer P, Valentine M, Jenkins-West T et al. Orally active endothelin receptor antagonist BMS-182874 suppresses neointimal development in balloon-injured rat carotid arteries. J Cardiovasc Pharmacol 1995; 26:908-915.
89. Fukuda K, Yanagida T, Okuda K et al. Role of endothelin as a mitogen in experimental glomerulonephritis in rats. Kidney Int 2001; 49:1320-1329.
90. Forbes JM, Leaker B, Hewitson TD et al. Macrophage and myofibroblast involvement in ischemic acute renal failure is attenuated by endothelin receptor antagonists. Kidney Int 1999; 55:198-208.
91. Forbes JM, Hewitson TD, Becker G et al. Simultaneous blockade of endothelin A and B receptors in ischaemic acute renal failure is detrimental to long term kidney function. Kidney Int 2001; 59:1333-1341.
92. Morrissey JJ, Ishidoya S, McCracken R et al. Nitric oxide generation ameliorates the tubulointerstitial fibrosis of obstructive nephropathy. J Am Soc Nephrol 1996; 7:2202-2212.
93. Haller H. Calcium antagonists and cellular mechanisms of glomerulosclerosis and atherosclerosis. Am J Kidney Dis 1993; 21 No.6 Suppl 3:26-31.
94. Ono T, Liu N, Kusano H et al. Broad antiproliferative effects of benidipine on cultured human mesangial cells in cell cycle phases. Am J Nephrol 2002; 22:581-586.
95. Epstein M. Calcium antagonists and renal disease. Kidney Int 1998; 54:1771-1784.
96. Akagi Y, Isaka Y, Arai M et al. Inhibition of TGF-beta 1 expression by antisense oligonucleotides suppressed extracellular matrix accumulation in experimental glomerulonephritis. Kidney Int 1996; 50:148-155.
97. Isaka Y, Tsujie M, Ando Y et al. Transforming growth factor-β1 antisense oligodeoxynucleotides block interstitial fibrosis in unilateral ureteral obstruction. Kidney Int 2000; 58:1885-1892.
98. Border WA, Okuda S, Languino LR et al. Suppression of experimental glomerulonephritis by antiserum against transforming growth factor β1. Nature 1990; 346:371-374.

99. Wolf YG, Rasmussen LM, Ruoslahti E. Antibodies against transforming growth factor -beta1 suppresses intimal hyperplasia in a rat model. J Clin Invest 1994; 93:1172-1178.
100. Shimizu T, Fukagawa M, Kuroda T et al. Pirfenidone prevents collagen accumulation in the remnant kidney in rats with partial nephrectomy. Kidney Int 1997; 52:S239-S243.
101. Border WA, Noble NA, Yamamoto T et al. Natural inhibitor of transforming growth factor-β protects against scarring in experimental kidney disease. Nature 1992; 360:361-364.
102. Shihab FS, Bennett WM, Tanner AM et al. Angiotensin II blockade decreases TGF-β1 and matrix proteins in cyclosporine nephropathy. Kidney Int 1997; 52:660-673.
103. Peters H, Border WA, Noble NA. Targeting TGF-β overexpression in renal disease: Maximizing the antifibrotic action of angiotensin II blockade. Kidney Int 1998; 54:1570-1580.
104. Li JH, Zhu H-J, Huang XR et al. Smad7 inhibits fibrotic effects of TGF-β on renal tubular epithelial cells by blocking Smad2 activation. J Am Soc Nephrol 2002; 13:1464-1472.
105. Clarkson MR, Gupta S, Murphy M et al. Connective tissue growth factor: A potential stimulus for glomerulosclerosis and tubulointerstitial fibrosis in progressive renal disease. Curr Opin Nephrol Hypertension 1999; 8:543-548.
106. Matsumoto K, Nakamura T. Hepatocyte growth factor: Renotropic role and potential therapeutics for renal diseases. Kidney Int 2001; 59:2023-2038.
107. Mizuno S, Matsumoto K, Nakamura T. Hepatocyte growth factor suppresses interstitial fibrosis in a mouse model of obstructive nephropathy. Kidney Int 2001; 59:1304-1314.
108. Morrissey J, Kruska K, Guo G et al. Bone morphogenetic protein-7 improves renal fibrosis and accelerates the return of renal function. J Am Soc Nephrol 2002; 13:S14-S21.
109. Ostendorf T, Kunter U, Grone HJ et al. Specific antagonism of PDGF prevents renal scarring in experimental glomerulonephritis. J Am Soc Nephrol 2001; 12:909-918.
110. Gilbert RE, Kelly DJ, McKay T et al. PDGF signal transduction ameliorates experimental mesangial proliferative glomerulonephritis. Kidney Int 2001; 59:1324-1332.
111. Ludewig D, Kosmehl H, Sommer M et al. PDGF receptor kinase blocker AG1295 attenuates interstitial fibrosis in rat kidney after unilateral obstruction. Cell Tissue Res 2000; 299:97-103.
112. Britton RS, Bacon BR. Intracellular signaling pathways in stellate cell activation. Alcohol Clin Exp Res 1999; 23:922-925.
113. Tomita N, Morishita R, Lan HY et al. In vivo administration of a nuclear transcription factor-kappaB decoy suppresses experimental crescentic gomerulonephritis. J Am Soc Nephrol 2000; 11:1244-1252.
114. Zheng F, Fornoni A, Elliot SJ et al. Upregulation of type I collagen by TGFβ in mesangial cella is blocked by PPARγ activation. Am J Physiol Renal Physiol 2002; 282:F639-F648.
115. McCarthy KJ, Routh RE, Shaw W et al. Troglitazone halts diabetic glomerulosclerosis by blockade of mesangial expansion. Kidney Int 2000; 58:2341-2350.
116. Ma L-J, Marcantoni C, Linton MF et al. Peroxisome proliferator-activated receptor-γ agonist troglitazone protects against nondiabetic glomerulosclerosis in rats. Kidney Int 2001; 59:1899-1910.
117. Johnson DW, Saunders HJ, Field MJ et al. In vitro effects of simvastatin on tubulointerstitial cells in a human model of cyclosporin nephrotoxicity. Am J Physiol 1999; 276:F467-F475
118. Sharpe CC, Dockrell ME, Scott R et al. Evidence of a role of Ki-RAS in the stimulated proliferation of renal fibroblasts. J Am Soc Nephrol 1999; 10:1186-1192.
119. O'Donnell MP. Renal tubulointerstitial fibrosis: New thoughts on its development and progression. Postgrad Med 2000; 108:159-172.
120. Tsai T-J, Lin R-H, Chang C-C et al. Vasodilator agents modulate rat glomerular mesangial cell growth and collagen synthesis. Nephron 1995; 70:91-99.
121. Hewitson TD, Martic M, Kelynack KJ et al. Pentoxifylline reduces in vitro renal myofibroblast proliferation and collagen secretion. Am J Nephrol 2000; 20:82-88.
122. Strutz F, Heeg M, Kochsiek T et al. Effects of pentoxifylline, pentifylline and gamma-interferon on proliferation, differentiation, and matrix synthesis of human renal fibroblasts. Nephrol Dial Transplant 2000; 15:1535-1546.
123. Lin S-L, Chen Y-M, Chien C-T et al. Pentoxifylline attenuated the renal disease progression in rats with remnant kidney. J Am Soc Nephrol 2002; 13:2916-2929.
124. Hillis GS, Duthie LA, MacLeod AM. Dipyridamole inhibits human mesangial cell proliferation. Nephron 1998; 78:172-178.

125. Himmelfarb J, Couper L. Dipyridamole inhibits PDGF- and bFGF-induced vascular smooth muscle cell proliferation. Kidney Int 1997; 52:1671-1677.
126. Hewitson TD, Tait MG, Martic M et al. Dipyridamole inhibits in vitro renal myofibroblast proliferation and collagen synthesis. J Lab Clin Med 2002; 140:199-208.
127. Olson JL. Role of heparin as a protective agent following reduction in renal mass. Kidney Int 1984; 25:376-382.
128. Castellot JJ, Hoover RL, Harper PA et al. Heparin and glomerular epithelial cell-secreted heparin like species inhibit mesangial cell proliferation. Am J Pathol 1985; 120:427-435.
129. Tiozzo R, Reggiani D, Cingi MR et al. Effect of heparin derived fractions on the proliferation and protein synthesis of cells in culture. Throm Res 1991; 62:177-188.
130. Hoover RL, Rosenberg RD, Haering W et al. Inhibition of rat arterial smooth muscle cell proliferation by heparin. II. in vivo studies. Circ Res 1980; 47:578-583.
131. Nagler A, Katz A, Aingorn H et al. Inhibition of glomerular mesangial cell proliferation and extracellular matrix deposition by halofuginone. Kidney Int 2001; 52:1561-1569.
132. Nagler A, Miao H-Q, Aimgorn H et al. Inhibition of collagen synthesis, smooth muscle cell proliferation and injury induced hyperplasia by halofuginone. Arterioscler Thromb Vasc Biol 1997; 17:194-202.
133. Gregory CR, Huang X, Pratt RE et al. Treatment with rapamycin and mycophenolic acid reduces arterial intimal thickening produced by mechanical injury and allows endothelial replacement. Transplantation 1995; 59:655-661.
134. Hauser IA, Renders L, Radeke H-H et al. Mycophenolate mofetil inhibits rat and human mesangial cell proliferation by guanosine depletion. Nephrol Dial Transplant 1999; 14:58-63.
135. Badid C, Vincent M, McGregor B et al. Mycophenolate mofetil reduces myofibroblast infiltration and collagen III deposition in rat remnant kidney. Kidney Int 2000; 58:51-61.
136. Franklin TJ. Therapeutic approaches to organ fibrosis. Int J Biochem Cell Biol 1997; 29:79-89.
137. Bani D. Relaxin: A pleiotropic hormone. Gen.Pharmac 1997; 28:13-22.
138. Garber SL, Mirochnik Y, Brecklin CS et al. Relaxin decreases renal interstitial fibrosis and slows progression of renal disease. Kidney Int 2001; 59:876-882.
139. Garber SL, Mirochnik Y, Brecklin C et al. Effect of relaxin in two models of renal mass reduction. Am J Nephrol 2002; 2003:8-12.
140. Unemori EN, Pickford LB, Salles AL et al. Relaxin induces an extracellular matrix degrading phenotype in human lung fibroblasts in vitro and inhibits lung fibrosis in a murine model in vivo. J Clin Invest 1996; 98:2739-2745.
141. Masterson R, Hewitson TD, Kelynack KJ et al. Relaxin downregulates renal fibroblast function and stimulates matrix remodelling in vitro. Nephrol Dial Transplant 2004; 19:544-52.
142. Shimizu F, Fukagawa M, Yamauchi S et al. Pirfenidone prevents the progression of irreversible glomerular sclerotic lesions in rats. Nephrology 1997; 3:315-322.
143. Shimizu T, Kuroda T, Hata S et al. Pirfenidone improves renal function and fibrosis in the post-obstructed kidney. Kidney Int 1998; 54:99-109.
144. Fukagawa M, Noda M, Shimizu T et al. Chronic progressive interstitial fibrosis in renal disease - are there novel pharmacological approaches? Nephrol Dial Transplant 1999; 14:2793-2795.
145. Kang DH, Hughes J, Mazzali M et al. Impaired angiogenesis in the remnant kidney model: II. Vascular endothelial growth factor administration reduces renal fibrosis and stabilizes renal function. J Am Soc Nephrol 2001; 12:1448-1457.
146. Msuda Y, Shimizu A, Mori T et al. Vascular endothelial growth factor enhances glomerular capillary repair and accelerates resolution of experimentally induced glomerulonephritis. Am J Pathol 2001; 159:599-608.
147. Mori T, Shimizu A, Masuda Y et al. Hepatocyte growth factor stimulates endothelial proliferation and accelerates angiogenic capillary repair in experimental progressive glomerulonephritis (abstract). J Am Soc Nephrol 2000; 11:495A.

Chapter 7

Myocardial Infarction and Cardiac Fibrogenesis

Shozo Kusachi and Yoshifumi Ninomiya

Abstract

Fibrogenesis is essential for infarct healing and affects ventricular remodeling, one of the most important prognostic factors after myocardial infarction. Fibrogenesis is initiated by a variety of cytokines and growth factors produced by activated macrophages and inflammatory cells during the initial inflammatory phase. Fibroblasts that proliferate and infiltrate into the infarct zone are transformed into myofibroblasts, which express a variety of extracellular matrix (ECM) components that reconstruct the ECM in the infarcted myocardium. Following the inflammatory phase, fibrogenesis occurs prominently in the granulation tissue around the necrotic myocardium. Fibrillar collagens play major structural roles in infarct fibrosis. In addition to fibrillar collagens, basement membrane components of the ECM, type IV collagen, perlecan proteoglycan and laminin, appear in the infarct zone and also contribute to infarct ECM reconstruction. Other glycoproteins and proteoglycans are also expressed in the infarct zone and function in ECM reconstruction through their biological activity. Matricellular proteins modulate ECM reconstruction through paracrine and autocrine processes. Among various mediators of ECM homeostasis, transforming growth factor-β1 (TGF-β1), connective tissue growth factor (CTGF) and angiotensin II function importantly in promoting infarct fibrogenesis. Stretching of the myocardial wall and hypoxia are physiological factors that are specific to myocardial infarction and stimulate infarct fibrogenic processes through enhancing the levels of the fibrogenic mediators. Intercellular fibrosis also occurs in the noninfarct zone and TGF-β1 and angiotensin II promote this fibrosis. Reperfusion and pharmacological intervention may modulate infarct fibrogenic processes and limit ventricular remodeling.

Introduction

Fibrogenesis is an essential process in the healing of infarcted myocardium. Infarct healing is an independent factor that, in addition to infarct size and ventricular wall stress, affects ventricular remodeling, that is, geometric changes of the ventricle occurring after acute myocardial infarction.[1,2] Ventricular remodeling has important effects on the function of the ventricle[2] and on the patient's prognosis.[3,4] In the fibrogenesis that occurs during the healing of myocardial infarcts, the extracellular matrix (ECM) plays a vital role not only in the ventricular remodeling but also in the ventricular function. Intervention in infarct fibrogenesis is thus important with respect to improvement of ventricular remodeling and patients' prognosis.

Fibrogenesis: Cellular and Molecular Basis, edited by Mohammed S. Razzaque.

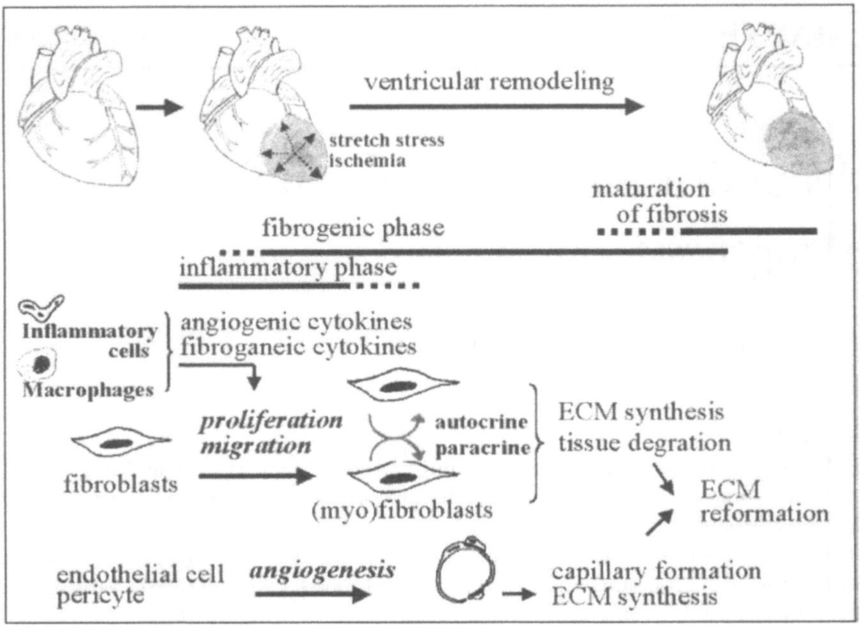

Figure 1. Schematic presentation of fibrosis in infarcted myocardium.

In the pathological course of myocardial infarction, fibrogenesis follows the inflammatory response to myocardial injury caused by the sudden cessation of blood supply. In the tissue repair process, ECM reformation, especially fibrillar collagen matrix reformation, is indispensable for the replacement of myocardial necrotic tissue. The migration and activation of cells contributing to these processes, especially macrophages and fibroblasts, ECM degradation and ECM reformation are essential for fibrogenesis. Many mediators, e.g. cytokines, growth factors and secreted ECM proteins, regulate these processes.

Characteristic pathophysiologic conditions associated with acute myocardial infarction are sudden ischemia/hypoxia of the infarcted myocardium and increased stretch stress to the myocardial wall evoked by ventricular enlargement.[5] These stimuli considerably affect infarct fibrogenesis.

In this chapter, processes of ECM degradation and reformation associated with myocardial infarction are reviewed, including our recent findings focused on the ECM and related molecules.

Cellular and Molecular Events after Induction of the Myocardial Infarction

Inflammatory Phase

Myocardial injury caused by sudden cessation of blood supply with acute coronary obstruction triggers all characteristics of inflammatory responses, that is essential processes for subsequent fibrogenesis for healing processes (Fig. 1). Various cytokines and growth factors are produced by activated macrophages and inflammatory cells. Most of them are chemoattractants and can regulate and promote fibroblast proliferation in addition to promoting angiogenesis,

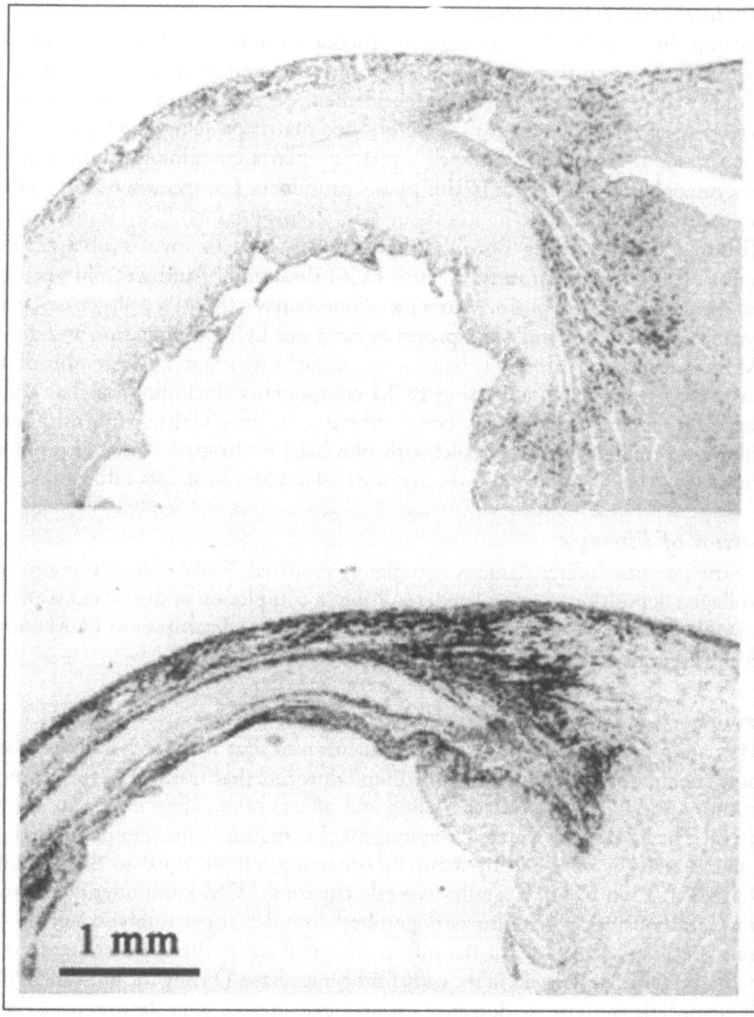

Figure 2. Digoxigenin-labelled in situ hybridization of RNA for type I collagen mRNA. Upper Panel) Type I collagen mRNA signals in the infarcted heart in experimental rat myocardial infarction on day 1. Positive signals for type I collagen mRNA can be seen in the infarct marginal zone. Lower Panel) Infarcted heart on day 7. Distinct positive signals for type I collagen can been seen in the infarct granulation tissue around necrotic infarct tissue.

which is essential for ECM reconstruction for infarct healing. The cytokines that contribute to fibrogenesis ("fibrogenic cytokines") include transforming growth factor (TGF), platelet derived growth factor (PDGF), fibroblast growth factor (FGF) and tumor necrosis factor (TNF).[6] Fibroblasts containing fibrillar collagen transcripts appears as early as the prominent inflammatory phase[7] in the infarct peripheral zone (Fig. 2, upper panel) in response to a functional cascade of fibrogenic cytokines and growth factors, and the fibrogenic process thus starts during this prominent inflammatory phase.

Granulation Tissue Formation

Following the cascade of inflammatory processes, the infarcted myocardium is reorganized to form granulation tissue. Macrophage infiltration and phagocytosis occur to debride necrotic infarct tissue. In addition to the debridement of necrotic infarct myocardium, macrophages release several cytokines, growth factors[8] and matrix proteins, including osteopontin.[9] Granulation tissue is arranged as loose or immature matrix containing fibronectin, laminin, collagens, proteoglycans and so on. In this phase, prominent fibrogenesis occurs in the granulation tissue around the necrotic infarct tissue (Fig. 2, lower panel).[10,11] Granulation tissue is rich in fibroblasts. Infiltrating fibroblasts are transformed to myofibroblasts, which are immunopositive for α-smooth muscle actin. ECM disassembly and assembly occur prominently and dynamically in this phase. Matrix metalloproteinases (MMPs; collagenases, gelatinases, proteoglycanases and so on) and other proteases carry out ECM degradation and disassembly. The ECM is reconstructed through enhancement of the biosynthesis and assembly of the ECM by activated (myo)fibroblasts expressing ECM components (including fibrillar collagen) as well as regulatory factors through autocrine and paracrine mechanisms. Fibrosis, detected by Azan-Mallory staining, occurs in parallel with fibroblast infiltration. Vascular proliferation is also prominent in this phase and also occurs in parallel with fibroblasts infiltration.

Maturation of Fibrosis

After the necrotic infarct tissue is completely removed, ECM reconstruction, including fibrillar collagen deposition, is completed, resulting in completion of the infarct scar. The levels of fibrillar collagen mRNA, however, still remains elevated, and turnover of ECM components occurs in the infarct scar.[12]

Disassembly of ECM of Infarcted Myocardium

MMPs are members of a family of metalloenzymes that can degrade injured structural ECM components in the infarcted myocardium, a process that is essential for subsequent reconstruction of the ECM for infarct healing and affects remodeling events in the infarcted myocardium. The MMPs are secreted as zymogens that require activation by limited proteolysis, and their activity is inhibited by naturally occurring inhibitors called tissue inhibitors of MMPs (TIMPs). Each MMP degrades several structural ECM components. In myocardial infarction, a study using in situ film zymography[13] found that gelatinolytic activity (MMP-9 and -2 activities) was detectable in the infarct marginal zone where inflammatory cells were abundant, and reached peak levels in the early fibrogenic phase. During the prominent fibrogenic phase, gelatinolytic activity is detected around the infarct granulation tissue mainly in (myo)fibroblasts. Increases in collagenase activity (MMPs -1, -8, -15 and -18) in the rat myocardium after infarction are observed by zymography.[14] The sequence of the increase in collagenase activity is similar to that in gelatinase activity in myocardial infarction. Transcripts for these MMPs are not concomitantly upregulated in association with their activity. Stromelysin-1 (MMP-3), which degrades many ECM components, including proteoglycan, laminin, fibronectin and other components, has been reported to be increased in infarcted myocardium as shown by zymography.[15] Membrane-type MMP-1 (MT1-MMP, MMP-14), a member of one of the more recently discovered subgroups of MMPs, activates pro-MMP-2 and pro-MMP-13.[16] MMP-14 is sensitive to inhibition by TIMP-2 and TIMP-3. Signals for MT1-MMP mRNA are observed in the infarct zone, especially around the granulation tissue during the prominent fibrogenic phase.[13] TIMP-1 expression has been reported to be down-regulated as shown by zymography in rabbit myocardial infarction, inverse of MMP-9 and MMP-3 expression.[15] In infarcted myocardium, activation of the latent form of MMPs leads to the degradation of the damaged ECM. The serum levels of MMP-1 and TIMP-1 are sequentially increased in patients with acute myocardial infarction.[17] The balance between collagenase activation and TIMP inhibition regulates the amount of collagenolysis in infarcted myocardial tissue. These lines of evidence indicate that disassembly of injured ECM components occurs dynamically and concomitantly with ECM reconstruction, especially in the infarct granulation tissue. Evidence is now accumulating that MMP inhibitors attenuate the ventricular enlargement induced by myocardial infarction.[18-20] The attenuation of ventricular enlargement in the infarcted heart by MMP inhibitors suggests that ECM disassembly may occur faster or to a greater extent than ECM reconstruction in the infarcted hearts. The balance between speed and the extent of ECM disassembly and assembly are thus important for limiting ventricular enlargement.

Regarding mediators of MMPs homeostasis, TGF-β1 has been found to upregulate TIMPs and to downregulate collagenases.[21,22] A recent study showed that TGF-β1 overexpression in transgenic mice decreases collagenase protein and activity levels and mRNA expression and increases TIMPs (TIMP-1, -2 and -4) in the hearts of the mice. TGF-β1 overexpression does not affect gelatinase (MMP-2 and -9) activity or expression. Angiotensin II, one of the fibrogenic mediators described below, has been reported to decrease collagenase activity in cultured adult cardiac fibroblasts and endothelial cells.[23,24] TIMP-1 synthesis is enhanced by angiotensin II.[25] Although the regulatory mechanism controlling the levels of MMPs and TIMPs have not been fully clarified, TGF-β1and angiotensin II are likely to modulate the disassembly of injured myocardium.

ECM Assembly

Infarct Zone

A diagram showing the sequential expression of notable ECM components in the infarcted myocardium is shown in Figure 3.

Fibrillar Collagens

Collagens are classified into several subtypes,[26,27] among which fibrillar collagens, types I and III collagens, are major components of the ECM. Among the collagen subtypes, type I collagen shows the greatest tensile strength and plays a major and essential role in infarct ECM reformation, e.g., replacement, fibrosis in response to tissue injury/necrosis, with the effect of preventing ventricular enlargement. Fibrillar collagen mRNAs do not appear within hours after coronary obstruction, rather, they appear after formation of the fibronectin scaffold[28] relatively early in the inflammatory phase in fibroblasts located in the infarct marginal zone. Immunopositive staining for type I collagen appears as a wavy pattern in the infarct marginal zone when the inflammatory response is declining or ending, a finding consistent with the results of Azan-Mallory staining.[29] Expressions of both types I and III collagen mRNAs in infarcted myocardium increases and reaches a maximal level when infarct ECM reformation, i.e., ventricular remodeling, is almost completed.[11,30] Types I and III mRNA signals are predominantly located in the infarct granulation tissue around necrotic infarct tissue. In the infarct scar several weeks after coronary ligation in rats, type I collagen mRNA is still maintained at an elevated level, and presumably contributes to turnover of the collagen matrix of the scar.[12] The time course of fibrillar collagen expression is almost paralleled by ventricular geometric changes.[11,30]

In situ hybridization has demonstrated that the cells producing fibrillar collagen mRNA are (myo)fibroblasts located in the infarct peripheral zone around necrotic infarct tissue, i.e., located in active infarct fibrogenic lesions.[30-32] The patterns of fibrillar collagen mRNA expression and localization are thus identical with the temporal and spatial changes in (myo)fibroblast distribution,[10] and (myo)fibroblasts play essential roles in infarct collagen matrix reformation.

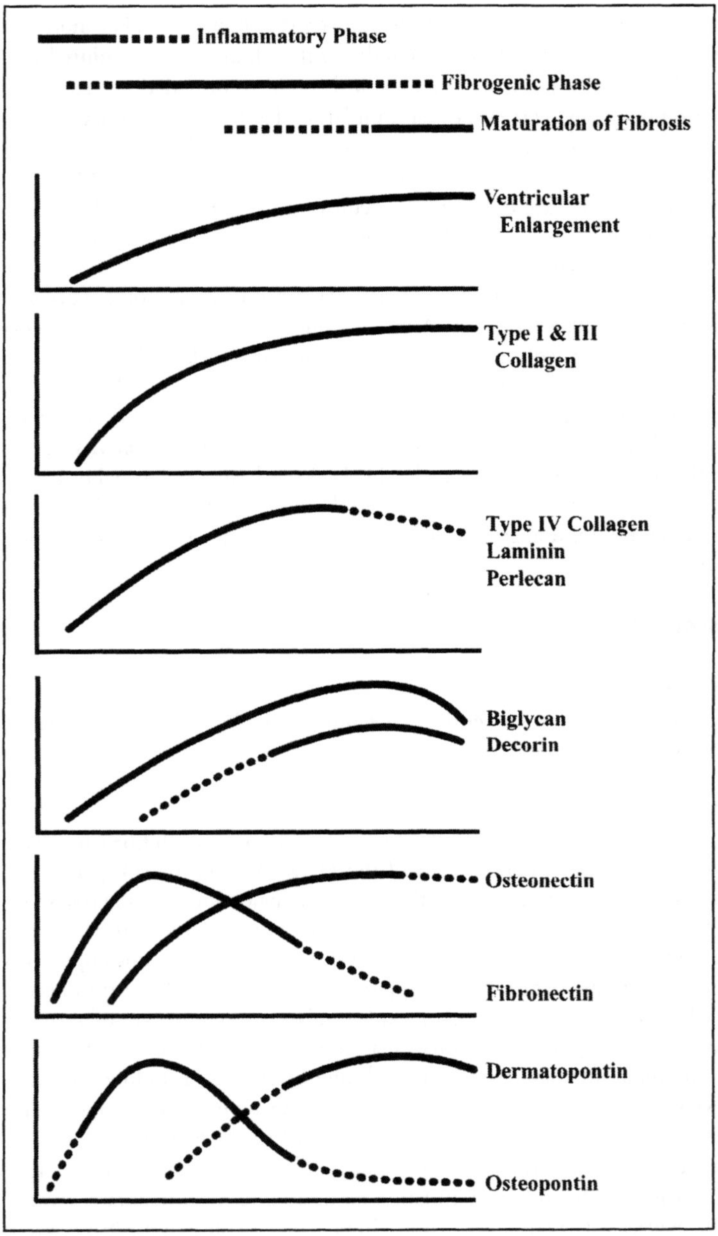

Figure 3. Diagrammatic presentation of sequential appearance of ECM and ECM-related substances in the infarcted myocardium.

Hypoxia increases the level of type I collagen mRNA in cultured human adult cardiac fibroblasts.[33] Enhancement of type I collagen synthesis by hypoxia in association with enhancement of the expression of TGF-β1, a potent stimulator of ECM formation, has also been demonstrated in human dermal fibroblasts.[34,35] Hypoxia also enhances connective tissue growth factor (CTGF) expression, which is selectively induced by TGF-β1 and a fibrillar collagen synthesis stimulator.[36] The findings indicate that hypoxia itself is one of the mediators that enhance fibrillar collagen synthesis in the infarcted myocardium.

Cyclic stretching increases the type III collagen mRNA level and the ratio of collagen type III to collagen type I in cardiac fibroblasts.[37] In other cells, stretch upregulates the transcription of TGF-β1 in smooth muscle derived from developing gut.[38] Mechanical stretch stimulates TGF-β1 mRNA expression in a time- and elongation-dependent manner in vascular smooth muscle.[39] Stretch also stimulates mRNA expression of ECM components, type I and type IV collagen, and fibronectin, in rat vascular smooth muscle cells.[39] Pressure stimuli also increase fibrillar collagen synthesis.[38] These lines of evidence indicate that increased mechanical stimulation of the ventricular wall associated with myocardial infarction is one of the important factors that regulates and enhances fibrillar collagen synthesis in the infarcted myocardium.

ECM Components of Basement Membrane

In normal tissues, basement membrane ECM components are restricted to the basement membrane structure. In fibrogenic lesions, several basement membrane ECM components, however, appear where basement membrane is not present.

Type IV Collagen

Type IV collagen is strictly distributed in the basement membrane of many tissues.[40-42] The major form of type IV collagen is a heterotrimer containing two α_1(IV) chains and one α_2(IV) chain, and this form appears to be ubiquitous in all basement membranes. In addition to the classical α_1(IV)[43] and α_2(IV) chains,[44] protein chemistry and molecular biological studies have identified the additional chains; α_3(IV), α_4(IV), α_5(IV) and α_6(IV).[45-48] Immunohistochemical studies have demonstrated the existence of the peculiar molecular form, α_3(IV) α_4(IV) α_5(IV). These studies indicated that α_3(IV) and α_4(IV) chains together with [α_5(IV)] chain formed a molecule containing triple-helix structure. Similarly, α_5(IV) and α_6(IV) make up a third molecular form $[\alpha_5(IV)]_2$ α_6(IV).[49,50]

In myocardial infarction, mRNA signals for both α_1(IV) and α_2(IV) chains begin to appear in (myo)fibroblasts located between the surviving myocytes in the infarct peripheral zone as early as the relatively early inflammatory phase.[51] Immunostaining for both α_1(IV) and α_2(IV) first appear when the inflammatory response is declining or ending, forming a wavy pattern in the infarct peripheral zone, and the staining is not restricted to the cell membrane.[52] Both α_1(IV) and α_2(IV) mRNA levels and protein expression then gradually increase. α_1(IV) and α_2(IV) are coexpressed in the infarct zone. Type IV collagen expression declines during the maturation of fibrosis. It thus appears that type IV collagen composed of α_1(IV) and α_2(IV) could contribute to matrix reformation from the early phase until the prominent fibrogenic phase of myocardial infarction.

Perlecan and Laminin

Perlecan is a major heparan sulphate proteoglycan (HSPG) that is a ubiquitous component of basement membranes.[53] The temporal changes in the perlecan mRNA signal distribution parallel those for type IV collagen mRNA. The mRNA signals for perlecan are located in (myo)fibroblasts in the infarct peripheral zone around necrotic infarct tissue. Some surviving myocytes in the infarct marginal zone also show positive signals for perlecan.

Laminin is a hetrotrimeric matrix protein. In myocardial infarction, the temporal pattern of laminin distribution is identical with that of the distribution of type IV collagen immunopositive staining.[54]

Type IV collagen binds laminin via a mechanism mediated by nidogen/entactin.[55,56] Type IV collagen also binds the heparan sulfate chain of perlecan.[57,58] Perlecan binds laminin and fibronectin through its heparan sulfate chain. Type IV collagen interacts with cells indirectly through laminin. Laminin has many biological functions[59,60] e.g., stimulation of cell attachment, migration, growth, and differentiation, which are essential for infarct fibrogenesis. Perlecan binds bFGF through its glycosaminoglycan side chain (domain I)[61-63] and functions as a receptor of bFGF. bFGF is known to have potent angiogenic function,[64,65] and has been shown to appear in the infarct zone.[66] Bound bFGF can be released from perlecan by stromelysin, collagenase, plasmin and heparanases.[62] Perlecan has been demonstrated to play a role in mitogenesis and angiogenesis via modulation of bFGF homeostasis.[61] Immunohistochemical staining showed that immunopositive staining for bFGF, which is observed around infarct granulation tissue, overlaps with perlecan staining.[30] Perlecan thus appears likely to play a role in infarct fibrogenesis via bFGF. Although the roles of these basement membrane components in infarct fibrogenesis have not yet been fully clarified, type IV collagen, laminin and perlecan are likely to form a fine multicomponent network structure and exert multifunctions that contribute to infarct ECM reconstruction.

Proteoglycans

Proteoglycans are macromolecules characterized by a core protein and one or more attached glycosaminoglycans. The solid ECM framework is made up primarily cross-linked fibers of collagens and needs coordination with the fine meshwork of proteoglycans. Proteoglycans are involved in maintaining tensile strength and in the storage of growth factors and cytokines.[67]

Biglycan

Among the proteoglycans, biglycan (previously referred to as PG I), one of the major small chondroitin sulphate/dermatan sulphate (CS/DS) proteoglycans,[68-70] has been postulated to interact with other matrix components, especially with type I collagen.[69,71-74] In myocardial infarction, the temporal pattern of biglycan mRNA expression parallels that of type I collagen mRNA expression[75] during the fibrogenic phase. The cells that express biglycan mRNA are (myo)fibroblasts and surviving myocytes located in the infarct peripheral zone around necrotic infarct tissue. Biglycan mRNA is coexpressed with type I collagen mRNA in the (myo)fibroblasts.

The concomitant expression or colocalization of biglycan with type I collagen has been reported previously in other tissues.[76] Biglycan binds type I collagen through its glycosaminoglycan chain.[72] Coordination of biglycan mRNA expression with fibrillar collagen mRNA expression in the myocardial infarct indicates that biglycan contributes to ECM framework assembly.

Like fibrillar collagen expression, biglycan expression is upregulated by TGF-β1 in cultured human skin fibroblasts.[77] Biglycan is also upregulated by bFGF in wounded tissue cultures.[78] In myocardial infarction in rats, increased TGF-β1 immunostaining is observed for at least 10 days after coronary ligation.[79] Similar concomitant increases in biglycan mRNA and TGF-β1 mRNA expression have been shown in fibrogenic processes of other pathological situations.[69,80] TGF-β1 is thus one of the mediators which enhance biglycan expression as well as fibrillar collagen expression in the infarct zone. Cyclic compression increases the biglycan biosynthesis in the tendon.[81] This process is similar to TGF-β1 induction of biglycan synthesis in the same tissue.[82,83] Hypoxia induces TGF-β1 expression in cultured dermal fibroblasts.[34] Mechanical stretching and hypoxia in the myocardial infarct zone affect biglycan synthesis similar to fibrillar collagen expression.

bFGF also upregulates biglycan synthesis.[78] In myocardial infarction, bFGF is increased during the inflammatory phase.[84] bFGF is thus thought to contribute to the initial increase in biglycan mRNA expression in the infarct zone.

Decorin

Decorin (previously referred to as PG2), is widely distributed in the ECM in association with collagen fibrils.[85] Decorin has a core protein of a size similar to that of biglycan and one CS/DS chain.[86] The leucine-rich-motif extracellular part of the core protein of decorin has been postulated to interact with other ECM molecules.[87] In myocardial infarction, in contrast to biglycan expression, decorin mRNA expression is not increased in the infarct zone during the inflammatory phase but is increased during the middle and late prominent fibrogenic phases. Decorin mRNA signals are observed in (myo)fibroblasts located in infarct granulation tissue around necrotic infarct tissue.

Biological studies have shown that decorin contributes to fiber stabilization,[88] tensile strength[89] and fiber thickness.[90] Since an increase of decorin is observed after the transition to the maturation of collagen matrix, decorin appears to contribute to the maturation of fibrillar collagen networks.

A study[91] using a collagen binding assay showed that the leucine-rich repeat of the core protein of decorin binds type I collagen. Another study revealed that decorin can also inhibit the fibrogenesis of types I and II collagens. Overexpression of decorin in decorin-gene-transfected rats attenuates the TGF-β1 mRNA expression in association with the suppression of fibrosis.[92] Furthermore, the administration of decorin inhibits the TGF-β1-mediated ECM production in experimentally induced glomerulonephritis in rats.[93] These findings suggest that decorin may be a component that counteracts the biological activity of TGF-β1, and that decorin thus contributes to ECM homeostasis via counteractive mechanism involving TGF-β1 in the maturation phase of infarct fibrosis.

Syndecan

Syndecans are cell-surface heparan sulphate proteoglycans.[94,95] In rat/mouse myocardial infarction, syndecans (syndecan-1 and -4) increase as early as 24 hours after the induction of myocardial infarction. The influx of macrophages from the bloodstream is responsible for the increase of syndecan in the infarct zone.

Biological studies have disclosed that syndecans interact with ECM proteins and cellular effectors, including fibronectin,[96] fibrillar collagen[97] and growth factors such as FGF, epidermal growth factor (EGF) and TGF,[98-100] through their heparan-sulphate chains and act as receptors and coreceptors. bFGF and heparin-binding EGF-like growth factor (HB-FGF) mRNAs increase in the infarct zone as early as within several hours in rat model of myocardial infarction.[84] Syndecans are therefore thought to play roles in regulating growth factors in the initial phase of infarct fibrogenesis.

Matrix Proteins

Fibronectin

Fibronectin is a high-molecular-weight ECM glycoprotein.[101,102] In the infarct zone of myocardial infarction, fibronectin mRNA increases relatively early in the inflammatory phase in the border zone.[103,104] Fibronectin protein begins to appear in the middle and late inflammatory phases and continues to increase during the prominent fibrogenic phase. Fibronectin protein is decreasing during maturation of infarct ECM reconstruction phase. Fibronectin thus exerts its functions from the initial until the prominent fibrogenic phase. Fibronectin EIIIA isoform is increased in myocardial infarcts.[104] Diffusions of plasma fibronectin through blood vessels and local production by fibroblasts are sources of infarct fibronectin.

A series of biological studies have demonstrated that fibronectin promotes cell migration and cell adhesion. Fibronectin is a chemoattractant for many cells and creates a scaffold to which other matrix components attach. Fibroblast migration and fibrillar collagen deposition are essential processes for infarct fibrogenesis. Fibronectin plays an indispensable role in the initiation and progression of infarct fibrogenesis.

Osteonectin

Osteonectin, an acidic secreted ECM glycoprotein, was first reported as a major noncollagenous constituent of bone.[105] Osteonectin is a Ca-binding glycoprotein that is secreted by many types of cells, including fibroblasts. In myocardial infarction, osteonectin mRNA expression does not markedly increase during the inflammatory phase.[106] Increased expression is observed in the prominent fibrogenic phase. Osteonectin mRNA signals are observed in (myo)fibroblasts and macrophages in the infarct granulation tissue around necrotic infarct tissue during the prominent fibrogenic phase.

Osteonectin has been reported to bind ECM components, including type I collagen, in a Ca-dependent manner.[107] Furthermore, coexpression of osteonectin and type I collagen has been demonstrated in cultured fibroblasts[108] and fibroblasts in rat dental tissues.[109] A study[110] on experimental wound healing in rats demonstrated that osteonectin protein and mRNA expression occur predominantly in parallel with the maturation of fibrotic changes. Osteonectin appears to play a role in mature infarct fibrosis.

TGF-β1 has been demonstrated to enhance osteonectin synthesis in addition to causing increases of type I collagen and fibronectin in cultured human fibroblasts.[111,112] Hypoxia and stretch stimulation of infarcted myocardium likely affect osteonectin expression through TGF-β1.

Osteopontin

Osteopontin, a highly acidic glycoprotein, is an ECM protein that was initially cloned from bone[113] and has been shown to bind integrin receptors though an arginine-glycine-aspartate motif;[114] these receptors function in the adhesion of many matrix molecules to the cell surface and promote cell adhesion, migration and activation.[115] Osteopontin is a multifunctional protein and evidence has been accumulated showing that osteopontin plays key roles in both granulomatous inflammation and the subsequent healing fibrosis during the pathological processes of tissue injury.[9]

In myocardial infarction, osteopontin mRNA expression increases rapidly during the inflammatory phase and then gradually decreases during the prominent fibrogenic phase.[106] Osteopontin mRNA signals are observed in macrophages in the infarct zone.

In granulomatous inflammation, osteopontin induces the migration and activation of inflammatory cells, including macrophages.[116,117] Osteopontin has been shown to negatively regulate MMP expression.[118,119] Through these functions, osteopontin is expected to contribute to the cell-mediated formation of granulation in infarcted myocardium, including the debridement of necrotic myocardium. Osteopontin also plays significant roles in the infarct healing phase that follows the inflammatory processes. In skin incisions of osteopontin-null mutant mice, a significantly decreased clearance of necrotic tissue, greater disorganization of the matrix, and alteration of collagen fibrillogenesis leading to small diameter collagen fibrils have been reported.[120] Osteopontin-null mutant mice also show less interstitial fibrosis in association with low macrophage infiltration and TGF-β1.[119] These findings suggest that osteopontin functions by regulating fibrogenic cytokine/growth factors and the recruitment of cells contributing to the regulation of fibrosis. Osteopontin is therefore thought to play an important role in the early inflammatory phase after myocardial infarction by enhancing the proliferation and migration of cells, including macrophages and fibroblasts, and likely contributes to the clearance of necrotic myocardial tissue and initiation of fibrogenesis.

Dermatopontin

Dermatopontin, a tyrosine-rich acidic matrix protein, has been recently found to be a 22-kDa secreted ECM protein.[121-125] Dermatopontin was recently demonstrated to bind several ECM components and to contribute to collagen fibril formation in vitro.[122] In myocardial infarction, dermatopontin mRNA expression does not markedly increase during the inflammatory phase but does increase during the prominent infarct fibrogenic phase.[126] Dermatopontin mRNA is found in macrophages and (myo)fibroblasts located in the infarct interior zone around necrotic infarct tissue. Dermatopontin mRNA is coexpressed with decorin and type I collagen mRNAs in (myo) fibroblasts in the infarct zone.

An in vitro study[127] examined collagen fibril formation by constructing fibrils from lathyritic rat skin type I collagen and found that dermatopontin accelerates collagen fibril formation. The study also revealed that dermatopontin stabilizes the collagen fibrils against dissociation by low temperature. Thus, dermatopontin is thought to contribute to collagen matrix maturation in the infarct zone.

Dermatopontin has been demonstrated to bind decorin and TGF-β1.[128] A study[129] used solid-phase binding assays to demonstrate that dermatopontin inhibited decorin-TGF-β1 complex formation. Decorin binds to TGF-β1 through the core protein[130] and attenuates the biological activity of TGF-β1.[130,131] The solid-phase binding study also revealed that dermatopontin enhanced the biological activity of TGF-β1. Dermatopontin is thus likely to contribute to the regulation of ECM reformation in the infarct through modifying the interaction between decorin and TGF-β1.

Mediators Affecting ECM Reformation

Many growth factors, cytokines and hormones contribute to the modulation of ECM reformation in myocardial infarction. Among mediators of ECM formation, angiotensin II and TGF-β1 play major roles in ECM reconstruction for infarct healing and have been extensively studied. The cellular response to these mediators occurs through cell surface receptors and intracellular signal transductions via multipathways with cross-talk. Because of the complexity of the intracellular signaling pathways, the signaling pathways are still poorly understood and are now under investigation (for reviews, see ref. 132, 133)

Angiotensin II

Several studies in cultured cardiac fibroblasts have indicated that angiotensin II enhances fibrillar collagen matrix formation.[134] Furthermore, angiotensin II has been reported to decrease MMP-1 activity[24] while TIMP-1 synthesis is enhanced by angiotensin II in rat heart endothelial cells.[25] These lines of evidence indicate that angiotensin II enhances fibrogenesis through an increase in fibrillar collagen synthesis and inhibition of collagen degradation.

Increases in angiotensinogen expression, angiotensin converting enzyme (ACE) binding and angiotensin II type 1 receptor (AT1 receptor) expression have been found in infarcted heart, and these findings indicate that angiotensin promotes infarct fibrogenesis. Angiotensinogen mRNA expression is increased in rat myocardial infarction, as shown by a solution hybridization assay.[135] An increase in the angiotensin II content in the scar of rat myocardial infarction has been reported.[136] ACE binding assays using quantitative autoradiography revealed increased ACE binding at the site of infarction during the prominent fibrogenic phase in rats.[137] (Myo)fibroblasts, macrophages and endothelial cells are responsible for ACE binding. ACE mRNA is increased in infarcted hearts.[138] AT1 receptor mRNA and protein are also upregulated in infarcted hearts[139] and (myo)fibroblasts expressing AT1 receptor. Angiotensin II, ACE and AT1 receptor are all locally upregulated in the infarcted heart. Angiotensin II has been reported to upregulate TGF-β1 mRNA and promote the conversion of the latent form to the active form of TGF-β1 in rat cardiac fibroblasts.[140] AT1 receptor blocker attenuates TGF-β1 and fibrillar collagen synthesis.[141] These recent lines of evidence indicate that locally induced angiotensin II promotes fibrillar collagen network reformation through TGF-β1.

TGF-β1

TGF-β1 is a key mediator of ECM formation. TGF-β1, like PDGF and bFGF, is a regulatory polypeptide for the growth of fibroblasts. TGF-β1 enhances fibrillar collagen synthesis by upregulation of collagen gene expression.[142-144] Increased expression of TGF-β1 has been reported in myocardial infarction.[141,145] The cells producing TGF-β1 in the infarct site are mainly (myo)fibroblasts. TGF-β1 mRNA is expressed in association with angiotensin II and ACE receptor binding. An AT1 receptor antagonist decreases TGF-β1 and fibrillar collagen mRNA expression at the site of infarct fibrosis. Mechanical stretching and hypoxia upregulate TGF-β1 expression.[34,38] TGF-β1 thus plays critical roles in ECM reconstruction, including fibrillar collagen matrix formation, in the infarct zone under the conditions of stretching and ischemia.

CTGF

Connective tissue growth factor (CTGF), a 36- to 38-kDa peptide, is selectively induced by TGF-β1 and has been suggested to contribute to tissue repair. In myocardial infarction, CTGF mRNA expression increases as early as the inflammatory phase and is maintained at high levels during the infarct prominent fibrogenic phase.[146] CTGF-mRNA producing cells are myocytes in the infarct marginal zone and (myo)fibroblasts. CTGF increased the expression of type I collagen.[147] CTGF has also been shown to stimulate fibronectin and integrin expression[147] and CTGF-specific receptors have recently been demonstrated in a human chondrocyte cell line.[148] A cascade of CTGF expression after TGF-β1 expression has been found in granulation tissue beds during cutaneous wound healing.[149] Thus, CTGF is likely to contribute to infarct fibrogenesis together with TGF-β1.

TNF-α

TNF-α has been reported to appear in the infarct border zone where fibrogenesis occurs.[150] In cultured cardiac fibroblasts, TNF-α stimulates the proliferation of fibroblasts and the expression of fibronectin. These facts indicate that TNF-α plays a role in early fibrogenesis in infarcted hearts.

Noninfarct Zone

The noninfarct zone, that is, zones remote from the infarct zone, also shows fibrosis, in which the total collagen content and fibrillar collagen are increased in association with increased TGF-β1.[151,152] The collagen volume in sites remote from myocardial infarction is reduced by an AT1 receptor blocker.[153] Both the TGF-β1 level and interstitial fibrosis of the noninfarct zone are attenuated by ACE inhibitors and AT1 receptor blockers.[154] Fibrosis in the noninfarct zone is thus regulated by TGF-β1 and angiotensin, as in the infarct zone. Diastolic stress is increased in the noninfarct ventricular wall as well as in the infarcted ventricular wall. Mechanical stress to the noninfarcted wall increases these mediators.

Intervention in Infarct Fibrogenesis

In theory, excessive fibrogenesis would increase ventricular stiffness, resulting in worsening of ventricular function, especially diastolic function, which would affect patients' prognosis.[155] Conversely, poor fibrogenesis causes ventricular dilation, which also affects patients' prognosis. An appropriate degree of fibrogenesis that prevents ventricular dilation without increasing ventricular stiffness is the goal of intervention in infarct fibrogenesis.

Reperfusion

Reperfusion of infarct-related artery is now applied widely in the treatment of acute myocardial infarction. An earlier study[156] using experimental myocardial infarction in rats demonstrated that 2-hour coronary ligation, i.e., too late reperfusion for myocyte salvage, produced transmural infarction and that the size of the infarct produced did not differ from that produced by permanent ligation, but infarct expansion was inhibited. Late reperfusion hastens fibrillar collagen mRNA and protein expression[7,29] without changing the distribution pattern of fibrillar collagen expression. Acceleration of infarct fibrogenic processes is expected to improve ventricular remodeling, which would improve the function of the ventricle.

Pharmacological Intervention

Evidence is now accumulating that inhibition of angiotensin II by either ACE inhibitors or AT1 receptor blockers affects ventricular remodeling and ECM deposition in myocardial infarction. ACE inhibitor limits ventricular remodeling[157,158] in association with a reduction of collagen deposition in infarcted hearts.[157] Ventricular remodeling is affected by 3 independent factors, namely, infarct size, ventricular wall stress and infarct healing.[1,2] Inhibition of angiotensin II metabolism decreases arterial pressure and total circulating volume, a resulting in a reduction of ventricular wall stress.[157,159] Reduction of ECM deposition in association with attenuation of ventricular loading appears to limit ventricular remodeling.

References

1. Weisman HF, Healy B. Myocardial infarct expansion, infarct extension, and reinfarction: pathophysiologic concepts. Prog Cardiovasc Dis 1987; 30:73-110.
2. Pfeffer MA, Braunwald E. Ventricular remodeling after myocardial infarction. Experimental observations and clinical implications. Circulation 1990; 81:1161-1172.
3. Kostuk WJ, Kazamias TM, Gander MP et al. Left ventricular size after acute myocardial infarction: serial changes and their prognostic significance. Circulation 1973; 47:1174-1179.
4. White HD, Norris RM, Brown MA et al. Left ventricular end-systolic volume as the major determinant of survival after recovery from myocardial infarction. Circulation 1987; 76:44-51.
5. Capasso JM, Li P, Zhang X et al. Heterogeneity of ventricular remodeling after acute myocardial infarction in rats. Am J Physiol 1992; 262(2 Pt 2):486-495.
6. Kovacs EJ, DiPietro LA. Fibrogenic cytokines and connective tissue production. FASEB J 1994; 8:854-861.
7. Moritani H, Kusachi S, Takeda K et al. Reperfusion accelerates the distribution of type I and III collagen messenger RNA expression after acute myocardial infarction: in situ hybridization in experimental infarction in rats. Coron Artery Dis 1999; 10:89-96.
8. Howell JM. Current and future trends in wound healing. Emerg Med Clin North Am 1992; 10:655-663.
9. Denhardt DT, Noda M, O'Regan AW et al. Osteopontin as a means to cope with environmental insults: regulation of inflammation, tissue remodeling, and cell survival. J Clin Invest 2001; 107:1055-1061.
10. Fishbein MC, Maclean D, Maroko PR. Experimental myocardial infarction in the rat: qualitative and quantitative changes during pathologic evolution. Am J Pathol 1978; 90:57-70.
11. Doi M, Kusachi S, Murakami T et al. Time-dependent changes of decorin in the infarct zone after experimentally induced myocardial infarction in rats: Comparison with biglycan. Pathol Res Pract 2000; 196:23-33.
12. Sun Y, Weber KT. Infarct scar: a dynamic tissue. Cardiovasc Res 2000; 46 250-256.
13. Suezawa C, Murakami T, Ayada Y et al. spatial changes of gelatinase activitys and membrane type 1-matrix metalloproteinase (MT1-MMP) mRNA expression in myocardial infarction in rats. Jpn Circ J 2001; 65 Supplement I-A:122.
14. Cleutjens JP, Kandala JC, Guarda E et al. Regulation of collagen degradation in the rat myocardium after infarction. J Mol Cell Cardiol 1995; 27:1281-1292.

15. Romanic AM, Burns-Kurtis CL, Gout B et al. Matrix metalloproteinase expression in cardiac myocytes following myocardial infarction in the rabbit. Life Sci 2001; 68:799-814.
16. Strnlicht MD, Werb Z. Membrane-type MMPs (MMPs 14, 15, 16 and 17). New York: Oxford University Press, 1999.
17. Hirohata S, Kusachi S, Murakami M et al. Time dependent alterations of serum matrix metalloproteinase-1 and metalloproteinase-1 tissue inhibitor after successful reperfusion of acute myocardial infarction. Heart 1997; 78:278-284.
18. Rohde LE, Ducharme A, Arroyo LH et al. Matrix metalloproteinase inhibition attenuates early left ventricular enlargement after experimental myocardial infarction in mice. Circulation 1999; 99:3063-3070.
19. Creemers EE, Cleutjens JP, Smits JF et al. Matrix metalloproteinase inhibition after myocardial infarction: a new approach to prevent heart failure. Circ Res 2001; 89:201-210.
20. Ducharme A, Frantz S, Aikawa M et al. Targeted deletion of matrix metalloproteinase-9 attenuates left ventricular enlargement and collagen accumulation after experimental myocardial infarction. J Clin Invest 2000; 106:55-62.
21. Massague J. The transforming growth factor-beta family. Annu Rev Cell Biol 1990; 6:597-641.
22. Slack JL, Liska DJ, Bornstein P. Regulation of expression of the type I collagen genes. Am J Med Genet 1993; 45:140-151.
23. Brilla CG, Zhou G, Matsubara L et al. Collagen metabolism in cultured adult rat cardiac fibroblasts: response to angiotensin II and aldosterone. J Mol Cell Cardiol 1994; 26:809-820.
24. Funck RC, Wilke A, Rupp H et al. Regulation and role of myocardial collagen matrix remodeling in hypertensive heart disease. Adv Exp Med Biol 1997; 432:35-44.
25. Chua CC, Hamdy RC, Chua BH. Angiotensin II induces TIMP-1 production in rat heart endothelial cells. Biochim Biophys Acta 1996; 28:175-180.
26. Myllyharju J, Kivirikko KI. Collagens and collagen-related diseases. Ann Med 2001; 33:7-21.
27. Trojanowska M, LeRoy EC, Eckes B et al. Pathogenesis of fibrosis: type 1 collagen and the skin. J Mol Med 1998; 76:266-274.
28. Casscells W, Kimura H, Sanchez JA et al. Immunohistochemical study of fibronectin in experimental myocardial infarction. Am J Pathol 1990; 137:801-810.
29. Yamasaki S, Kusachi S, Moritani H et al. Reperfusion hastens appearance and extent of distribution of type I collagen in infarct zone: immunohistochemical study in rat experimental infarction. Cardiovasc Res 1995; 30:763-768.
30. Nakahama M, Murakami T, Kusachi S et al. Expression of perlecan proteoglycan in the infarct zone of mouse myocardial infarction. J Mol Cell Cardiol 2000; 32:1087-1100.
31. Sun Y, Weber KT. Angiotensin converting enzyme and myofibroblasts during tissue repair in the rat heart. J Mol Cell Cardiol 1996; 28:851-858.
32. Sappino A, Schurch W, Gabbiani G. Differentiation repertoire of fibroblastic cells: expression of cytoskeletal proteins as marker of phenotypic modulations. Lab Invest 1990; 63:144-161.
33. Agocha A, Lee HW, Eghbali-Webb M. Hypoxia regulates basal and induced DNA synthesis and collagen type I production in human cardiac fibroblasts: effects of transforming growth factor beta 1, thyroid hormone, angiotensin II and basic fibroblast growth factor. J Mol Cell Cardiol 1997;:2233-2244.
34. Falanga V, Zhou L, Yufit T. Low oxygen tension stimulates collagen synthesis and COL1A1 transcription through the action of TGF-beta1. J Cell Physiol 2002; 191:42-50.
35. Falanga V, Martin TA, Takagi H et al. Low oxygen tension increases mRNA levels of alpha 1 (I) procollagen in human dermal fibroblasts. J Cell Physiol 1993; 157:408-412.
36. Kondo S, Kubota S, Shimo T et al. Connective tissue growth factor increased by hypoxia may initiate angiogenesis in collaboration with matrix metalloproteinases. Carcinogenesis 2002; 23:769-776.
37. Carver W, Nagpal ML, Nachtigal M et al. Collagen expression in mechanically stimulated cardiac fibroblasts. Circ Res 1991; 69:116-122.
38. Gutierrez JA, Perr HA. Mechanical stretch modulates TGF-beta1 and alpha1(I) collagen expression in fetal human intestinal smooth muscle cells. Am J Physiol 1999; 277:1074-1080.
39. Joki N, Kaname S, Hirakata M et al. Tyrosine-kinase dependent TGF-beta and extracellular matrix expression by mechanical stretch in vascular smooth muscle cells. Hypertens Res 2000; 23:92-99.

40. Furthmayr H. Basement membrane collagen: Structure, assembly, and biosynthesis. In: Reid LM, ed. Extracellular matrix. New York: Marcel Dekker, 1993:149-183.
41. Razzaque MS, Koji T, Taguchi T et al. In situ localization of type III and type IV collagen-expressing cells in human diabetic nephropathy. J Pathol 1994; 174:131-138.
42. Olsen BR, Ninomiya Y. Basement membrane collagen. In: Vale R, ed. Extracellular matrix, anchor and adhesion proteins. 2 ed. New York: Oxford Unversity Press, 1999:395-399.
43. Kuhn K, Wiedemann H, Timpl R et al. Macromolecular sturucture of basement membrane collagens. Identification of 7s collagen as a cross-linking domain of type IV collagen. FEBS Lett 1981; 125:123-128.
44. Bachinger HP, Fessler LI, Fessler LH. Mouse procollagen IV. Characterization and supramolecular association. J Biol Chem 1982; 257:9796-9803.
45. Butkowski RJ, Wieslander J, Wilson B et al. Properties of the globular doamin of type IV collagen and its relationship to the Goodpasture antigen. J Biol Chem 1985; 260:3739-3745.
46. Butkowski R, .J., Langeveld JPM et al. Localization of the Goodpasture epitope to a novel chain of basement membrane collagen. J Biol Chem 1987; 262:7874-7877.
47. Hostikka SL, Eddy RL, Byers MG et al. Identification of a distinct type IV collagen chain with restricted kidney distrubution and assignment of its gene to the locus of X chromosome-linked Alport syndrome. Proc Natl Acad Sci USA 1990; 87:1606-1610.
48. Zhou J, Mochizuki T, Smeets H et al. Deletion of the paired 5(IV) and 3(IV) collagen genes in inherited smooth muscle tumors. Science 1993; 261:1167-1169.
49. Ninomiya Y, Kagawa M, Iyama K et al. Differential expression of two basement membrane collagen genes, COL4A6 and COL4A5, demonstrated by immunofluorescence staining using peptide-specific monoclonal antibodies. J Cell Biol 1995; 130:1219-1229.
50. Sado Y, Kagawa M, Kishiro Y et al. Establishment by the rat lymph node method of epitope-defined monoclonal antibodies recognizing the six different alpha chains of human type IV collagen. Histochem Cell Biol 1995; 104:267-275.
51. Murakami M, Kusachi S, Nakahama M et al. Expression of the alpha 1 and alpha 2 chains of type IV collagen in the infarct zone of rat myocardial infarction. J Mol Cell Cardiol 1998; 30:1191-1202.
52. Yamanish A, Kusachi S, Nakahama M et al. Sequential changes in the localization of the type IV collagen alpha chain in the infarct zone: immunohistochemical study of experimental myocardial infarction in the rat. Pathol Res Pract 1998; 194:413-422.
53. Iozzo RV, Cohen IR, Grassel S et al. The biology of perlecan: the multifaceted heparan sulphate proteoglycan of basement membranes membranes and pericellular matrices. Biochem J 1994; 302:625-639.
54. Morishita N, Kusachi S, Yamasaki S et al. Sequential changes in laminin and type IV collagen in the infarct zone—Immunohistochemical study in rat myocardial infarction. Jpn Circ J 1996; 60:108-114.
55. Timpl R, Dziadek M, Fujiwara S et al. Nidogen: a new, self-aggregating basement membrane protein. Eur J Biochem 1983; 137:455-465.
56. Carlin B, Jaffe R, Bender B et al. Entactin, a novel basal lamina-associated sulfated glycoprotein. J Biol Chem 1981; 256:5209-5214.
57. Fujiwara S, Wiedemann H, Timpl R et al. Structure and interactions of heparan sulfate proteoglycans from a mouse tumor basement membrane. Eur J Biochem 1984; 143:143-157.
58. Battaglia C, Mayer U, Aumailley M et al. Basement-membrane heparan sulfate proteoglycan binds to laminin by its heparan sulfate chains and to nidogen by sites in the protein core. Eur J Biochem 1992; 208:359-366.
59. Beck K, Hunter I, Engel J. Structure and function of Laminin: anatomy of multidomain glycoprotein. FASEB J 1990; 4:148-160.
60. Martin GR. Laminin and other basement membrane components. Ann Rev Cell Biol 1987; 3:87-85.
61. Aviezer D, Hecht D, Safran M et al. Perlecan, basal lamina proteoglycan, promotes basic fibroblast growth factor-receptor binding, mitogenesis, and angiogenesis. Cell 1994; 79:1005-1013.
62. Whitelock JM, Murdoch AD, Iozzo RV et al. The degradation of human endothelial cell-derived perlecan and release of bound basic fibroblast growth factor by stromelysin, collagenase, plasmin, and heparanases. J Biol Chem 1996; 271:10079-10086.

63. Groffen AJ, Bukens CAF, Tryggvason K et al. Expression and characterization of human perlecan domains I and II synthesized by baculovirus-infected insect cells. Eur J Biochem 1996; 241:827-834.
64. Sellke FW, Laham RJ, Edelman ER et al. Therapeutic angiogenesis with basic fibroblast growth factor: technique and early results. Ann Thorac Surg 1998; 65:1540-1544.
65. Watanabe E, Smith DM, Sun JS et al. Effect of basic fibroblast growth factor on angiogenesis in the infarcted porcine heart. Basic Res Cardiol 1998; 93:30-37.
66. Obama H, Biro S, Tashiro T et al. Myocardial infarction induces expression of midkine, a heparin-binding growth factor with reparative activity. Anticancer Res 1998; 18:145-152.
67. Iozzo RV. Proteoglycans: Structure, biology, and molecular interactions. New York: Marcel Dekker, 2000.
68. Krusius T, Ruoslahti E. Primary structure of an extracellular matrix proteoglycan core protein deduced from cloned cDNA. Proc Natl Acad Sci USA 1986; 86:7683-7687.
69. Scott JE. Proteoglycan-fibrillar collagen interactions. Biochem J 1988; 252:313-323.
70. Fisher LW, Termine JD, Young MF. Deduced protein sequence of bone small proteoglycan I (biglycan) shows homology with proteoglycan II (decorin) and several nonconnective tissue proteins in a variety of species. J Biol Chem 1989; 264:4571-4576.
71. Oldberg A, Franzen A, Heinegard D. Cloning and sequence analysis of rat bone sialoprotein (osteopontin) cDNA reveals an Arg-Gly-Asp cell-binding sequence. J Biol Chem 1988; 25:19430-19432.
72. Pogany G, Hernandez DJ, Vogel KG. The in vitro interaction of proteoglycans with type I collagen is modulated by phosphate. Arch Biochem Biophys 1994; 313:102-111.
73. Kresse H, Hausser H, Schonherr E et al. Biosynthesis and interactions of small chondroitin/dermatan sulphate proteoglycans. Eur J Clin Chem Clin Biochem 1994; 32:259-264.
74. Schonherr E, Witsch-Prehm P, Harrach B et al. Interaction of biglycan with type I collagen. J Biol Chem 1995; 270:2776-2783.
75. Yamamoto K, Kusachi S, Ninomiya Y et al. Increase in the expression of biglycan mRNA expression Colocalized closely with that of type I collagen mRNA in the infarct zone after experimentally-induced myocardial infarction in rats. J Mol Cell Cardiol 1998; 30:1749-1756.
76. Riessen R, Isner JM, Blessing E et al. Regional differences in the distribution of the proteoglycans biglycan and decorin in the extracellular matrix of atherosclerotic and restenotic human coronary arteries. Am J Pathol 1994; 144:962-974.
77. Breuer B, Schmidt G, Kresse H. Nonuniform influence of transforming growth factor-beta on the biosynthesis of different forms of small chondroitin sulphate/dermatan sulphate proteoglycan. Biochem J 1990; 269:551-554.
78. Kinsella MG, Tsoi CK, Jarvelainen HT et al. Selective expression and processing of biglycan during migration of bovine aortic endothelial cells. The role of endogenous basic fibroblast growth factor. J Biol Chem 1997; 272:318-325.
79. Casscells W, Bazoberry F, Speir E et al. Transforming growth factor-beta 1 in normal heart and in myocardial infarction. Ann NY Acad Sci 1990; 593:148-160.
80. Westergren-Thorsson G, Hernnas J, Sarnstrand B et al. Altered expression of small proteoglycans, collagen, and transforming growth factor-beta 1 in developing bleomycin-induced pulmonary fibrosis in rats. J Clin Invest 1993; 92:632-637.
81. Evanko SP, Vogel KG. Proteoglycan synthesis in fetal tendon is differentially regulated by cyclic compression in vitro. Arch Biochem Biophys 1993; 307:153-164.
82. Vogel KG, Hernandez DJ. The effects of transforming growth factor-beta and serum on proteoglycan synthesis by tendon fibrocartilage. Eur J Cell Biol 1992; 59:304-313.
83. Robbins JR, Evanko SP, Vogel KG. Mechanical loading and TGF-beta regulate proteoglycan synthesis in tendon. Arch Biochem Biophys 1997; 342:203-211.
84. Iwabu A, Murakami T, Kusachi S et al. Concomitant expression of heparin-binding epidermal growth factor-like growth factor mRNA and basic fibroblast growth factor mRNA in myocardial infarction in rats. Basic Res Cardiol 2002; 97:217-222.
85. Kjellen L, Lindahl U. Proteoglycans: structures and interactions. Annu Rev Biochem 1991; 60:443-475.

86. Oldberg A, Antonsson P, Lindblom K et al. A collagen-binding 59-kd protein(fibromodulin) is structurally related to the small interstitial proteoglycans PG-S1 and PG-S2(decorin). EMBO J 1989; 8:2601-2604.
87. Krull NB, Zimmermann T, Gressner AM. Spatial and temporal patterns of gene expression for the proteoglycans biglycan and decorin and for transforming growth factor-beta1 revealed by in situ hybridization during experimentally induced liver fibrosis in the rat. Hepatology 1993; 18:581-589.
88. Scott JE. Proteodermatan and proteokeratan sulfate (decorin, lumican/fibromodulin) proteins are horseshoe shaped. Implications for their interactions with collagen. Biochemistry 1996; 35:8795-8799.
89. Pins GD, Christiansen DL, Patel R et al. Self-assembly of collagen fibers. Influence of fibrillar alignment and decorin on mechanical properties. Biophys J 1997; 73:2164-2172.
90. Kuc IM, Scott PG. Increased diameters of collagen fibrils precipitated in vitro in the presence of decorin from various connective tissues. Connect Tissue Res 1997; 36:287-296.
91. Svensson L, Heinegard D, Oldberg A. Decorin-binding sites for collagen type I are mainly located in leucine-rich repeats 4-5. J Biol Chem 1995; 270:20712-20716.
92. Isaka Y, Brees DK, Ikegaya K et al. Gene therapy by skeletal muscle expression of decorin prevents fibrotic disease in rat kidney. Nat Med 1996; 2:418-423.
93. Border WA, Noble NA, Yamamoto T et al. Natural inhibitor of transforming growth factor- protects against scarring in experimental kidney disease. Nature 1992; 360:361-364.
94. Kojima T, Shworak NW, Rosenberg RD. Molecular cloning and expression of two distinct cDNA-encoding heparan sulfate proteoglycan core proteins from a rat endothelial cell line. J Biol Chem 1992; 267:4870-4877.
95. Kojima T, Leone CW, Marchildon GA et al. Isolation and characterization of heparan sulfate proteoglycans produced by cloned rat microvascular endothelial cells. J Biol Chem 1992; 267:4859-4869.
96. Saunders S, Bernfield M. Cell surface proteoglycan binds mouse mammary epithelial cells to fibronectin and behaves as a receptor for interstitial matrix. J Cell Biol 1988; 106:423-430.
97. Koda JE, Rapraeger A, Bernfield M. Heparan sulfate proteoglycans from mouse mammary epithelial cells. Cell surface proteoglycan as a receptor for interstitial collagens. J Biol Chem 1985; 260:8157-8162.
98. Kiefer MC, Stephans JC, Crawford K et al. Ligand-affinity cloning and structure of a cell surface heparan sulfate proteoglycan that binds basic fibroblast growth factor. Proc Natl Acad Sci USA 1990; 87:6985-6989.
99. Subramanian SV, Fitzgerald ML, Bernfield M. Regulated shedding of syndecan-1 and -4 ectodomains by thrombin and growth factor receptor activation. J Biol Chem 1997; 272:14713-14720.
100. Elenius K, Maatta A, Salmivirta M et al. Growth factors induce 3T3 cells to express bFGF-binding syndecan. J Biol Chem 1992; 267:6435-6441.
101. Hynes RO. Fibronectins. New York: Springer, 1990.
102. Mosher DF. Fibronectin. New York: Academic Press, 1989.
103. Ulrichm MM, Janssen AM, Daemen MJ et al. Increased expression of fibronectin isoforms after myocardial infarction in rats. J Mol Cell Cardiol 1997; 29:2533-2543.
104. Knowlton AA, Connelly CM, Romo GM et al. Rapid expression of fibronectin in the rabbit heart after myocardial infarction with and without reperfusion. J Clin Invest 1992; 89:1060-1068.
105. Termine JD, Kleinman HK, Whitson SW et al. Osteonectin, a bone-specific protein linking mineral to collagen. Cell 1981; 26:99-105.
106. Sezaki S, Komatsubara I, Ayada Y et al. Spatially and temporally different expression of osteonectin and osteopontin in the infarct zone of myocardial infarction in rats. Jpn Circ J 2001; 65 Supplement I-A:278.
107. Sage EH, Vernon RB, Decker J et al. Distribution of the calcium-binding protein SPARC in tissues of embryonic and adult mice. J Histochem Cytochem 1989; 37:819-829.
108. Vuorio T, Kahari VM, Black C et al. Expression of osteonectin, decorin, and transforming growth factor-beta 1 genes in fibroblasts cultured from patients with systemic sclerosis and morphea. J Rheumatol 1991; 18:247-251.
109. Salonen J, Uitto VJ, Pan YM et al. Proliferating oral epithelial cells in culture are capable of both extracellular and intracellular degradation of interstitial collagen. Matrix 1991; 11:43-55.

110. Reed MJ, Poulakkainen P, Lane TF et al. Differential expression of SPARC and thrombospondin 1 in wound repair: Immunolocalization and in situ hybridization. J Histochem Cytochem 1993; 41:1467-1477.
111. Wrana JL, Overall CM, Sodek J. Regulation of the expression of a secreted acidic protein rich in cysteine (SPARC) in human fibroblasts by transforming growth factor-beta. Eur J Biochem 1991; 197:519-528.
112. Reed MJ, Vernon RB, Abrass IB et al. TGF-beta 1 induces the expression of type I collagen and SPARC, and enhances contraction of collagen gels, by fibroblasts from young and aged donors. J Cell Physiol 1994; 158:169-179.
113. Pytela R, Pierschbacher MD, Ginsberg MH et al. Platelet membrane glycoprotein IIb/IIIa: member of a family of Arg-Gly-Asp—Specific adhesion receptors. Science 1986; 231:1559-1562.
114. Miyauchi A, Alvarez J, Greenfield EM et al. Recognition of osteopontin and related peptides by an $\alpha v\beta 3$ integrin stimulates immediate cell signals in osteoclasts. J Biol Chem 1991; 260:20369-20374.
115. Denhardt DT, Guo X. Osteopontin: a protein with diverse functions. FASEB J 1993; 7:1475-1482.
116. Ashkar S, Weber GF, Panoutsakopoulou V et al. Eta-1 (osteopontin): an early component of type-1 (cell-mediated) immunity. Science 2000; 287:860-864.
117. Yamamoto S, Nasu K, Ishida T et al. Effect of recombinant osteopontin on adhesion and migration of P388D1 cells. Ann N Y Acad Sci 1995; 760:378-380.
118. Cowan KN, Jones PL, Rabinovitch M. Elastase and matrix metalloproteinase inhibitors induce regression, and tenascin-C antisense prevents progression, of vascular disease. J Clin Invest 2000; 105:21-34.
119. Nemir M, Bhattacharyya D, Li X et al. Targeted inhibition of osteopontin expression in the mammary gland causes abnormal morphogenesis and lactation deficiency. J Biol Chem 2000; 275:969-976.
120. Liaw L, Birk DE, Ballas CB et al. Altered wound healing in mice lacking a functional osteopontin gene (suppl). J Clin Invest 1998; 101:1468-1478.
121. Cronshow AD, MacBeath JR, Shackleton DR et al. TRAMP (tyrosine rich acidic matrix protein), a protein that copurifies with lysyl oxidase from porcine skin. Identification of TRAMP as the dermatan sulphate proteoglycan-associated 22K extracellular matrix protein. Matrix 1993; 13:255-266.
122. MacBeath JR, Shackleton DR, Hulmes DJ. Tyrosine-rich acidic matrix protein (TRAMP) accelerates collagen fibril formation in vitro. J Biol Chem 1993; 268:19826-19832.
123. Neame PJ, Choi HU, Rosenberg LC. The isolation and primary structure of a 22-kDa extracellular matrix protein from bovine skin. J Biol Chem 1989; 264:5474-5479.
124. Okamoto O, Suzuki Y, Kimura S et al. Extracellular matrix 22-kDa protein interacts with decorin core protein and is expressed in cutaneous fibrosis. J Biochem 1996; 132:106-114.
125. Supeeti-Fuga A, Rocchi M, Schafer BW et al. Complementary DNA sequence and chromosomal mapping of a human proteoglycan-binding cell-adhesion protein (dermatopontin). Genomics 1993; 17:463-467.
126. Takemoto S, Murakami T, Kusachi S et al. Increased expression of dermatopontin mRNA in the infarct zone of experimentally induced myocardial infarction in rats: Comparison with decorin and type I collagen mRNAs. Basic Res Cardiol 2002; 97:461-468.
127. MacBeath JR, Shackleton DR, Hulmes DJ et al. Tyrosine-rich acidic matrix protein (TRAMP) accelerates collagen fibril formation in vitro. J Biol Chem 1993; 268:19826-19832.
128. Takeuchi Y, Kodama Y, Matsumoto T. Bone matrix decorin binds transforming growth factor-beta and enhances its bioactivity. J Biol Chem 1994; 269:32634-32638.
129. Okamoto O, Fujiwara S, Abe M et al. Dermatopontin interacts with transforming growth factor beta and enhances its biological activity. Biochem J 1999; 337:537-541.
130. Yamaguchi Y, Mann DM, Ruoslahti E. Negative regulation of transforming growth factor-beta by the proteoglycan decorin. Nature 1990; 346:281-284.
131. Markmann AHH, Schonherr E, Kresse H. Influence of decorin expression on transforming growth factor-beta-mediated collagen gel retraction and biglycan induction. Matrix Biol 2000; 19:631-636.
132. Booz GW, Baker KM. Molecular signalling mechanisms controlling growth and function of cardiac fibroblasts. Cardiovasc Res 1995; 30:537-543.

133. Raghow R. The role of extracellular matrix in postinflammatory wound healing and fibrosis. FASEB J 1994; 8:823-831.
134. Campbell SE, Katwa LC. Angiotensin II stimulated expression of transforming growth factor-beta1 in cardiac fibroblasts and myofibroblasts. J Mol Cell Cardiol 1997; 29:1947-1958.
135. Lindpaintner K, Lu W, Neidermajer N et al. Selective activation of cardiac angiotensinogen gene expression in post-infarction ventricular remodeling in the rat. J Mol Cell Cardiol 1993; 25:133-143.
136. Yamagishi H, Kim S, Nishikimi T et al. Contribution of cardiac renin-angiotensin system to ventricular remodelling in myocardial-infarcted rats. J Mol Cell Cardiol 1998; 25:1369-1380.
137. Sun Y, Cleutjens JP, Diaz-Arias AA et al. Cardiac angiotensin converting enzyme and myocardial fibrosis in the rat. Cardiovasc Res 1995; 28:1423-1432.
138. Passier RC, Smits JF, Verluyten MJ et al. Activation of angiotensin-converting enzyme expression in infarct zone following myocardial infarction. Am J Physiol 1995; 269:H1268-1276.
139. Sun Y, Weber KT. Cells expressing angiotensin II receptors in fibrous tissue of rat heart. Cardiovasc Res 1996; 31:518-525.
140. Lee AA, Dillmann WH, McCulloch AD et al. Angiotensin II stimulates the autocrine production of transforming growth factor-beta 1 in adult rat cardiac fibroblasts. J Mol Cell Cardiol 1995; 27:2347-2357.
141. Sun Y, Zhang JQ, Zhang J et al. Angiotensin II, transforming growth factor-beta1 and repair in the infarcted heart. J Mol Cell Cardiol 1998; 30:1559-1569.
142. Raghow R, Postlethwaite AE, Keski-Oja J et al. Transforming growth factor-beta increases steady state levels of type I procollagen and fibronectin messenger RNAs posttranscriptionally in cultured human dermal fibroblasts. J Clin Invest 1987; 79:1285-1288.
143. Rossi P, Karsenty G, Roberts AB et al. A nuclear factor 1 binding site mediates the transcriptional activation of a type I collagen promoter by transforming growth factor-beta. Cell 1988; 52:405-414.
144. Varga J, Rosenbloom J, Jimenez SA. Transforming growth factor beta (TGF beta) causes a persistent increase in steady-state amounts of type I and type III collagen and fibronectin mRNAs in normal human dermal fibroblasts. Biochem J 1987; 247:597-604.
145. Thompson NL, Bazoberry F, Speir EH et al. Transforming growth factor beta-1 in acute myocardial infarction in rats. Growth Factors 1988; 1:91-99.
146. Ohnishi H, Oka T, Kusachi S et al. Increased expression of connective tissue growth factor in the infarct zone of experimentally induced myocardial infarction in rats. J Mol Cell Cardiol 1998; 30:2411-2422.
147. Frazier K, Williams S, Kothapalli D et al. Stimulation of fibroblast cell growth, matrix production, and granulation tissue formation by connective tissue growth factor. J Invest Dermatol 1996; 107:404-411.
148. Nishida T, Nakanishi T, Shimo T et al. Demonstration of receptors specific for connectve tissue growth factor on a human chondrocyte cell line (HCS-2/8). Biochem Biophys Res Commun 1998; 247:905-909.
149. Igarashi A, Okochi H, Bradham DM et al. Regulation of connective tissue growth factor gene expression in human skin fibroblasts and during wound repair. Mol Biol Cell 1993; 4:637-645.
150. Jacobs M, Staufenberger S, Gergs U et al. Tumor necrosis factor-alpha at acute myocardial infarction in rats and effects on cardiac fibroblasts. J Mol Cell Cardiol 1999; 31:1949-1959.
151. Yu CM, Tipoe GL, Wing-Hon Lai K et al. Effects of combination of angiotensin-converting enzyme inhibitor and angiotensin receptor antagonist on inflammatory cellular infiltration and myocardial interstitial fibrosis after acute myocardial infarction. J Am Coll Cardiol 2001; 38:1207-1215.
152. Sun Y, Zhang JQ, Zhang J et al. Cardiac remodeling by fibrous tissue after infarction in rats. J Lab Clin Med 2000; 135:316-323.
153. De Carvalho Frimm C, Sun Y, Weber KT. Angiotensin II receptor blockade and myocardial fibrosis of the infarcted rat heart. J Lab Clin Med 1997; 129:439-446.
154. Youn TJ, Kim HS, Oh BH. Ventricular remodeling and transforming growth factor-beta 1 mRNA expression after nontransmural myocardial infarction in rats: effects of angiotensin converting enzyme inhibition and angiotensin II type 1 receptor blockade. Basic Res Cardiol 1999; 94:246-253.
155. Zile MR, Brutsaert DL. New concepts in diastolic dysfunction and diastolic heart failure: Part II: causal mechanisms and treatment. Circulation 2002; 105:1503-1508.

156. Hochman JS, Choo H. Limitation of myocardial infarct expansion by reperfusion independent of myocardial salvage. Circulation 1987; 75:299-306.
157. Jugdutt BI, Menon V. Beneficial effects of therapy on the progression of structural remodeling during healing after reperfused and nonreperfused myocardial infarction: different effects on different parameters. J Cardiovasc Pharmacol Ther 2002; 7:95.
158. Ali SM, Brown Jr EJ, Nallapati SR et al. Early angiotensin converting enzyme inhibitor therapy after experimental myocardial infarction prevents left ventricular dilation by reducing infarct expansion: a possible mechanism of clinical benefits. Coron Artery Dis 1998; 9:815-821.
159. Litwin SE, Raya TE, Warner A et al. Effects of captopril on contractility after myocardial infarction: experimental observations. Am J Cardiol 1991; 68:26D-34D.

Chapter 8

Cardiac Fibrosis and Aging

Serge Masson, Roberto Latini, Monica Salio and Fabio Fiordaliso

Abstract

Excessive deposition of extracellular matrix proteins during aging leads to a progressive reduction of myocardial and arterial compliances. The increased cardiovascular stiffness may in turn determine a reduced capacity of the aged heart to respond to stressful situations. Our knowledge on the biology of extracellular matrix during aging derives mainly from observations made in animal models. The age-dependent accumulation of collagen in the heart is not related to an increased synthesis, but rather to a reduction of its degradation by specific matrix metalloproteinases and to an increased cross-linking of mature collagen that renders it more resistant to degradation. Specific pharmacological treatments are currently in use to limit or even reverse the progressive stiffening of the arteries and myocardium.

Age-Dependent Cardiac Fibrosis

The incidence and severity of cardiovascular diseases, and ultimately of heart failure, increases with age.[1-3] In apparently healthy individuals, mild changes in cardiac structure and function develop with advancing age, in part as an adaptive response to a stiffening of the arterial system.[2] Though left ventricular ejection fraction and contractility are generally well preserved, myocardial and arterial compliance, baroreceptor sensitivity and response to beta-adrenergic agonists may be compromised. There is a progressive reduction of ventricular myocyte number in the physiological aging heart of rodents and human.[4,5] This phenomenon triggers in turn a reactive hypertrophy of the remaining myocytes and, often associated with necrotic cell death, and deposition of interstitial collagen. Interestingly, there is a striking gender difference related to cardiac aging: whereas the number of myocytes in the left and right ventricles progressively decreases in men (45 and 19 million myocyte loss/year, respectively), no such loss and compensatory hypertrophy is observed in women's heart during aging.[6] The amount of replacement and interstitial fibrosis in these hearts was however not reported in this study.

Ventricular function is influenced by several factors, including molecular characteristics of the interstitium, and more precisely by the amount, phenotype and degree of cross-linking of collagen network. Extracellular matrix in the myocardium provides a scaffold for myocytes and blood vessels, serves to maintain tissue architecture and coordinates the delivery of force generated by myocytes. Myocytes, being large cells, account for 80% of the left ventricular volume, but the fibroblasts, responsible for collagen production,[7] represent as much as two-thirds of the cell population in the heart. Fibroblasts synthesize different hormones and cyotokines, but also the two prevailing forms of fibrillar collagen in the extracellular matrix, i.e., type I (85%) and

Fibrogenesis: Cellular and Molecular Basis, edited by Mohammed S. Razzaque.

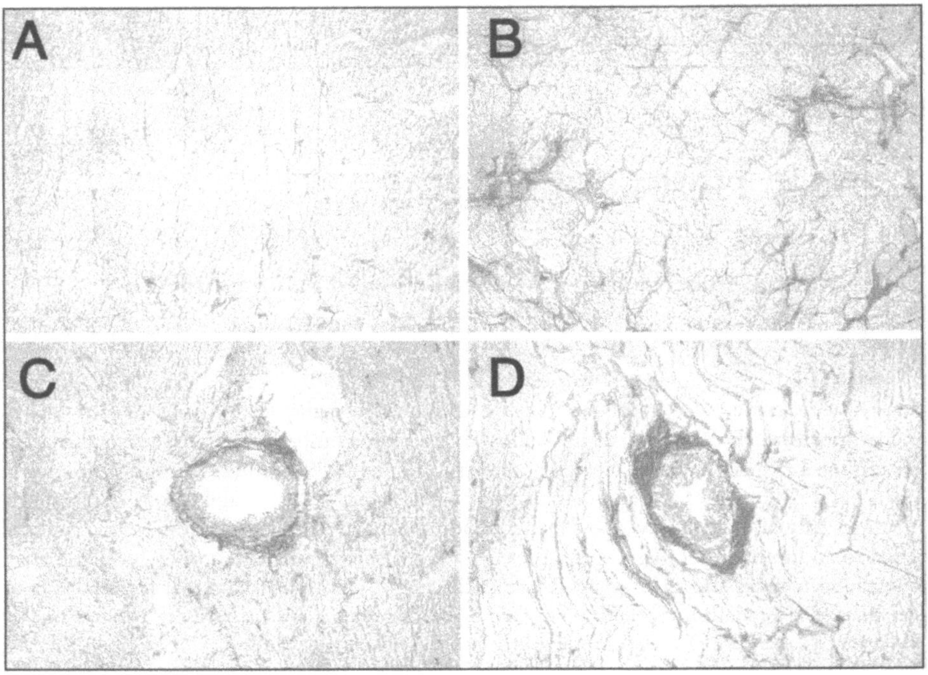

Figure 1. Microphotographs of 5-μm histological sections of the left ventricular myocardium from 3-month (A,C) and a 18-month old rats (B,D) stained with picrosirius red, a dye specific for collagen. Interstitial fibrosis (A,B) and perivascular (C,D) collagen deposition around intramural coronary arteries are significantly elevated in the aged animals.

III (11%) collagens.[8,9] While type I collagen fibers have a high tensile strength, type III fibers are more distensible and play an important role in all the processes of tissue remodeling after acute or chronic injuries. In the rat, a generally accepted model of cardiac aging, the amount of collagen remains essentially unchanged in the adult and augments more steadily in old and senescent animals.[10] For instance, the amount of collagen increases by 2.3-2.8 times in the left and right ventricles of old (26 months) Fischer 344 rats compared to younger animals (1 month). Comparable results are reported with different analytical techniques for measuring collagen (histology, immunohistology) in different strains of rats (Figs. 1,2).[11-15] The relative proportion of types I and III collagen also changes with age, being one-third at 2 weeks of age but reaching more than 50% after 1 year in the heart.[16] Collagen content increases with age in the normal human heart.[17] Likewise, hydroxyproline content increases with age in human heart muscle, but more in males than in females.[18] In human subjects free of pathologies, a study on necroptic cardiac tissue indicates that collagen content (mainly type I and III) increases by almost 50% between classes of ages ranging from 20-25 to 67-87 years.[19] The diameter of collagen fibrils is also greater in the old hearts.

Mature collagen fibrils are degraded by interstitial collagenases yielding stable telopeptides that can be readily measurable in plasma or serum. There is now accumulating evidence to show that circulating levels of these procollagen peptide fragments can be used as markers for cardiac collagen turnover. For instance, procollagen type I C-terminal propeptide (PIP) provides indirect diagnostic information on the extent of myocardial fibrosis and the efficacy of

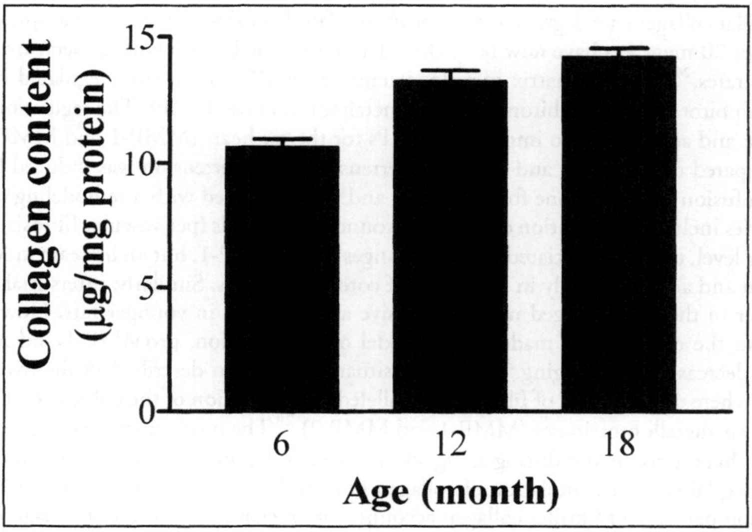

Figure 2. Quantitative accumulation of total collagen in the myocardium of aged rats. Collagen content was determined with a biochemical method using Sirius red dye in formalin-fixed histological sections.[44] Results are expressed as mean ± SEM, relative to total protein content.

hypertensive drugs in reducing collagen synthesis and fibrosis.[20] The high baseline level of procollagen type III amino-terminal peptide (PIIINP) is associated to poor outcome in patients with severe heart failure, with a mean age of 69 years, and decreases in patients receiving an aldosterone receptor antagonist enrolled to the RALES clinical trial.[21] Even more interesting is the suggestion that limitation of excessive ECM turnover might contribute to the reduction of mortality observed in the patients randomized to receive spironolactone. Circulating levels of matrix metalloproteinases, the enzymes that degrades collagen, may also be used as markers of its turnover. A word of caution should be however put forward since most observations on the potential role of makers of ECM turnover in cardiac diseases were made in a context of relatively selected patient populations. The significance of these markers in the aged population, where the incidence of comorbidities associated to elevated collagen turnover is high (pulmonary and liver fibrosis, bone remodeling), needed to be defined.

Why Does Collagen Accumulate in the Aged Myocardium?

Extracellular matrix is not an inert structure but rather a dynamic and complex system regulated at several pre and post-transcriptional levels, including post-translational processing. Collagen levels are determined by the balance between synthesis and degradation.[22] Experimental studies measuring the incorporation of radiolabeled proline into newly formed collagen have shown that collagen fractional synthesis amounts to 19% per day in the heart of 1-month old rat, significantly more than what happens in other organs like the lungs, skeletal muscles or skin.[23] Synthesis remains high in rats aged 15 months (they have a lifespan of 24-28 months), but falls sharply in senescent animals (2% per day at the age of 24 months). However, most of the newly synthesized collagen is rapidly degraded, especially in the adult and aged rats, where ~ 90% of this pool of collagen is degraded every day in the myocardium.[23] On the other hand, the expression of myocardial fibrillar collagens I and III messenger RNA is reduced during aging.[12,14] Therefore, alterations in the catabolism of collagen or in its stability are required to explain the age-dependent accumulation of myocardial collagen.

Fibrillar collagens are degraded in the heart by a family of zinc-containing endoproteinases (more than 20 members have now been cloned and identified), each with a spectrum of specific substrates.[24-26] These matrix metalloproteinases (MMP) are in turn regulated by their natural inhibitors (tissue inhibitors of matrix metalloproteinase, TIMP). The regulation of the expression and activity of two important MMPs for the rat heart (MMP-1 and MMP-2) has been compared during aging and during hypertension.[27] Hypertension was induced by continuous infusion of aldosterone for 2 months, and was associated with a remodeling of coronary arteries including deposition of collagen around these vessels (perivascular fibrosis). At the molecular level, this was associated with no changes in proMMP-1, but an increase in MMP-2 expression and activity, mainly in the media of coronary arteries. Similarly, interstitial fibrosis was higher in the heart of aged nonhypertensive animals than in younger rats. However, in contrast to the observations made in the model of hypertension, proMMP-1 and MMP-2 activities decreased during aging.[27] A similar situation has been described in the liver of the aging rat where development of fibrosis is paralleled by a reduction of the collagenolytic activity of matrix metalloproteinases (MMP-1 and MMP-2).[28] The mechanisms leading to cardiac fibrosis in hypertension and during aging are therefore different: whereas increase in collagen synthesis explains the accumulation of collagen in a model of hypertension, reduction of the degradation pathway of fibrillar collagen accounts for excessive deposition of collagen during aging.

Biosynthesis of fully mature and functional fibrillar collagen includes an extremely complex post-translational processing.[29] Briefly, extracellular fibrils align themselves head-to-tail and specific lysine and hydroxylysine residues are deaminated oxidatively by the enzyme lysyl oxidase. The resulting aldehydes functions undergo a nonenzymatic condensation, creating covalent cross-links that stabilize fibrillar collagen. In a further step of maturation, condensation of two reducible divalent keto-imine cross-links yields a mature and nonreducible trivalent cross-link that increases markedly the tensile strength of collagen and limits its degradation.[30] The degree of maturation (cross-link) may be measured in tissue by assaying hydroxylysylpyridinoline (HP) residues after hydrolysis. The concentration of ventricular HP increases approximately fivefold in senescent Fischer 344 rats (23 months) compared to sedentary young adults (5 months).[31] HP cross-link concentration is significantly elevated in the left ventricle free wall and septum, but not in the right ventricle myocardium of old sedentary rats.[31] Interestingly, training of the aged animals (treadmill walking) over a week period completely abrogates the age-associated increase in collagen cross-link seen in the left ventricle.[31,32] In conclusion, the excessive fibrosis observed in the aged myocardium is determined by a decreased activity of the enzymes of the degradative pathway of mature collagen and an increased cross-linking, and therefore resistance to degradation of extracellular collagen. As a consequence, myocardial stiffness increases in the aged myocardium as ventricular fibrosis progresses, as seen in the hypertensive heart.[33] However, the degree of cross-linking, rather than the total amount of collagen or the ratio between types I and III, seems to be determinant for myocardial stiffness.[34]

Attenuation of Myocardial Stiffness During Aging

Long-lived proteins (including vascular and myocardial collagen) undergo continuous cross-linking through reaction with reducing sugars. These reactions slowly form advanced glycosylation end products (AGEs). Mounting evidences indicate that AGEs can negatively influence vascular elasticity and myocardial compliance through cross linking with ECM collagen, or other proteins. Inhibitors of AGEs have therefore been proposed in different experimental settings to reduce this detrimental influence on organ function. Two therapeutic approaches are available, one inhibiting formation and one breaking established AGE cross-links.

Aminoguanidine is one of the first available inhibitor of AGEs formation. Chronic administration of aminoguanidine in rats treated from 6 to 24 months reduces the age-related cardiac hypertrophy and the decrease in vasodilatory response observed in old untreated animals.[35] In 30-month old rats treated for 6 months, aminoguanidine can also prevent cardiac hypertrophy and the increase in aortic impedance observed in the aged animals.[36] This reduction of age-related arterial stiffening by aminoguanidine occurs in absence of modifications of the arterial wall: neither collagen or elastin content, nor media thickness or the number of smooth muscle cells are altered by the treatment.

AGE cross-link breakers is a class of thiazolium derivatives that are able to break established AGE cross-links. Their efficacy has been tested in aged dogs, and after 1 month of treatment, the age-related left ventricular stiffness is significantly reduced in the group of animals receiving the AGE breaker.[37] This is accompanied by an improvement in cardiac function and underlines how a brief treatment is apparently able to reverse left ventricular stiffness. The same compound (ALT-711) not only improves arterial and left ventricular chamber compliance, but also cardiac output in old rhesus monkeys.[38] In aged human subjects with arterial wall stiffening, this AGE breaker reduces significantly pulse pressure and improves arterial compliance, compared to placebo.[39]

Synthesis of collagen by cardiac fibroblasts is under neurohormonal control. In particular, hormones of the renin-angiotensin-aldosterone system (angiotensin II and aldosterone) and endothelin are believed to stimulate collagen production. Consequently, drugs reducing the activity of these stimulators should contribute to limit the accumulation of collagen in the myocardium. It is however difficult to assess the impact of these drugs on age-dependent fibrosis per se or on its reversal, in absence of other cardiac pathologies known to induce collagen deposition (hypertension, myocardial infarction, heart failure). For instance, although there are experimental and clinical evidences that myocardial fibrosis is increased in hypertensive heart disease, the respective contribution of hemodynamic overload and hormonal stimulation to collagen synthesis in hypertension is still debated.[40] In an experimental study, cardiovascular changes elicited by long-term inhibition of the renin-angiotensin system has been evaluated, in mice treated with an ACE inhibitor from weaning to 24 months of age.[41] Myocardial fibrosis that developed during natural aging in these mice was completely prevented by the life-long ACE inhibition, as well as the thickening of arterial tunica media from the aorta and arteries in the lungs, brain, kidney and heart. Such long-term therapies are however of limited interest in humans during natural aging of otherwise healthy individuals. Age-related cardiac fibrosis being merely due to a diminished degradation rather than an augmented biosynthesis, further observations on the reversal of myocardial fibrosis in aged human subjects by drugs reducing neurohormonal stimulation of collagen synthesis are required.

Diastolic Dysfunction and Failure

This has been an area of great debate, mainly due to the difficulty in establishing the existence of a diastolic dysfunction. The improvement of noninvasive methodologies, in particular Doppler assessment of diastolic transmitral flow velocities, have allowed a better characterization of the entity. Increased collagen deposition and/or its cross-linking is a major determinant of increased myocardial stiffness, which together with altered relaxation create the basis for development of diastolic dysfunction. Diastolic dysfunction eventually leads to diastolic heart failure.[42] It has been shown that the prevalence of diastolic heart failure increases with age, being around 15% in patients below 50 years of age, and as high as 50% in those aged more than 70.[43] Diastolic heart failure carries a bad prognosis with a 5-year morbidity and a mortality around 50%. There is now an increasing interest in designing new therapeutic strategies to treat this disease, affecting mainly the elderly. Blockade of the AT_1 receptor for angiotensin II is being tested in 2 randomized clinical trials, CHARM (Candesartan Cilexetil in Heart Failure Assessment of Reduction in Mortality and Morbidity) and Wake Forest. The rationale of these trials is that selectively blocking angiotensin II type 1 receptor will decrease collagen deposition in the myocardium and by this way, will improve diastolic heart failure and outcome. In case angiotensin receptor blockers will prove beneficial, they will become an important therapeutic option, for a disease very frequent in the elderly.

References

1. McKee PA, Castelli WP, McNamara PM et al. The natural history of congestive heart failure: The Framingham study. New Engl J Med 1971; 285:1441-1446.
2. Lakatta EG. Cardiovascular aging research: The next horizons. J Am Geriat Soc 1999; 47:613-625.
3. Sharpe N, Doughty R. Epidemiology of heart failure and ventricular dysfunction. Lancet 1998; 352:3-7.
4. Kajstura J, Cheng W, Sarangarajan R et al. Necrotic and apoptotic myocyte cell death in the aging heart of Fischer 344 rats. Am J Physiol 1996; 271:H1215-H1228.
5. Olivetti G, Melissari M, Capasso JM et al. Cardiomyopathy of the aging human heart. Myocyte loss and reactive cellular hypertrophy. Circ Res 1991; 68:1560-1568.
6. Olivetti G, Giordano G, Corradi D et al. Gender differences and aging: Effects of the human heart. J Am Coll Cardiol 1995; 26:1068-1079.
7. Eghbali M, Czaja MJ, Zeydel M et al. Collagen mRNAs in isolated adult heart cells. J Mol Cell Cardiol 1988; 20:267-276.
8. Medugorac I, Jacob R. Characterization of left ventricular collagen in the rat. Cardiovasc Res 1983; 17:15-21.
9. Bishop J, Greenbaum J, Gibson D et al. Enhanced deposition of predominantly type I collagen in myocardial disease. J Mol Cell Cardiol 1990; 22:1157-1165.
10. Eghbali M, Robinson TF Seifter S. et al. Collagen accumulation in heart ventricles as a function of growth and aging. Cardiovasc Res 1989; 23:723-9729.
11. Capasso JM, Palackal T, Olivetti G et al. Severe myocardial dysfunction induced by ventricular remodeling in the aging rat hearts. Am J Physiol 1990; 259:H1086-H1096.
12. Besse S, Robert V, Assayag P et al. Nonsynchronous changes in myocardial collagen mRNA and protein during aging: Effect of DOCA-salt hypertension. Am J Physiol 1994; 267:H2237-H2244.
13. Thomas DP, McCormick RJ, Zimmerman SD et al. Aging- and training-induced alterations in collagen characteristics of rat left ventricle and papillary muscle. Am J Physiol 1992; H778-H783.
14. Annoni G, Luvarà G, Arosio B et al. Age-dependent expression of fibrosis-related genes and collagen deposition in the rat myocardium. Mech Ageing Dev 1998; 101:57-72.
15. Masson S, Arosio B, Fiordaliso F et al. Left ventricular response to β-adrenergic stimulation in aging rats. J Gerontol 2000; 55A:B35-B41.
16. Mays PK, Bishop JE, Laurent GJ. Age-related changes in the proportion of types I and III collagen. Mech Ageing Dev 1988; 45:203-212.
17. Burns TR, Klima M, Teasdale TA et al. Morphometry of the aging heart. Mod Pathol 1990; 3:336-342.
18. Ito A, Yamagiwa H, Sasaki R. Effects of aging on hydroxyproline in human heart muscle. J Am Geriatr Soc 1980; 28:398-404.
19. Gazoti Debessa CR, Mesiano Maifrino LB, Rodriguez de Souza R. Age related changes of the collagen network of the human heart. Mech Ageing Dev 2001; 122:1049-1058.
20. López B, Gonzàlez A, Varo N et al. Biochemical assessment of myocardial fibrosis in hypertensive heart disease. Hypertension 2001; 38:1222-1226.
21. Zannad F, Alla F, Dousset B et al. On behalf of the RALES Investigators. Limitation of excessive extracellular matrix turnover may contribute to survival benefit of spironolactone therapy in patients with congestive heart failure. Insights from the Randomized Aldactone Evaluation Study (RALES). Circulation 2000; 102:2700-2706.
22. Laurent GJ. Dynamic state of collagen: Pathways of collagen degradation in vivo and their possible role in regulation of cardiac mass. Am J Physiol 1987; 252:C1-C9.

23. Mays PK, McAnulty RJ, Campa JS et al. Age-related changes in collagen synthesis and degradation in rat tissues. Importance of degradation of newly synthesized collagen in regulating collagen production. Biochem J 1991; 276:307-313.
24. Dollery CM, McEwan JR, Henney AM. Matrix metalloproteinases and cardiovascular disease. Circ Res 1995; 77:863-868.
25. Creemers EEJM, Cleutjens JPM, Smits JFM et al. Matrix metalloproteinase inhibition after myocardial infarction. A new approach to prevent heart failure? Circ Res 2001; 89:201-210.
26. Spinale FG. Matrix metalloproteinases. Regulation and dysregulation in the failing heart. Circ Res 2002; 90:520-530.
27. Robert V, Besse S, Sabri A et al. Differential regulation of matrix metalloproteinases associated with aging and hypertension in the rat heart. Lab Invest 1997; 76:729-738.
28. Gagliano N, Arosiso B, Grizzi F et al. Reduced collagenolytic activity of matrix metalloproteinases and development of liver fibrosis in the aging rat. Mech Ageing Dev 2002; 123:413-425.
29. Van der Rest M, Garrone R. Collagen family of proteins. FASEB J 1991; 5:2814-2823.
30. Reiser KM, McCormick RJ, Rucker RB. Enzymatic and nonenzymatic cross-linking of collagen and elastin. FASEB J 1992; 6:2439-2449.
31. Thomas DP, Cotter TA, Li X et al. Exercise training attenuates aging-associated increases in collagen and collagen crosslinking of the left but not right ventricle in the rat. Eur J Appl Physiol 2001; 85:164-169.
32. Thomas DP, McCormick RJ, Zimmerman SD et al. Aging- and training-induced alterations in collagen characteristics of rat left ventricle and papillary muscle. Am J Physiol 1992; 263:H778-H783.
33. Yamamoto K, Masuyama T, Sakata Y et al. Myocardial stiffness is determined by ventricular fibrosis, but not by compensatory or excessive hypertrophy in hypertensive heart. Cardiovasc Res 2002; 55:76-82.
34. Norton GR, Tsotetsi J, Trifunovic B et al. Myocardial stiffness is attributed to alterations in cross-linked collagen rather than total collagen or phenotypes in spontaneously hypertensive rats. Circulation 1997; 96:1991-1998.
35. Li YM, Steffes M, Donnelly T et al. Prevention of cardiovascular and renal pathology of aging by the advanced glycation inhibitor aminoguanidine. Proc Natl Acad Sci USA 1996; 93:3902-3907.
36. Corman B, Duriez M, Poitevin P et al. Aminoguanidine prevents age-related arterial stiffening and cardiac hypertrophy. Proc Natl Acad Sci USA 1998; 95:1301-1306.
37. Asif M, Egan J, Vasan S et al. An advanced glycation endproduct cross-link breaker can reverse age-related increases in myocardial stiffness. Proc Natl Acad Sci USA 1999; 97:2809-2813.
38. Vaitkevicius PV, Lane M, Spurgeon H et al. A cross-link breaker has sustained effects on arterial and ventricular properties in older rhesus monkeys. Proc Natl Acad Sci USA 2000; 98:1171-1175.
39. Kass DA, Shapiro EP, Kawaguchi M et al. Improved arterial compliance by a novel advanced glycation end-product crosslink breaker. Circulation 2001; 104:1464-1470.
40. Weber KT, Sun Y, Guarda E. Structural remodeling in hypertensive heart disease and the role of hormones. Hypertension 1994; 23:869-877.
41. Inserra F, Romano L, Ercole L et al. Cardiovascular changes by long-term inhibition of the renin-angiotensin system in aging. Hypertension 1995; 25:437-442.
42. Zile MR, Brutsaert DL. New concepts in diastolic dysfunction and diastolic heart failure: Part I. Diagnosis, prognosis, and measurements of diastolic function. Circulation 2002; 105:1387-1393.
43. Zile MR, Brutsaert DL. New concepts in diastolic dysfunction and diastolic heart failure: Part II. Causal mechanism and treatment. Circulation 2002; 105:1503-1508.
44. López-de León A, Rojkind M. A simple micromethod for collagen and total protein determination in formalin-fixed paraffin-embedded sections. J Histochem Cytochem 1985; 33:737-743.

Chapter 9

Matrix Remodeling and Atherosclerosis Effect of Age

Ladislas Robert

Abstract

Athero-arteriosclerosis is the most common age-related pathology. Frequent manifestations are thrombo-embolic incidents, mostly of the coronary circu-lation and heart insufficiency resulting from an increased work load on the heart as a result of progressive increase of peripheral resistance and loss of contractile cardiac muscle fibers. Major advances made during the second half of the XXth century in the cell-biological-biochemical description of the underlying mechanisms led to increasing success of prevention, treatment and, displacing the pathological consequences to later years of life. This will result in the XXIst century in a new emphasis on the importance of biochemical and structural alterations of the large vessel wall, mainly responsible for the increasing risk of heart failure due to overload. This process, arteriosclerosis, can best be described by the progressive modification of the macromolecular composition of the vascular extracellular matrix. This is partially the result of a progressive shift in the phenotype of vascular smooth muscle cells and partially the result of postsynthetic modifications of matrix macromolecules. We shall summarize some of the most important findings made over the years in a number of laboratories concerning this process.

Introduction

Age-dependent remodeling of vascular extracellular matrix (ECM) is a complex process and plays a very important role in a number of age-related cardiovascular pathologies such as athero-arteriosclerosis, heart insufficiency, cerebrovascular disease and diabetic micro- and macro-angiopathies, to mention only the most frequent pathologies. After a transient optimism, sustained mainly by the important results of Alvar Svanborg's investigations in Goteborg,[1,2] it became clear that "dying of old age" means essentially to die of cardiovascular failure. Those who escape age-related dementias (vascular factors are important in this respect also) and cancer, have to face the fact that heart failure, and heart arrest often preceded by arythmia and fibrillation will represent the fatal end-point. Autopsy studies on centenarians reported already at the middle of the last century by L. Haranghy[3,4] and cited in a previous review[5] clearly showed the prevalence of cardiovascular pathologies at this advanced age. This was even more so surprising because the autopsied persons in the above-cited study had a variety of socio-professional conditions. But these autopsies as well as a number of other reports on persons autopsied at advanced ages (above 90 years) also confirmed the presence of several possibly fatal pathologies (the so-called polypathology of old age) with cardiovascular alterations as a

Fibrogenesis: Cellular and Molecular Basis, edited by Mohammed S. Razzaque.

constant feature.[3,5] It appears therefore of importance to have a closer look on the underlying mechanisms of these age-related cardiovascular diseases. As will be shown, progressive remodeling of the vascular wall, essentially of its extracellular matrix is a central feature of all the described modifications.

The Genesis of the Disease

Considered for long as a "degenerative disease", athero-arteriosclerosis owes this double designation to the observations that two apparently independent processes can be distinguished in the vascular wall. One which was first described by pathologists of the early XIXth century as Lobstein in Strasbourg emphasize the hardening of the vessel wall which was sometimes compared to calcified pipes. This process, arteriosclerosis, is clearly age-dependent. The other morphological manifestation which attracted the attention of most of the scientists since the second half of the XXth century concerns the formation of lipidic plaques, as observed also in rabbits fed a cholesterol-rich diet. This model rendered much easier the study of atherogenesis since the description by Anitchkoff of his rabbit-cholesterol model experiments (reviewed in ref. 6). The discovery of serum-lipoproteins by M. Macheboeuf in Paris at the middle of the XXth century, followed by ever increasing precision of their detailed description largely facilitated the understanding of the lipid-initiated, local disease, atheromatosis. However pathologists infrequently encountered one form without the other, the term coined by the German physician Marchand, athero-arteriosclerosis became legitimate. This legitimacy is however questionable for at least two reasons. One is the observation of lipidic plaques in young children, on perfectly elastic vascular walls, as shown for instance by the important epidemiological study of G. Berenson (the Bogalusa study: ref. 7). On the other hand it was reported by Max Burger[8] that human (and also horse) aorta and femoral arteries accumulate progressively and diffusely calcium and cholesterol (Fig. 1). These data suggest that profound and diffuse alterations of the vessel wall are present in most persons above 60-70 years even in the absence of lipidic deposits in the subintima. The above-cited and many other studies have shown the importance of dietary factors in the early formation of lipidic plaques.[7,9]

Impact of Cell Biology

Since the availability of cell culture techniques, great advances were made in this field on the study of vascular smooth muscle cells (SMC) in culture.[10-12] The Hayflick-paradigm, is the study of cell-aging by serial passages of cells in culture[11] applied to vascular SMC-s yielded important results. It was reported by July and Gordon Campbell in Australia that during successive passages there is a progressive shift of SMC-phenotype.[12] From sessile and contractile, these cells become mitotically and biosynthetically activated. This modified SMC-phenotype would lead to the uncoordinated deposition of matrix macromolecules not only in the lipidic plaques but also diffusely in the vascular wall. The resulting fibrous plaques, rich in collagen fibers could eventually rupture resulting in thrombo-embolic events with mostly fatal consequences. One important consequence of this phenotypic shift of SMC-s is the ensuing modification of matrix synthesis and degradation. There is however good reason to believe that this process does not remain localized but spreads over most of the vascular wall. This would suggest a relationship between progressive shift of SMC-phenotype and the age-dependent rigidification of the vascular wall attributed to a change in the "programme" of matrix biosynthesis by SMC-s. No more deposition of concentric elastic lamellae can be observed in adults although elastin biosynthesis is not completely suppressed as shown by our studies on aorta explant cultures. Collagen synthesis is maintained and the overall result is a loss of elasticity and hardening of the vessel wall.

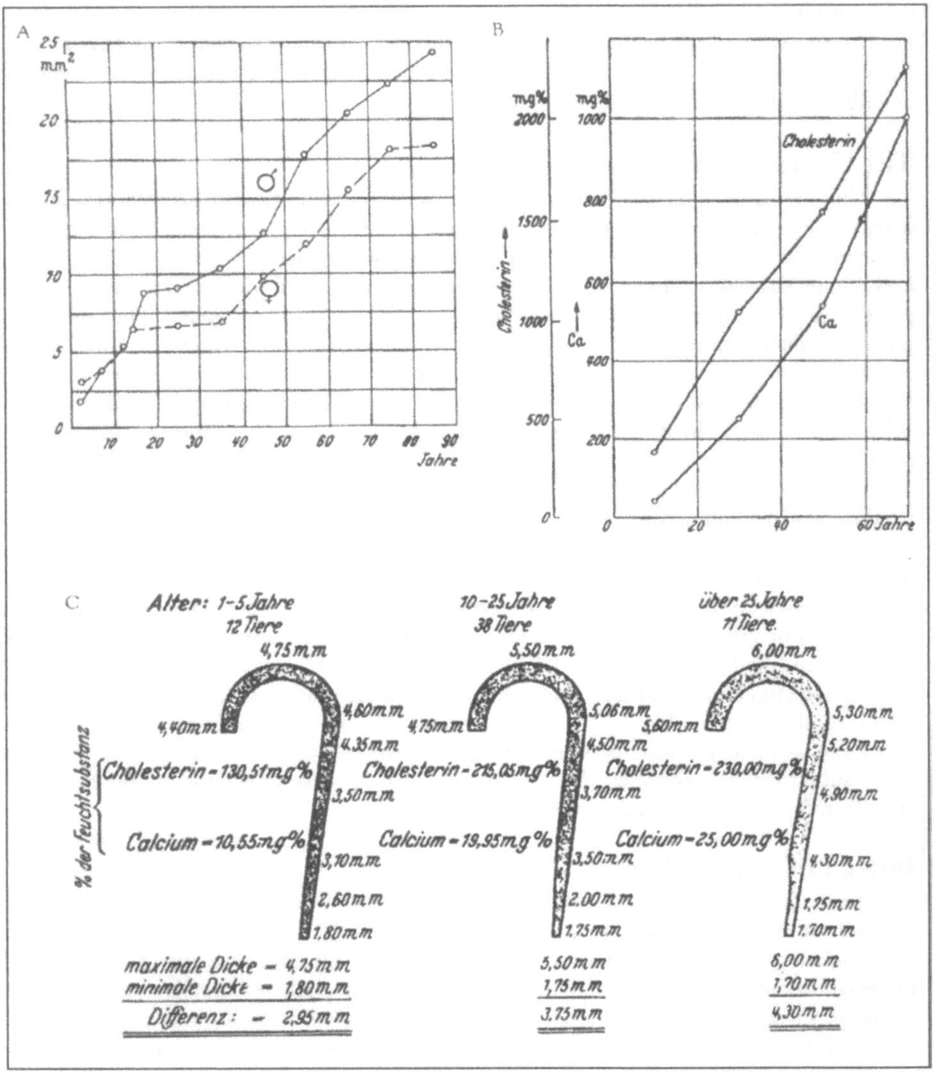

Figure 1. Diffuse calcium and cholesterol deposition in human aorta and femoral artery and in horse aorta, according to M. Burger (ref. 8) A) and B) Abscissa -age-years, ordinates: A) wall thickness of femoral artery, B) cholesterol and Ca-content of human aorta. C: wall thickness, cholesterol and Ca-content of horse aorta at 1-5 years of age, 25 years of age and over 25 years of age.

Other Etiological Factors

Since the early XXth century several propositions were made in order to account for the major anatomical and functional changes of the larger elastic vessels. These early interpretations were well reviewed by J. Balo[6] and others.[13] Let us mention one important proposition, made by Virchow in the middle of XIXth century and reemphasized by the Viennese pathologist Rokitansky implying an initial inflammatory process at the origin of the arteriosclerotic alterations. The isolation of acid-soluble collagen by Nageotte in Paris in the 1920s suggested a correlation to these pathologists concerning the "infiltration" of soluble collagen from the circulation in the vessel wall, accounting for its "fibrosis" and progressive hardening (for review, see ref. 6). Inflammation, "insudation" of collagen appears quite early in the pathological literature. Most recent knowledge on inflammatory mediators, cytokines, white blood cells among others enabled the elaboration of modernized versions of this early recognized pathogenetic process. Another observation of these early workers deserves mention. This is the recognition of the importance of neuro-humoral factors as shown by the induction of arteriosclerotic alterations in rabbits by repeated adrenalin injections (reviewed in ref. 6). A more complete description of these early tentatives of pathogenetic description of these diseases can be found in the cited reviews.

Role of Immune Factors

With the advent of cellular immunology, together with well-established methods of "humoral" immunology it became tempting to study the role of both these factors in the genesis of vascular diseases. Several laboratories described the successful production of typical arteriosclerotic lesions in rabbits by immunisation with vascular homogenates. Szigeti in Hungary,[14] Scebat and Renais in Paris[15] reported such experiments. These were however performed with total vascular wall homogenates rendering difficult to identify the most important vascular antigen(s) responsible for the described modifications. We carried out similar experiments with Y. Grogogeat in Paris, using purified vascular extracts. We could show that highly purified elastin was the most active lesion-initiating antigen.[16] As shown in Figure 2, the lesions observed on the rabbit aorta were mostly of the arteriosclerotic type with heavy calcifications. Further ultrastructural studies showed that the elastic lamellae of the immunised rabbits were disrupted, and strongly modified in their detailed morphology.[16,17] These modifications were obtained without adding any lipid to the rabbit diet. No lipid deposits were observed either.[16-18] In collaboration with Dr. Worcel at the Roussel-UCLAF laboratories, we could measure radioactive Calcium influx in control and immunized aortas.[18] It appeared that immunisation strongly increased Ca-influx in the otherwise apparently intact vascular fragments. Previous treatment with Calcitonin (porcine, injectable) could partially prevent these modifications.[18,19] These experiments tended to confirm the original pathogenetic hypothesis of J. Balo, attributing a primary role to the degradation of elastic fibers in the development of the athero-arteriosclerotic disease.[6]

Role of Elastolysis

The essential features of the progressive modifications of vascular ECM during atherogenesis concern primarily the loss of elastic fibers and the progressive preponderance of collagen fibers. During the development of the vascular wall the ratio of elastin to collagen is tightly controlled. In the human aorta the thoracic segment contains approximately twice as much elastin as collagen, in the abdominal segment this ratio is inversed. This enables the upper part of the aorta to dilate at the reception of the systolic blood volume pumped by the left ventricle in the thoracic aorta. As a result of its elasticity this upper segment will efficiently pump the blood towards the abdominal segment and to the iliac bifurcation. The progressive loss of vascular elasticity will result in a loss of this pumping (or "secondary heart") effect. The progressively dilated upper segment will passively accomodate the systolic blood volume without being able to pump it towards the periphery because of the loss of elasticity and rigidification of the vascular wall. This will increase the workload on the heart which loses contractile myofilaments during aging (see ref. 20 for review). Its conducting system also becomes less efficient because of the "fibrosis" of aging heart tissue. Collagen fibers form basket-like structures around the myofibrils. These collagen fibrils increase in thickness and density and progressively dislocate the contractile fibers with ensuing disturbances of the propagation of the contractile impulses. This process can be considered as the most plausible explanation of heart failure in old age.

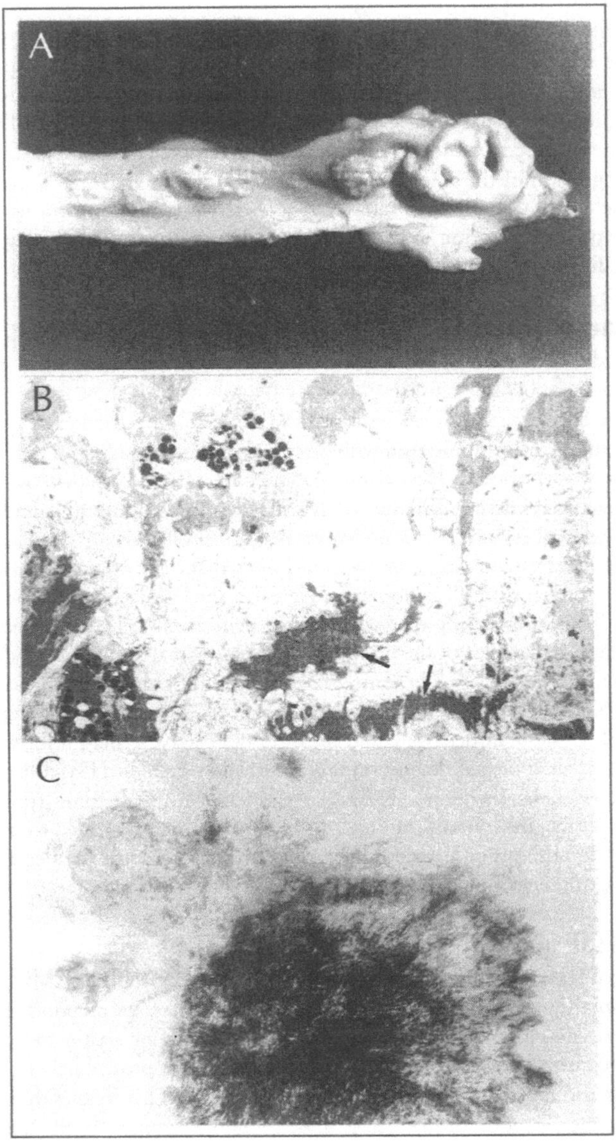

Figure 2. A) Arteriosclerotic modifications of a rabbit aorta immunized with purified elastin, as described in ref. 16. B) heavy Calcium deposits in the endothelial cells, disruption and calcification of elastic fibers (➔) C) crystalline Calcium needles in the subintima.

Mechanism of Elastolysis

The first confirmation of Balo's hypothesis attributing a primary role to elastolysis in the atherogenic process came from the isolation of a potent elastolytic protease from the pancreas (see ref. 21 for review). The other proteases known at that time as trypsin and chymotrypsin did not attack elastin. It appeared therefore logical to attribute an important role to pancreatic

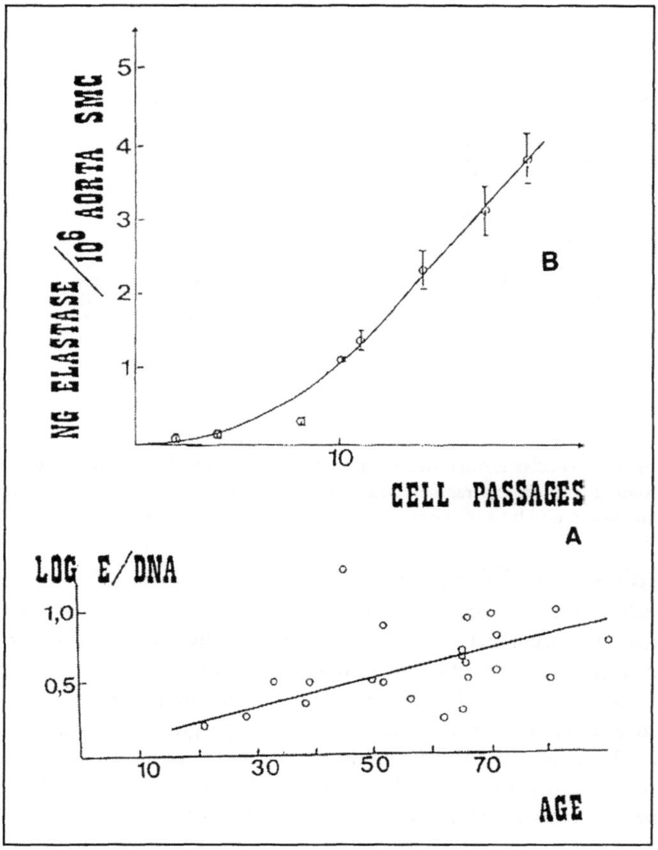

Figure 3. A) Exponential increase of elastase-type endopeptidase activity in human aorta extracts. Log Elastase activity in arbitrary units per μg DNA determined in lesion-free areas of human aortas. B) Elastase activity in equivalents of pancreatic elastase per 10^6 smooth muscle cells, determined in successive passages of SMC (from refs. 26, 29).

elastase in the destruction of vascular elastin. Logic is however the best way to go wrong with confidence. Although trace amounts of pancreatic elastase could be identified in the circulation, its role in the destruction of vascular elastin had to be abandoned. A major shift in the conceptual approach to this problem came from the recognition that fast-acting enzymes such as the pancreatic elastase could not be involved in such a slow process as the progression of the modifications with age of vascular matrix. On the contrary, only slowly acting enzymes could be involved in a process which takes decades to develop. This was hard to accept for scientists trained in enzyme kinetics but was well received by pathologists. The first demonstration of a slowly acting vascular elastase was reported from our laboratory by B. Robert et al in 1974,[22,23] followed by its characterization, localization on SMC-membrane[24,25] and most importantly the demonstration of its age-dependent increase in human aorta[26] and its progressive upregulation during successive passages in cell cultures[27] (Fig. 3). These results were obtained with very sensitive methods, worked out especially for this purpose[28] and fitted well with the observed progressive fragmentation of vascular elastic fibers (Fig. 4). Even the action of PMN-elastase, isolated by Janoff about at the same time as platelet elastase in our laboratory[29] acted too fast to be involved in such a slow process.

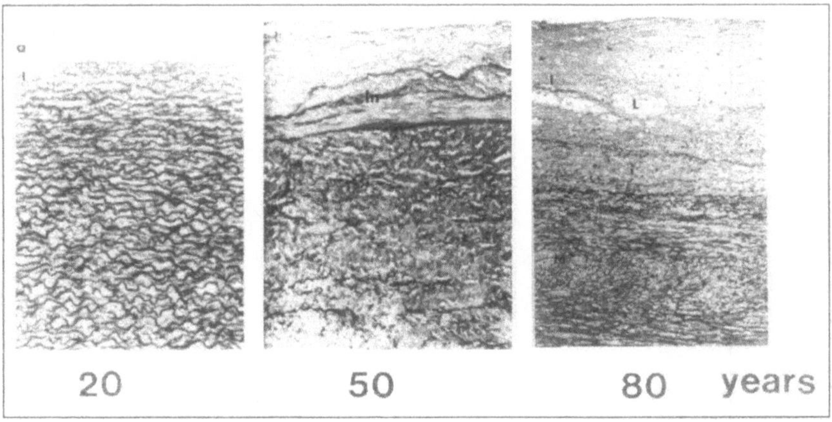

Figure 4. Fragmentation of elastic fibers in human aorta with age and atherosclerosis. Left: intact aorta from a young adult; middle: elastic fibers fragmented and lysed in the subintima and inner part of the media; right: advanced fragmentation and lysis of the elastic fibers (modified after Bouissou, from ref. 27).

Role of Lipids and of Calcium

But elastolysis is not the only postsynthetic modification which concerns vascular elastic fibers. The progressive accumulation of calcium was shown by Lansing and others in the middle of the XXth century,[30] confirmed in our laboratory. We could show that part of this accumulated calcium is associated with the microfibrillar glycoproteins part of the elastic "tissue" (Fig. 5). It was also recognized that lipids also accumulate progressively in elastic fibers.[31,32] As shown in Table 1, the most salient accumulation concerns free fatty acids, cholesterol and its esters. The studies carried out in collaboration with B. Jacotot's team at the Henri Mondor Hospital at the University Paris 12 in Creteil definitively confirmed for human aorta the results

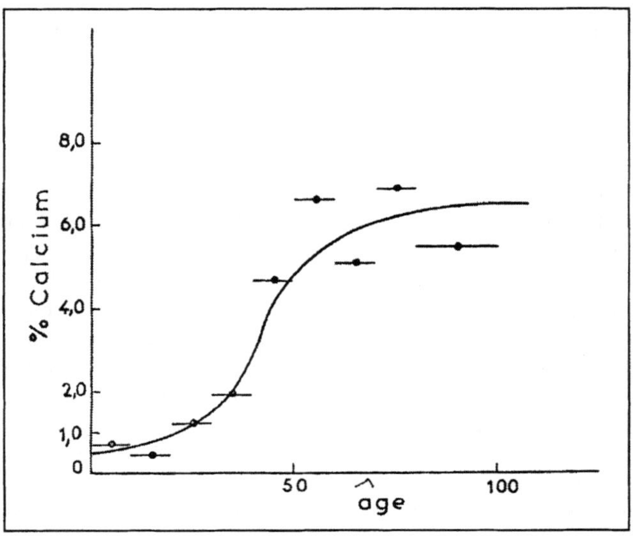

Figure 5. Calcium deposition in human aorta as a function of age (years). % Calcium determined in purified aorta elastin (45 min. boiling in 0.1 NaOH) on a dry weight basis.[30]

Table 1. Lipid content of the elastin of human aortas

Neutral Lipids	Lipids Extracted Before Heating in NaOH		Lipids Extracted After Heating in NaOH	
	Group I	**Group II**	**Group I**	**Group II**
Free Cholesterol	8.6	16.0	0.95	1.53
Esterified Cholesterol	30.9	30.6	4.28	6 12
Triglycerides (as µg fatty acids)	78	154	7.1	6.9
Free fatty acids	48	204	10.2	20.1
PhospholipidS	46	124	5.2	2.8
Total lipids	211.5	528.6	27.73	37.45

Group I: No or very early atherosclerosis. Group II: advanced atherosclerosis. Determinations were done of collagen-free elastic fibers before and after beating in 0.1 N NaoH for 45 min. at 100°. Data taken from reference 32.

of previous observations carried out on rats fed a labeled cholesterol containing diet showing its accumulation in several tissues and especially in the aorta wall.[31,32] Experiments in Partridge's laboratory[33] and in ours[34] showed that the uptake of calcium and of lipids is a mutually potentiated process. The presence of calcium enhances the uptake of labeled cholesterol by elastic fibers.[34] Although the studies of D. Urry suggested that some of the typical β-turns of elastin[35] might act as calcium coordination sites, the detailed molecular mechanism for the mutually potentiated lipid and calcium uptake by elastin remains to be elucidated. It appears however that these post synthetic processes are primarily responsible, even without ensuing elastolysis, for the loss of elasticity of elastic fibers. Lipid- and calcium-loaded elastic fibers are apparently faster degraded by elastase-type proteases than native fibers.[36]

Nature of Vascular Fibrosis

Fibrotic process is mostly related to an increased synthesis of fibrous collagens. This process was however not convincingly demonstrated in aging vascular wall except in the local development of fibrotic plaques as described above. Increasing cross-linking of vascular collagen(s) by the Maillard reaction was however demonstrated.[37,38] Without going in detail for the description of the Maillard reaction (see refs. 37,38), it should be remembered that this is the result of the nonenzymatic interaction between reducing sugars (glucose and others) and available free amino groups on proteins. First described by Fritz Verzar[39] for rat tail tendon collagen, the age-dependent increase of this process was further confirmed by several investigators. It cannot be excluded however that biosynthetic modifications contribute to the stiffening of the vascular wall. An intriguing experiment by Moczar et al[40] showed that the addition of hyperlipidemic human sera to rabbit aorta explant cultures inhibited the incorporation of ^{3}H-proline in all vascular macromolecular fractions except in collagen where it increased. This experiment suggested that hyperlipidemic environment might stimulate the biosynthesis of collagen(s). A more recent epidemiological observation reinforces this concept. Gariepi et al[41] measured the thickness of the common carotid artery by ultrasound echography in normo-lipidemic and hyperlipidemic individuals as a function of age. They found a significantly more rapid age-dependent increase of wall thickness in hyperlipidemic individuals.[41] The more recent identification of at least 26 types of collagen in vertebrates with more than half of them represented in the vascular wall[42,43] indicate the necessity of a novel appraisal of the biosynthetic modification of vascular collagens as a function of age and atherosclerosis.

Role of Proteoglycans

A number of different proteoglycans were identified in the vascular wall (for a review, see Wight, et al).[44] These macromolecules are associated mostly with other matrix components and play a crucial role in the fibrillogenetic process. According to J. Scott, the smaller proteoglycans as biglycan and decorin play an important role in the organisation of fibril size and orientation.[45] Other, larger proteoglycans and mainly hyaluronan are the principal regulators of water and electrolyte balance and traffic in tissues. Hyaluronan of ECM is one of the important factors regulating its rheological properties. Proteoglycans were also shown to be involved in the regulation of the cell cycle. Cell-membrane bound heparan sulphate proteoglycans were shown to bind growth factors as bFGF. The cleavage of the GAG-chains, and their internalisation appear also to be involved in the regulation of the cell cycle of vascular SMCs.[46] Our experiments showed that short chain heparan sulphate preparations by the Choay Laboratories in Paris, added to mesenchymal cells in culture produced modifications of the biosynthesis of several matrix components and down-regulated fibronectin biosynthesis which is increasingly synthesized with age and diabetes.[47,48]

Role of Receptors

Interactions between cells or cells and ECM-components are mediated by a number of receptors, such as integrins[49] or the elastin-laminin receptors (ELR). Here we shall describe the role of ELR in the progressive age-dependent modifications of vascular ECM. But before doing so we have to remember the crucial role played by the identification of the apoE-B-receptor by Brown and Goldstein in the elaboration of the lipid-induction theory of atherosclerosis (review in ref. 13). Most of the receptors involved in the above-mentioned cell-cell and cell-matrix interactions were described in detail,[49-51] therefore here we shall only mention some of the most important aspects of the ELR in vascular modifications. ELR was first described as a result of its involvement in cell-elastin interactions.[52] Its agonists are elastin peptides, some of the more closely studied sequences are shown in Table 2. Such short sequences as the tripeptide VGV can already activate this receptor on vascular endothelial cells as shown by the triggering of NO-dependent vasodilation (Faury et al).[53,54] This NO-dependent vasodilation of noradrenalin precontracted rat aorta rings enabled Faury et al to study the age-dependent behavior of this effect.[55] As shown in Figure 6 the ELR-mediated vasodilation is hardly detectable in the newborn animal, reaches its maximal effect in the young adult and decreases progressively with age. This behavior of the ELR-mediated vasodilation is reminiscent of the long-described age-dependent increase of systolic blood pressure in humans. With a number of other comparable observations it suggested an age-dependent "uncoupling" of the ELR from its normal transmission pathway which was studied on several cell types such as endothelial cells[53-55] and circulating white blood cells.[56,57] In respect to the topic of this review one of the most relevant effects of the stimulation of ELR was the increase of elastase-type protease synthesis (Table 3). This effect was dose-dependent and increased with increasing elastin peptide concentration added to the culture medium of fibroblasts or SMC-s.[58] Another effect of the stimulation of ELR by elastin peptides was the increase of cell proliferation.[59] These two processes, increased proliferation and increased lytic enzyme production are the most characteristic features of the modulation of SMC-phenotype as described by the Campbells.[10,12] Another relevant feature of ELR is the activation of release of reactive oxygen species (ROS) after stimulation by elastin peptides. Superoxide release was shown to increase with age but no more inhibition with pertussis toxin could be obtained in cells from elderly persons showing its "uncoupling" from the normal transmission pathway which involves a pertussis toxin sensitive G-protein in cells from

Table 2. Peptide sequences which are active or inactive on the elastin-laminin receptor, determined by NO⁻ -dependent vasodilation[54]

Sequences		
Active		**Inactive**
VGV		VGVGVA
PGV	dose-dependent	PGVGVA
VGVAPG		
PVGV		
VGVA	at 10^{-11}M	VPVGGA

young donors. It has to be reminded that superoxide combines with NO to form the highly toxic peroxynitrite anion ($ONOO^-$). This $ONOO^-$ anion can be neutralized by reduced glutathione (GSH), but its production was reported to decrease with age. It appears from this succinct summary of the properties of the ELR that it might well be involved in the age-dependent modifications of the vascular wall. Oligo- or polysaccharides with a galactose configuration at the nonreducing end-position were shown to act as efficient antagonists of the ELR. As shown in Table 3, melibiose is one of these efficient antagonists as shown in our laboratory. Elastin peptides were shown to be present in the circulating blood.[60] An epidemiological study on more than 1000 individuals, men and women, between 59 and 71 years indicated an average value of about 10μg /ml, saturating for the ELR with a Kd in the nanomole range. Using immunofluorescence the ELR could be demonstrated on most cell types studied, also on SMC-s, fibroblasts and most importantly on mononuclear cells in freshly excised human plaques.[61] These findings further emphasize the physio-pathological importance of the

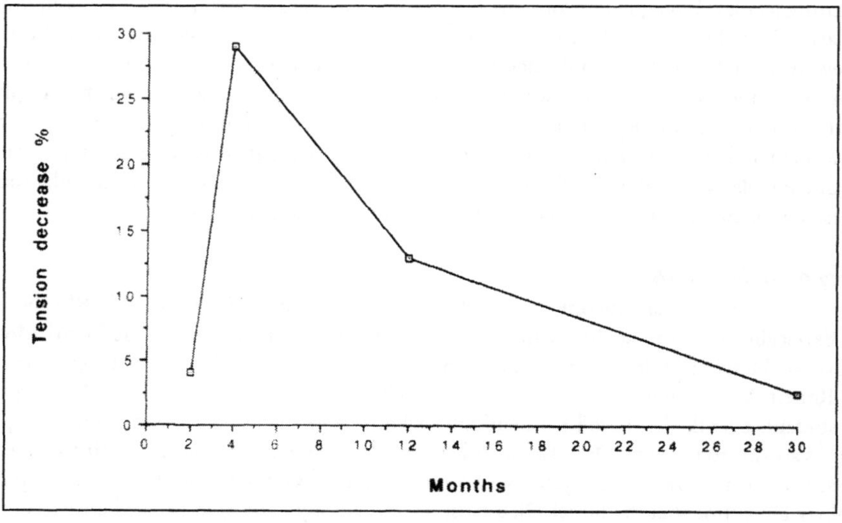

Figure 6. Effect of the age of rats on the vaso-dilation obtained by adding elastin peptides (1 μg/ml) to noradrenalin precontracted aorta rings, as described by Faury et al.[55] Abscissa: age in months, ordinates: decrease of wall tension. Maximal effect in young adults followed by an age-dependent decrease.

Table 3. Increase of elastase-type activity of fibroblasts in presence of elastin peptides (k-elastin) with a synthetic substrate (succinyl-tri ala p NA)

Conc. of k-Elastin Added	Enzyme Activity (μg/ml)	
	No Melibiose	+ Melibiose(10μg/ml)
0	2600	2000
10	3330	2030
100	6070	1840

Results expressed as μM substrate hydrolyzed/ 10^6 cells.

above-summarized properties of the ELR. They also show that elastin is not only a passive player in vascular remodeling but plays an active role by the action of its degradation products on the remodeling processes, mediated by the ELR. A number of other properties of this receptor were described elsewhere, as for instance its involvement in cancer progression and metastases formation.[62]

Conclusions

Age-dependent remodeling of vascular extracellular matrix is a key process in the development of vascular diseases such as atheromatous plaque formation, arteriosclerosis and other age-related vascular pathologies. We summarized in this review the central role of the matrix-synthetic properties of vascular smooth muscle cells and its progressive modifications with age and at the site of atherosclerotic plaques. Lipidic plaque formation can start early in life as influenced by nutritional factors. Progressive rigidification of the vascular wall is age-dependent and results from the progressive modulation of the biosynthetic activity of vascular smooth muscle cells as well as from age-dependent postsynthetic modifications of matrix macromolecules. Upregulation of matrix degrading enzymes as elastases and cross-linking of collagen fibers by the Maillard reaction are the most important postsynthetic modifications. They are preceded by lipid- and calcium uptake by vascular elastic fibers as a major contribution to this loss of elasticity followed by their degradation. These effects are further amplified by the action of elastin degradation products on the elastin-laminin receptor. Among the reactions mediated by this receptor, increased cell proliferation and increased elastase-type protease production play a crucial role in the age-dependent modifications of vascular extracellular matrix. These modifications appear to play a crucial role in heart failure of old age.

Acknowledgements

Most of the original experiments described and cited were carried out in our CNRS-laboratory on Extracellular Matrix at the Medical Faculty of University Paris 12 and the Henri Mondor University Hospital, in close collaboration with Bernard Jacotot, Jacqueline Labat-Robert, Alexandre M. Robert, Marie-Paule Jacob, William Hornebeck, Tamas Fülöp and a number of other graduate students over the last decades. I thankfully acknowledge their valuable contributions. Special thanks are due to my wife, Dr. J. Labat-Robert, for discussing, correcting and editing this review. 1 also want to acknowledge the generous hospitality of Prof. Gilles Renard, Head of the Ophthalmology Department at the Hôtel-Dieu Hospital (University Paris 5) in Paris where we continue our research on Extracellular Matrix, Aging and related Pathologies.

References

1. Svanborg A. Health and vitality. Gotheborg MFR, ed. 1988.
2. Svanborg A. Seventy years old people in Gotheborg. II: General presentation of social and medical conditions. Acta Med Scand 1977; 611:3.
3. Haranghy L, Füredi E. Post mortem examination of five persons died above 100 years of age. Acta Morphol Acad Sci Hung 1970; 18:91-94.
4. Haranghy L. Gerontological studies on hungarian centenarians. Budapest: Akademiai Kiado, 1965.
5. Robert L. Aging of the vascular wall and atherosclerosis. Exp Gerontol 1999; 34:491-501.
6. Balo J. Connective tissue changes in atherosclerosis. Int Rev Connect Tisssue Res 1963:1:241-306.
7. Berenson GS. Cardiovascular risk-factors in children. New York: Oxford University Press, 1980.
8. Bürger M. Altern und krankheit. Leipzig, Georg Thieme: 1947.
9. Jacotot B. Epidemiologie et facteur de risque. In: Jacotot B, ed. Atherosclérose. Sandoz, Paris: 1993:27-45.
10. Campbell JH, Campbell GR. Vascular Smooth Muscle in Culture. Vol. I - II. Boca Raton: CRC Press, 1987.
11. Hayflick L. The cellular basis for biological aging. In: Finch CE, Hayflick L, eds. Handbook of the Biology of Aging. London, New York: Van Nostrand Reinhold Co, 1977:159-188.
12. Campbell JH, Campbell GR. Recent advances in molecular pathology. Smooth muscle phenotypic changes in arterial wall homeostasis: Implications for the pathogenesis of atherosclerosis. Exp Mol Pathol 1985; 42:139-162.
13. Olsson AG, ed. Atherosclerosis, Biology and Clinical Science. Churchill, Livingstone, Edinburgh, London: 1967.
14. Szigeti I, Jako I, Doman J. Study of the antigenic and immunopathological effects of structural proteins of the mammalian arterial wall and their role in the pathogenesis of atherosclerosis. In: Le rôle de la paroi artérielle dans l'athérogénèse. Paris: CNRS Editions, 1968:388-392.
15. Renais J, Groult N, Scebat L et al. Pouvoir antigénique et pathogène du tissu artériel. In: Le rôle de la paroi artérielle dans l'athérogénèse. Paris: CNRS Editions, 1968:388-392.
16. Robert AM, Grogogeat Y, Reverdy V et al. Lésions artérielles produites chez le lapin par immunisation avec l'élastine et les glycoprotéines de structure de l'aorte. Etudes biochimiques et morphologiques. Atherosclerosis 1971; 13:427-449
17. Jacob MP, Hornebeck W, Lafuma C et al. Ultrastructural and biochemical modifications of rabbit arteries induced by immunization with soluble elastin peptides. Exp Mol Pathol 1984; 41:171-190.
18. Jacob MP, Moura AM, Tixier JM et al. Prevention by calcitonin of the pathological modifications of the rabbit arterial wall induced by immunization with elastin peptides: Effect on vascular smooth muscle permeability. Exp Mol Pathol 1987; 46:345-356.
19. Robert AM, Miskulin M, Godeau G et al. Action of calcitonine on the atherosclerotic modifications of brain microvessels induced in rabbits by cholesterol feeding. Exp Mol Pathol 1982; 37:67-73.
20. Wissler RW, Robert L. Aging and cardiovascular disease. Circulation 1996; 93:1608-1612.
21. Robert L, Hornebeck W. Elastin and Elastases. Vol. I and II. Boca Raton: CRC Press, Inc., 1989.
22. Robert B, Derouette JC, Robert L. Mise en évidence d'une protéase à activité élastolytique dans les extraits d'aortes humaines et animales. CR Acad Sci 1974; 278:3251-3254.
23. Robert B, Robert L, Robert AM. Elastine, élastase et artériosclérose. Pathol Biol 1974; 22:661-669.
24. Leake DS, Hornebeck W, Bréchemier D et al. Elastase-like activities of arterial smooth muscle cells in culture are located on the plasma membrane. Clin Sci 1982; 63:13-14.
25. Hornebeck W, Derouette JC, Robert L. Isolation, purification and properties of aortic elastase. FEBS Letters 1975; 58:66-70.
26. Hornebeck W, Derouette JC, Roland J et al. Corrélation entre l'âge, l'artériosclérose et l'activité élastinolytique de la paroi aortique humaine. CR Acad Sci 1976; 292:2003-2006.
27. Robert L, Labat-Robert J, Hornebeck W. Aging and atherosclerosis. In: Gotto AM, Paoletti R, eds. "Atherosclerosis reviews". New York: Raven Press, 1986:14:143-170.
28. Robert B, Robert L. Determination of elastolytic activity with ^{125}I and ^{131}I-labelled elastin. Eur J Biochem 1969; 11:62-67.
29. Robert B, Szigeti M, Robert L et al. Release of elastolytic activity from blood platelets. Nature 1970; 227:1248-1249.

30. Lansing AI. The arterial wall. Baltimore: Williams Wilkins Co, 1959.
31. Szigeti M, Monnier G, Jacotot b et al. Distribution of ingested ^{14}C-cholesterol in the macromolecular fractions of rat connective tissues. Connect Tissue Res 1972; 1:145-152.
32. Claire M, Jacotot B, Robert L. Characterization of lipids associated with macromolecules of the intercellular matrix of human aorta. Connect Tiss Res 1976; 4:61-71.
33. Hornebeck W, Partridge SM. Conformation changes in fibrous elastin due to calcium ions. Eur J Biochem 1975; 51:73-78.
34. Jacob MP, Hornebeck W, Robert L. Studies on the interaction of cholesterol woth soluble and insoluble elastins. Int J Biol Macromol 1983; 5:275-278.
35. Urry DW. Sequential polypeptides of Elastin: Structural properties and molecular pathologies. In: Robert AM, Robert L, eds. Front Matrix Biol. Karger, Basel: 1980:8:78-103.
36. Jacob MP, Bréchemier D, Robert L et al. Variation of elastase-type protease activity and elastin biosynthesis in rabbit aorta induced by cholesterol diet. Artery 1982; 10:310-316.
37. Colaco C ALS. The glycation hypothesis of atherosclerosis. Springer: Landes Bioscience, 1997.
38. Odetti P, Aragno I, Garibaldi S et al. Role of advanced glycation end products in aging collagen. Gerontology 1998; 44:187-191.
39. Verzar F. Aging of the collagen fiber. Int Rev Connect Tissue Res 1964; 2:243-300.
40. Moczar M, Robert L. Action of human hyperlipidemic sera on the biosynthesis of intercellular matrix macromolecules in aorta organ cultures. Paroi Artérielle 1976; 3:105-113
41. Gariepy J, Simon A, Massonneau M et al of the PCVMETRA Group. Wall thickening of carotid and femoral arteries in male subjects with isolated hypercholesterolemia. Atherosclerosis 1995; 113:141-151.
42. Wight TN. Arterial wall. In: Comper WD, ed. Extracellular matrix. Tissue Function. Harwood: Academic Publishers, 1996:1:175-202.
43. Franco CD, Hou G, Bendeck MP. Collagens, integrins and the discoidin domain receptor in arterial occlusive disease. Trends Cardiovsc Med 2002; 12:142-148.
44. Wight TN, Lark MW, Kinsell MG. Blood vessel proteoglycans. In: Wight TN, Mecham RP, eds. Biology of proteoglycans. Academic Press Inc., 1987:267-300.
45. Scott JE. Structure and function in extracellular matrices depend on interactions between anionic proteoglycans. Pathol Biol 2001; 49:284-289
46. Marcum JA, Reilly CF, Rosenberg RD. Heparan sulfate species and blood vessel wall function. In: Wight TN, Mecham RP, eds. Biology of proteoglycans. Academic Press Inc., 1987:301-343.
47. Asselot-Chapel C, Kern P, Labat-Robert J. Biosyntheses of interstitial collagens and fibronectin by porcine aorta smooth muscle cells. Modulation by low molecular weight heparin fragments. Biochim Biophys Acta 1989; 993:240-244.
48. Asselot-Chapel C, Combacau L, Labat-Robert J et al. Expression of fibronectin and interstitial collagen genes in smooth muscle cells : Modulation by low molecular weight heparin fragments and serum. Biochem Pharmacol 1995; 49:653-659.
49. Eble JA, Kühn K. Integrin-ligand interaction. New York: Springer, 1997.
50. Robert L, Jacob MP, Fülöp Jr T et al. Elastonectin and the elastin receptor. Pathol Biol 1989; 37:736-741.
51. Labat-Robert J. Cell-matrix interactions, alteration with aging and age associated diseases. Pathol Biol 2001; 49:349-352.
52. Hornebeck W, Tixier JM, Robert L. Inducible adhesion of mesenchymal cells to elastic fibers : Elastonectin. Proc Natl Acad Sci 1986; 83: 5517-5520.
53. Faury G, Usson Y, Robert-Nicoud M et al.Nuclear and cytoplasmic free calcium level changes induced by elastin peptides in human endothelial cells. Proc Natl Acad Sci 1998; 95:2967-2972.
54. Faury G, Garnier S, Weiss AS et al. Action of tropoelastin and synthetic elastin sequences on vascular tone and on free calcium level in human vascular endothelial cells. Circ Res 1998; 82:328-336.
55. Faury G, Chabaud A, Ristori MT et al. Effect of age on the vasodilatory action of elastin peptides. Mech Age Dev 1997; 95: 31-42.
56. Varga Zs, Jacob MP, Robert L et al. Identification and signal transduction mechanism of elastin peptide receptor in human leukocytes. FEBS Lett 1989; 258:5-8.

57. Péterszegi G, Texier S, Robert L. Human helper and memory lymphocytes exhibit an inducible elastin-laminin receptor. Int Arch Allergy Immunol 1997; 114:218-223.'
58. Archilla-Marcos M, Robert L. Control of the biosynthesis and excretion of the elastase-type protease of human skin fibroblasts by the elastin receptor. Clin Physiol Biochem 1993; 10:86-91.
59. Ghuysen-Itard AF, Robert L, Jacob MP. Effet des peptides d'élastine sur la prolifération cellulaire. CR Acad Sci 1992; 315:473-478.
60. Bizbiz L, Alpérovitch A, Robert L et al. Aging of the vascular wall : Serum concentration of elastin peptides and elastase inhibitors in relation with cardiovascular risk factor. The EVA study. Atherosclerosis 1997; 131:73-78.
61. Péterszgi G, Mandet C, Texier S et al. Lymphocytes in human atherosclerotic plaque exhibit the elastin-laminin receptor: Potential role in atherogenesis. Atherosclerosis 1997; 135:103-107.
62. Robert L, ed. Cell-matrix interactions in cancer spreading: Effect of aging. Semin Cancer Biol 2002; 12:157-163.

Chapter 10

Molecular and Cellular Aspects of Liver Fibrosis

Norifumi Kawada

Abstract

Hepatic stellate cells play essential roles in the pathogenesis of liver fibrosis. Transformation of stellate cells from the vitamin A-storing pheno-type to the "myofibroblastic" one closely correlates to hepatic fibrogenesis during chronic liver diseases. Understanding the molecular mechanisms of stellate cell activation have made a great progress, in particular, in the field of intracellular signal transduction of transforming growth factor-β and platelet-derived growth factor and collagen gene expression. Accumulation of the information on the stellate cell activation would shed light on the establishment of a novel therapeutic strategy against fibrotic liver diseases.

Stellate Cell As a Principal Player of Liver Fibrosis

Hepatic stellate cells, which reside in the space of Disse in close contact with both sinusoidal endothelial cells and hepatocytes, play multiple roles in the pathophysiology of the liver.[1] Quiescent stellate cells represent a retinol-storing phenotype and metabolize a small amount of basement membrane-forming substrata such as laminine and type IV collagen. When liver injury occurs, they undergo transformation into myofibroblasts that actively proliferate in response to platelet-derived growth factor-BB (PDGF-BB) and insulin-like growth factor-I (IGF-I), produce increased amount of extracellular matrix (ECM) material, show augmented contractility accompanied by the expression of smooth muscle α-actin and the production of endothelin-1 (ET-1), secrete transforming growth factor-β (TGF-β) and monocyte chemotactic protein-1 (MCP-1), lose the retinoid , and exhibit active apoptosis. Stellate cell activation is initiated by oxidative stress of lipid hydroperoxide and reactive aldehyde generated in and released from damaged hepatocytes, by paracrine stimulation of PDGF-BB, IGF-I and TGF-β derived from activated Kupffer cells, endothelial cells, platelets and infiltrating leukocytes, and by early ECM changes including the production of a splice variant of cellular fibronectin (EIIIA isoform).[2-5] Transcriptional activation by a zinc finger gene KLF6/Zf9, which is induced at the very early stage of liver injury, AP-1 and CCAAT/enhancer binding protein (C/EBP) enhances gene expression regulating ECM accumulation.[6]

Activated stellate cells are highly responsive to PDGF-BB and IGF-1 through the induction of receptors for individual growth factors. For instance, the autophosphorylation induced after dimerization of PDGF receptors activates transduction molecules containing SH2 domains, such as phosphatidylinositol 3-kinase (PI3-kinase), phospholipase C-γ (PLC-γ), the Src

Fibrogenesis: Cellular and Molecular Basis, edited by Mohammed S. Razzaque.

family of tyrosine kinases, the tyrosine phosphatase SHP-2, and a GTPase activating protein (GAP) for Ras. In human stellate cell, activation of the extracellular-signal regulated kinase (ERK) pathway followed by an increased expression of *c-fos* in response to PDGF has been demonstrated.[7,8]

TGF-β is a key regulatory molecule for ECM metabolism and functions as an autocrine and a paracrine mediator.[9] Cellular sources of TGF-β1 are multiple, including hepatic stellate cell, Kupffer cell, hepatocyte, sinusoidal endothelial cell and platelet. Proteolytic cleavage of latent TGF-β binding protein (LTBP) is supposed to be a prerequisite for the release and generation of bioactive (mature) TGF-β, which is induced by urokinase plasminogen activator (uPA) or tissue PA (tPA). KLF6/Zf9 transcriptionally activates uPA, resulting in the increase of bioactive TGF-β.[10] The impact of TGF-β1 on liver fibrosis is well documented in a TGF-β1 knockout mouse model,[11] in the remarkable attenuation of the development of liver fibrosis by using soluble type II TGF-β receptor,[12] and in adenoviral delivery of dominant-negative TGF-β receptor.[13] Role of Smad cascade in TGF-β signaling is characterized in stellate cells.[14]

Contraction of stellate cells, particularly induced by ET-1, causes constriction of sinusoids, leading to a persistent disturbance of intrahepatic microcirculation and portal hypertension.[12,13] ET-1 synthesis by stellate cells is regulated by endothelin-converting enzyme-1 (ECE-1) during repair of hepatic injuries. ET-1 release is increased in stellate cells, whereas markedly decreased in endothelial cells after liver injury, depending on the ECE-1 mRNA expression and its stability. Blockade of ET receptors using specific ET receptor antagonists markedly modulates the development of liver fibrosis and portal hypertension in rat models. Figure 1 summarizes some important molecules those contribute to the hepatic fibrogenesis.

Novel Therapeutic Strategy for Preventing Liver Fibrosis

Increasing numbers of studies have shown a variety of therapeutic approaches for liver fibrosis based on the molecular inhibition of stellate cell activation. Among them, IFNα has a clinical potential to treat liver fibrosis of chronic hepatitis type C by eliminating HCV virus from patients.[19] IFNγ inhibits the cell activation and type I collagen mRNA expression in rats and the collagen transcription through directly acting onto IFN-responsive element, a proximal element within the human α2(I) collagen (COL1A2) promotor.[20] IFNγ inhibits TGF-β/Smad signaling through STAT pathway.

An antioxidative agent, *N*-acetyl-L-cysteine, triggers the degradation of PDGF receptor β mediated by cathepsin B and inhibits PDGF signaling and PDGF-dependent DNA synthesis. The agent affects additionally the expression of TGF-β receptor type II in stellate cells. *N*-acetyl-L-cysteine dramatically attenuates liver fibrosis developed in rats.[21] Natural flavonoids included in herb have also antioxidative action and are promising agents that protect the liver from developing fibrosis.

Hepatocyte growth factor (HGF) gene therapy is an another candidate for the treatment of liver fibrosis. Repeated transfection of the human HGF gene into skeletal muscle suppress the increase of TGF-β1, inhibit fibrogenesis and hepatocyte apoptosis, and resolve fibrosis in the cirrhotic rat liver.[22] Continuous administration of recombinant human HGF (rhHGF) is also effective for liver fibrosis. Although rhHGF reduces mRNA levels of procollagen α2(I), α1(IV) and TGF-β1, the detailed molecular mechanism of these anti-fibrotic effects of HGF remains to be elucidated.[23]

Peroxisome proliferator-activated receptor γ (PPARγ) ligands, such as 15-deoxy-Delta-prostaglandin J2 and troglitazone, inhibit proliferation, collagen synthesis and chemotaxis induced by PDGF of stellate cells.[24] Furthermore, thiazolidinediones reduced ECM deposition and stellate cell activation in toxic model of liver fibrosis induce by dimethylnitrosamine or CCl_4 or in a bile-duct ligation model.[25]

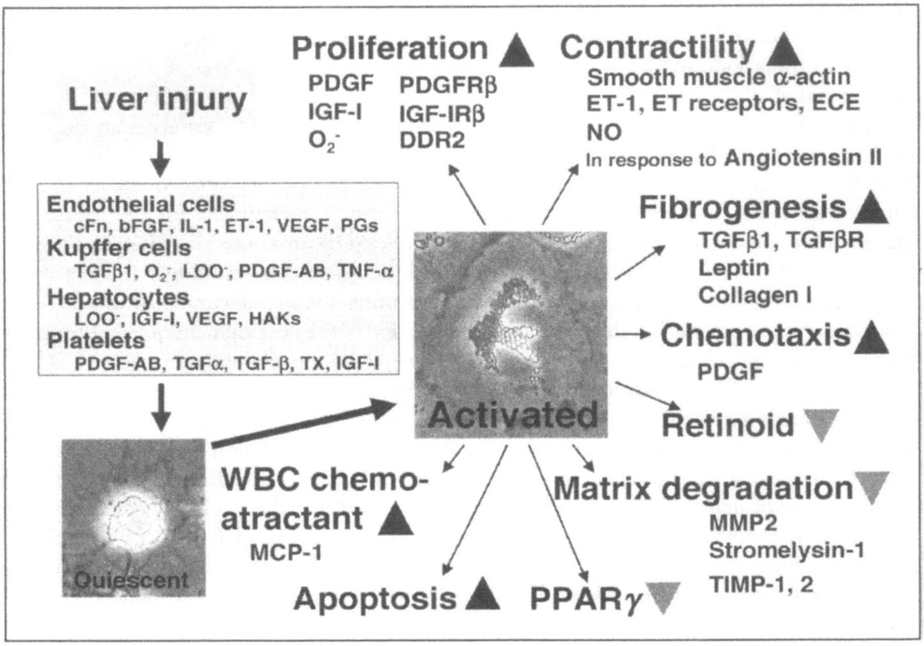

Figure 1. Schematic diagram of stellate cell activation. In response to liver injury, hepatic constituent cells including endothelial cells, Kupffer cells, hepatocytes as well as platelets generate multiple bioactive substances, such as growth factors, endothelins, prostanoids and reactive oxygen species. When exposed to these mediators, stellate cell undergoes transdifferentiation into activated myofibroblast. The latter phenotype exhibits active proliferation, increased contractility, collagen production, etc., which play pivotal roles in the development of liver fibrosis.

Recently, leptin has been suggested to play important roles in liver fibrogenesis. Supplementation of recombinant murine leptin to mice received CCl_4 or thioacetamide augments liver fibrosis accompanied by the increased expression of mRNAs for collagen α1(I), TGF-β1, and smooth muscle α-actin[26] Zucker (fa/fa) rats and ob/ob mice, that are insensitive to leptin, are protected from toxin-induced liver fibrosis.[27] Leptin receptor (Ob-Rb) is expressed more in sinusoidal endothelial cells and Kupffer cells than stellate cells, and thus the former cell population may be the target to circulating leptin.[28] Further studies will determine the effectiveness of anti-leptin therapy in the treatment of liver fibrosis.

References

1. Friedman SL. The cellular basis of hepatic fibrosis. N Engl J Med 1993; 328:1828-35.
2. Gressner AM, Bachem MG. Molecular mechanism of liver fibrosis-a homage to the role of activated fat-storing cells. Digestion 1995; 56:335-46.
3. Olaso E, Friedman SL. Molecular regulation of hepatic fibrogenesis. J Hepatol 1998; 29:836-47.
4. Kawada N. The hepatic perisiunusoidal stellate cell. Histol Histopathol 1997; 12:1069-80.
5. Okuyama H, Shimahara Y, Kawada N. The hepatic stellate cells in the post-genomic ear. Histol Histopathol 2002; 17:487-95.

6. Kim Y, Ratziu V, Choi SG et al. Transcriptional activation of transforming growth factor beta1 and its receptors by the Kruppel-like factor Zf9/core promoter-binding protein and Sp1. Potential mechanisms for autocrine fibrogenesis in response to injury. J Biol Chem 1998; 273:33750-8.
7. Marra F, Gentilini A, Pinzani M et al. Phosphatidylinositol 3-kinase is required for platelet-derived growth factor's actions on hepatic stellate cells. Gastroenterology 1997; 112:1297-306.
8. Kawada N, Ikeda K, Seki S et al. Expression of cyclins D1, D2 and E correlates with proliferation of rat stellate cells in culture. J Hepatol 1999; 30:1057-64.
9. Friedman SL. Cytokines and fibrogenesis. Semin Liver Dis 1999; 19:129-40.
10. Kojima S, Hayashi S, Shimokado K et al. Transcriptional activation of urokinase by the Kruppel-like factor Zf9/COPEB activates latent TGF-beta 1 in vascular endothelial cells. Blood 2000; 95:1309-16.
11. Hellerbrand C, Stefanovic B, Giordano F et al. The role of TGFbeta1 in initiating hepatic stellate cell activation in vivo. J Hepatol 1999; 30:77-87.
12. George J, Roulot D, Koteliansky VE et al. In vivo inhibition of rat stellate cell activation by soluble transforming growth factor beta type II receptor: A potential new therapy for hepatic fibrosis. Proc Natl Acad Sci USA 1999; 96:12719-24.
13. Qi Z, Atsuchi N, Ooshima A et al. Blockade of type beta transforming growth factor signaling prevents liver fibrosis and dysfunction in the rat. Proc Natl Acad Sci USA 1999; 96:2345-9.
14. Inagaki Y, Nemoto T, Nakao A et al. Interaction between GC box binding factors and Smad proteins modulates cell lineage-specific alpha 2(I) collagen gene transcription. J Biol Chem 2001; 276:16573-9.
15. Kawada N, Tran-Thi TA, Klein H et al. The contraction of hepatic stellate (Ito) cells stimulated with vasoactive substances. Possible involvement of endothelin 1 and nitric oxide in the regulation of the sinusoidal tonus. Eur J Biochem 1993; 213,:815-23.
16. Rockey DC, Weisiger RA. Endothelin induced contractility of stellate cells from normal and cirrhotic rat liver: implications for regulation of portal pressure and resistance. Hepatology 1996; 24:233-40.
17. Shao R, Yan W, Rockey DC. Regulation of endothelin-1 synthesis by endothelin-converting enzyme-1 during wound healing. J Biol Chem 1999; 274:3228-34.
18. Rockey DC, Chung JJ. Endothelin antagonism in experimental hepatic fibrosis. Implications for endothelin in the pathogenesis of wound healing. J Clin Invest 1996; 98:1381-8.
19. Shiratori Y, Imazeki F, Moriyama M et al. Histologic improvement of fibrosis in patients with hepatitis C who have sustained response to interferon therapy. Ann Intern Med 2000; 132:517-24.
20. Higashi K, Kouba DJ, Song YJ et al. A proximal element within the human alpha 2(I) collagen (COL1A2) promoter, distinct from the tumor necrosis factor-alpha response element, mediates transcriptional repression by interferon-gamma. Matrix Biol 1998; 16:447-56.
21. Okuyama H, Shimahara Y, Kawada N et al. Regulation of cell growth by redox-mediated extracellular proteolysis of platelet-derived growth factor receptor beta. J Biol Chem 2001; 276:28274-80.
22. Ueki T, Kaneda Y, Tsutsui H et al. Hepatocyte growth factor gene therapy of liver cirrhosis in rats. Nat Med 1999; 5:226-30.
23. Sato M, Kakubari M, Kawamura M et al. The decrease in total collagen fibers in the liver by hepatocyte growth factor after formation of cirrhosis induced by thioacetamide. Biochem Pharmacol 2000; 59:681-90.
24. Miyahara T, Schrum L, Rippe R et al. Peroxisome proliferator-activated receptors and hepatic stellate cell activation. J Biol Chem 2000; 275:35715-22.
25. Galli A, Crabb DW, Ceni E et al. Antidiabetic thiazolidinediones inhibit collagen synthesis and hepatic stellate cell activation in vivo and in vitro. Gastroenterology 2002; 122:1924-40.
26. Ikejima K, Takei Y, Honda H et al. Leptin receptor-mediated signaling regulates hepatic fibrogenesis and remodeling of extracellular matrix in the rat. Gastroenterology 2002; 122:1399-410.
27. Saxena NK, Ikeda K, Rockey DC et al. Leptin in hepatic fibrosis: evidence for increased collagen production in stellate cells and lean littermates of ob/ob mice. Hepatology 2002; 35:762-71.
28. Ikejima K, Honda H, Yoshikawa M et al. Leptin augments inflammatory and profibrogenic responses in the murine liver induced by hepatotoxic chemicals. Hepatology 2001; 34:288-97.

Chapter 11

Recent Therapeutic Developments in Hepatic Fibrosis

Ichiro Shimizu

Abstract

Currently, hepatic fibrogenesis is viewed as a dynamic process strictly related to the extent and duration of liver injury. Activated hepatic stellate cells (HSCs) are identified as the major source of extracellular matrix components in the injured liver, and are regarded as a target for antifibrogenic therapy. In addition, transforming growth factor-β (TGF-β) is recognized as a key fibrogenic cytokine produced by Kupffer cells and HSCs. There are several approaches to inhibit TGF-β; use of soluble receptors and gene therapy approaches. Hepatocyte growth factor is a hepatotrophic factor for liver regeneration and appears to suppress hepatic fibrogenesis in animals. HOE 77 and S 4682 are inhibitors of prolyl 4-hydroxylase, which is essential for collagen formation. α-Tocopherol and silybinin reduce lipid peroxidation and attenuate HSC activation in experimental models. Silymarin is extracted from milk thistle, the principle component of which is silybinin. Unfortunately, they have had mixed effects in human liver diseases. A Japanese herbal medicine Sho-saiko-to functions as a potent fibrosuppressant via the inhibition of oxidative stress in hepatocytes and HSCs. Its active components are baicalin and baicalein of flavonoids with chemical structures very similar to silybinin.

Understanding the cellular and molecular mechanisms underlying hepatofibrogenesis provides valuable information on the search for effective antifibrogenic therapies.

Introduction

Hepatic fibrosis, or the deposition of extracellular matrix (ECM), is associated with inflammation and cell death, which accompanies the repair processes, and is a consequence of severe liver damage that occurs in many patients with chronic liver injury of any etiology. The term hepatic fibrogenesis describes a progressive process that ranges from mild fibrosis to fully advanced cirrhosis. The origin of the abnormal ECM proteins is a cell known as the hepatic stellate cell (HSC) (also known as the fat-storing cell, lipocyte, or the Ito cell). HSCs are located in the space of Disse in close contact with hepatocytes and sinusoidal endothelial cells. Their three-dimensional structure consists of the cell body and several long and branching cytoplasmic processes (Fig. 1).[1] HSCs undergo a proliferation and transformation under inflammatory and peroxidative stimuli into myofibroblast-like cells, which are the origin of much of the collagen hypersecretion and nodule formation that occurs during hepatofibrogenesis.[2,2]

Fibrogenesis: Cellular and Molecular Basis, edited by Mohammed S. Razzaque.

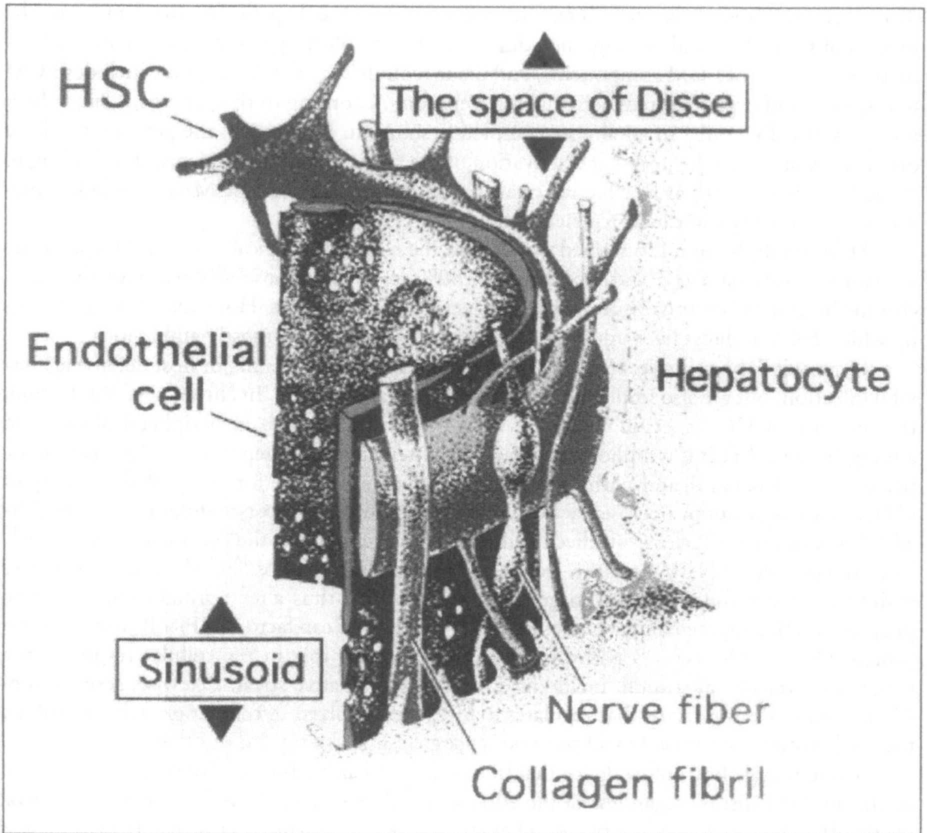

Figure 1. Schematic representation of the sinusoidal wall of the liver, based on findings reported by Wake.[1] Solid triangles indicate space directions. Hepatic stellate cells (HSCs) contact both endothelial cells and hepatocytes. Collagen fibrils course through the space of Disse.

The most common cause of hepatic fibrosis is currently chronic hepatitis C virus (HCV) infection. In fact, the World Health Organization reported that up to 3% of the world's population was infected with HCV, suggesting that more than 170 million chronic carriers were at risk for developing hepatic fibrosis to cirrhosis and hepatocellular carcinoma.[3] The likelihood that HCV infection will lead to cirrhosis is significant and varies with the duration of infection as well as the mechanism of infection. It should be noted that interferon (IFN) and a combination of IFN and ribavirin are the only currently available therapies for chronic HCV infection, but a sustained response to the treatment has been unsatisfactory. Hepatic fibrosis and its final common result, cirrhosis, appear to be much more common than clinically appreciated. Therefore, understanding the cellular and molecular mechanisms underlying the hepatofibrogenesis provides valuable information in the search for effective antifibrogenic therapies.

Hepatocyte Injury and Oxidative Stress

Parenchymal cell membrane damage could result in the release of oxygen-derived free radicals and other reactive oxygen species (ROS) derived from lipid peroxidative processes, which represent a general feature of sustained inflammatory response and liver injury, leading to necrosis and apoptosis. In general, the process leading to hepatic fibrosis, similar to the process of normal wound healing, includes three phases following tissue injury: acute inflammation, synthesis of ECM components, and tissue remodeling. This latter phase includes ECM remodeling and organized parenchymal regeneration. According to this scheme, a single liver injury eventually results in an almost complete resolution. In contrast, the persistence of the original insult causes the prolonged activation of the tissue repair mechanisms, thus leading to tissue fibrosis rather than to effective tissue repair. In other words, ECM accumulation predominates over hepatocellular regeneration.[4]

HCV multiplication is sustained throughout the course of a typical infection.[5] Hepatocytes are continuously damaged and replicated. In 80% to 85% of those infected with the virus, chronic hepatitis C eventually develops, which can lead to cirrhosis. However, the mechanisms by which HCV induces liver injury and hepatocyte death remain mostly ambiguous.

Cytotoxic T lymphocytes (CTLs) are not only thought to be a major host defense against viral infection, but are also implicated in the immunopathogenesis. In chronic HCV infection, the presence of HCV-specific CTLs has been demonstrated both in peripheral blood[6] and among liver-infiltrating lymphocytes[7] of patients with chronic hepatitis C. Two pathways, perforin and Fas/Fas ligand pathways, were reported to account for all cytolytic activity of CTLs,[8] although tumor necrosis factor (TNF)-α, a proinflammatory cytokine, is released by all HCV-specific CTL clones studied[9] and may also contribute to the cytotoxicity of CTLs.[10] The core protein of HCV has been recognized as a target for CTLs.[11,12] Viral proteins posses various accessory functions that target host proteins and thus alter normal cellular growth properties, affecting signaling molecules, such as the nuclear factor (NF)-κB and activator protein (AP)-1. The NF-κB pathway plays an important role in the cellular response to a variety of extracellular stimuli, including TNF-α and oxidative stress. Likewise, activation of AP-1, a transcription factor that regulates many genes involved in cellular growth control, by the viral proteins presumably will promote oncogenesis and a myriad of other effects.

There is large body of evidence that in humans and animals oxidative stress is implicated in chronic liver disease regardless of the etiology (viral infection, alcohol consumption, metal overload)[13] and serves as a link between hepatic injury and fibrosis.[14] Fortunately, cells possess antioxidant protective mechanisms, including enzymatic defense molecules such as superoxide dismutase (SOD) and glutathione peroxidase (GPx), which can react with ROS and neutralize them before they inflict damage on vital components. Overexpression of Bcl-2 suppresses lipid peroxidation and prevents apoptosis. Bcl-2, Bcl-X_L, Bad, and Bax are members of the Bcl-2 family proteins that play an important role in regulating cell survival and apoptosis. In addition to Bcl-2, Bcl-X_L also prevents apoptosis in response to a wide variety of stimuli. In contrast, proapoptotic proteins, Bad and Bax can accelerate death and in some situations are sufficient to cause apoptosis. Our preliminary studies showed that oxidative stress induced ROS generation, lipid peroxidation, the activation of AP-1 and NF-κB, the loss of SOD and GPx activities, and down-regulation of Bcl-2 and Bcl-X_L expression and up-regulation of Bad expression in cultured rat hepatocytes (Fig. 2).[15,16]

In particular, histopathological studies of chronic HCV infection showed fatty changes in 31% to 72% of patients,[17] indicating that hepatic steatosis is a characteristic feature of chronic HCV infection. It was suggested that hepatic steatosis may reflect a direct cytopathic effect of HCV and play a role in the progression of the disease. In support of these findings, a transgenic mouse model, which expressed the HCV core gene, was observed to develop progressive hepatic steatosis.[18] It is conceivable that following hepatocyte injury, hepatic steatosis leads to an increase in lipid peroxidation, which might contribute to HSC activation by releasing soluble mediators,[19,20] and, thus, induce hepatic fibrosis. In addition, chronic infection of HCV is associated with excess iron in the liver of subjects who are neither alcoholics nor recipients of blood transfusions.[21] Hepatic iron is a cofactor involved in inducing fibrosis in patients with

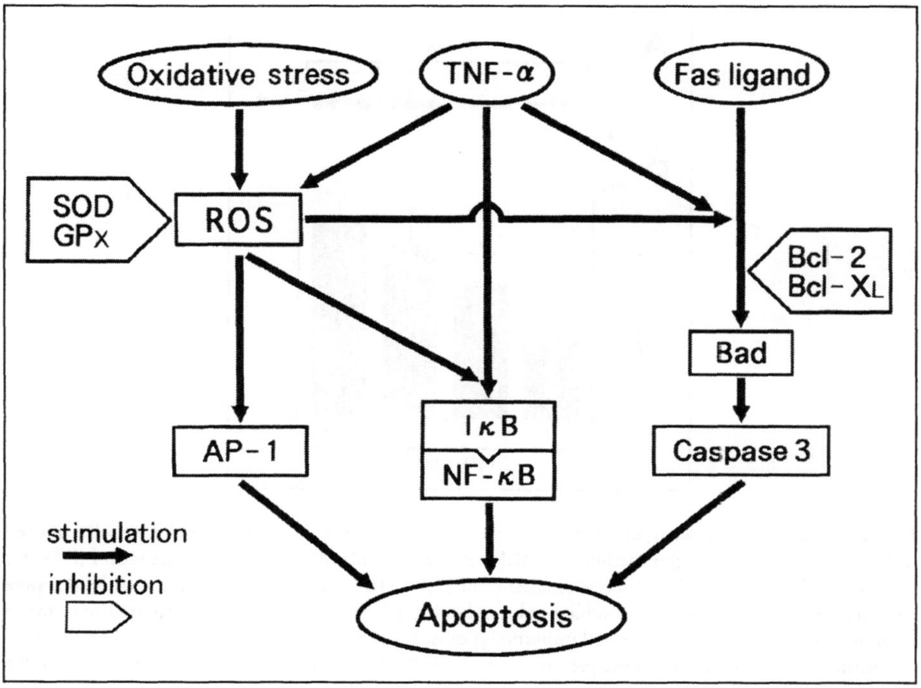

Figure 2. Working hypothesis regarding the role of reactive oxygen species (ROS) in inducing apoptosis in the liver. SOD: superoxide dismutase; GPx: glutathione peroxidase; AP-1: activator protein-1; TNF-α: tumor necrosis factor-α; NF-κB: nuclear factor-κB.

chronic hepatitis C.[22,23] Iron is known to be a potent in vivo factor in the production of hydroxy radicals, which induce lipid peroxidation and DNA cleavage,[24] leading to DNA mutagenesis.[25]

Increased lipid peroxidation was also observed to be common to alcoholic liver disease and nonalcoholic steatohepatitis as evident in studies of human alcohol-related liver injury[26] and animal models of diet-induced hepatic steatosis[27] and drug-induced steatohepatitis.[28]

HSC Activation and Growth Factors

In the injured liver, HSCs are regarded as the primary target cells for inflammatory and peroxidative stimuli, and are transformed into myofibroblast-like cells. These HSCs are referred to as activated cells, and this activation is accompanied by a loss of cellular retinoid, and the synthesis of α-smooth muscle actin (α-SMA) and large quantities of the major components of the ECM, including collagen types I, III, and IV, fibronectin, and laminin. The large majority of collagen III, IV and laminin are synthesized by HSCs and sinusoidal endothelial cells, whereas all cell types synthesize small amounts of collagen type I. During active hepatofibrogenesis, however, HSCs become the major ECM producing cell type, with a predominant production of collagen type I.[29] Following cell activation in vivo, HSCs express the genes encoding the key components required for matrix degradation such as matrix metalloproteinase (MMP)-1 and -2. However, through the activation of tissue inhibitor of metalloproteinase (TIMP)-1 and -2, activated HSCs also inhibit the activity of interstitial collagenases, which degrade fibrillar collagen.[30] This hypothesis is supported by studies of experimental hepatic fibrosis and in human liver disease, in which TIMP-1 expression is significantly

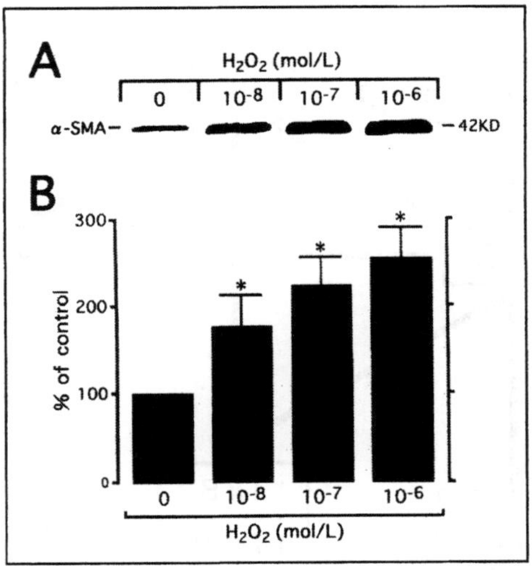

Figure 3. A) Effect of hydrogen peroxide (H_2O_2) on α-SMA expression in cultured rat HSCs. Cells cultured in Dulbecco's modified Eagle's medium (DMEM) supplemented with 10% fetal bovine serum (FBS) for 5 days were incubated in the presence or absence of H_2O_2 (10^{-8}-10^{-6} mol/L) for 24 h. Cell lysates were subjected to 12% sodium dodecyl sulfate-polyacrylamide gel electrophoresis and transferred onto nylon membranes, and α-SMA was detected immunologically. B) Densitometric analysis of α-SMA expression in cultured rat HSCs induced in the presence of H_2O_2 (10^{-8}-10^{-6} mol/L) shown in (A). The results of the densitometric analysis are expressed as mean percentages (± SD) of the control values (n=5). *$P < 0.05$ compared with controls.

up-regulated in cirrhotic liver compared with normal liver, while expression of collagenases remains unchanged.[31] In addition, transgenic mice overexpressing TIMP rapidly develop hepatic fibrosis following injury.[32] It is clear that fibrogenesis involves a dynamic interplay of matrix synthesis and degradation. In the HSC-mediated response to cell injury, this interplay is disrupted and leads to fibrosis.

There is evidence to show that the products of lipid peroxidation modulate collagen gene expression in HSCs.[33,34] It was reported that paracrine stimuli derived from hepatocytes undergoing oxidative stress induce HSC proliferation and collagen synthesis.[35] Although hepatocytes are the major site for oxygen utilization, nonparenchymal cells also are sources of ROS, and may thereby contribute to hepatocyte necrosis and/or HSC activation. Kupffer cells, fibroblasts, and invading mononuclear cells are capable of releasing ROS. Kupffer cells are activated by a variety of stimuli to produce ROS, such as opsonized zymosan, phorbol esters, lipopolysaccarides, and Ca^{2+} ionophores.[36,37] ROS include hydrogen peroxide (H_2O_2), singlet molecular oxygen ($^1O_2$), hydroxyl ($OH^·$), superoxide (O_2^-), alkoxyl ($RO^·$), peroxyl ($ROO^·$), and nitric oxide ($NO^·$) free radicals, highly heterogeneous in terms of reactivity against cellular targets.[38] HSCs have been shown to be activated by the generation of free radicals with Fe^{2+}/ascorbate,[20] by H_2O_2 (Fig. 3), and by malondialdehyde (MDA)[39] and 4-hydroxynonenal (HNE) (Fig. 4),[33,39] aldehydic products of lipid peroxidation, and antioxidants were observed to inhibit HSC activation.[13,14]

Inflammatory cells, such as Kupffer cells and invading mononuclear cells, which release cytokines, transforming growth factor (TGF)-β1 and platelet-derived growth factor (PDGF) may also contribute to the fibrogenic response to liver injury. It has been shown that TGF-β1 and PDGF activate cultured HSCs, and act as paracrine and autocrine (i.e., from HSCs) mediators which trigger and activate the transformation of HSCs in vivo. Importantly, TGF-β1 is

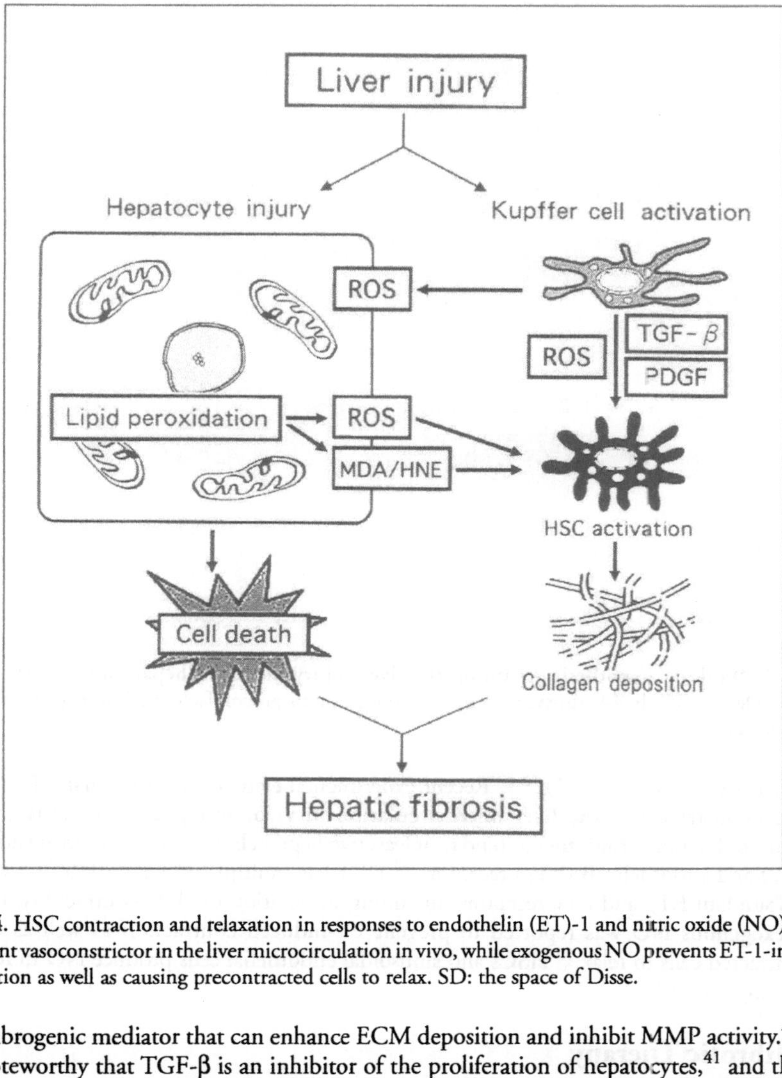

Figure 4. HSC contraction and relaxation in responses to endothelin (ET)-1 and nitric oxide (NO). ET-1 is a potent vasoconstrictor in the liver microcirculation in vivo, while exogenous NO prevents ET-1-induced contraction as well as causing precontracted cells to relax. SD: the space of Disse.

a key fibrogenic mediator that can enhance ECM deposition and inhibit MMP activity.[40] It is also noteworthy that TGF-β is an inhibitor of the proliferation of hepatocytes,[41] and that, at higher concentrations, TGF-β induces oxidative stress leading to hepatocyte apoptosis.[41] Thus, studies are underway to develop therapies that neutralize this cytokine. In addition, HSC proliferation contributes indirectly but significantly to fibrogenesis. An increase in HSC number has been well documented after both experimental and human liver injury.[42] Proliferation is an important component of the activation cascade because it amplifies the HSC-mediated response to injury. A number of mitogens appear to be important in the stimulation of HSC proliferation and include PDGF,[43] epidermal growth factor,[44] and endothelin (ET)-1.[45] The major mitogen driving HSC proliferation is PDGF, a cytokine that also plays a key role in smooth muscle cell proliferation during other forms of injury.

Due to their anatomical location, their ultrastructural features and similarities with pericytes that regulate blood flow in other organs, HSCs function as liver-specific pericytes.[46] Previous studies have shown that the contraction and relaxation of HSCs regulate hepatic sinusoidal blood flow.[46,47] Two vasoregulatory compounds with obvious effects on HSCs include ET-1

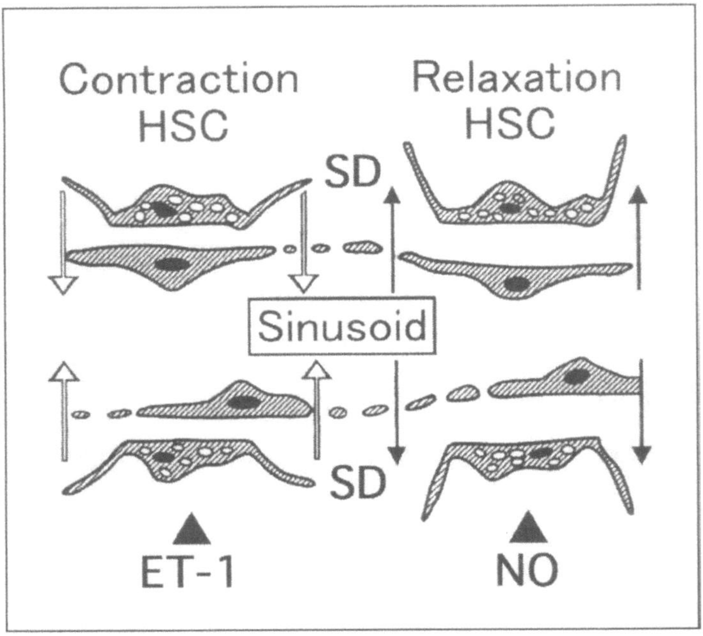

Figure 5. Working hypothesis regarding the liver injury-mediated hepatofibrogenesis. MDA: malondialdehyde; HNE: 4-hydroxynonenal; TGF: transforming growth factor β; PDGF: platelet-derived growth factor.

and nitric oxide (NO) (Fig. 5).[48-50] Recent experimental evidence suggests that ET-1 is a potent vasoconstrictor in the liver microcirculation in vivo, acting at both sinusoidal and extrasinusoidal sites.[51] Endothelins bind to at least two heptahelical receptors, termed endothelin A (ET_A) and endothelin B (ET_B) receptors.[52] ET-1 has multiple effects on HSCs, which express abundant ET_A and ET_B receptors, including stimulation of HSC contractility. In addition, exogenous NO was reported to prevent ET-induced contraction as well as causing precontracted cells to relax.[53] HSCs and sinusoidal endothelial cells produce NO in response to various stimuli.[54,55]

Antifibrotic Therapy

Removal of Causative Agent

Removing the causative agent is currently the most effective way to treat patients with hepatic fibrosis. This approach is effective in chronic hepatitis B virus or HCV infection, alcohol consumption, drug-induced liver disease, and iron and copper overload. Most of these agents induce hepatocellular injury and subsequently hepatic inflammation, finally resulting in HSC activation and collagen deposition.[30] However, some factors, including alcohol metabolites[56] and ferritin,[57] may directly activate HSCs and enhance their fibrogenic potential.

Based on our current understanding of the cellular and molecular basis of hepatofibrogenesis, antifibrotic agents can be categorized as follows:

1. Agents for removal of causative agent (i.e., eradication of HCV RNA).
2. Agents for cytoprotection. Reduction of inflammation and cell death.
3. Agents for inhibition of HSC activation.

4. Agents for fibrolysis. Promotion of matrix degradation.

Most compounds in each of these categories have been identified in cell cultures and in various animal models of hepatic fibrosis, and suggested as treatments for patients with hepatic fibrosis (Table 1). However, therapeutic application in clinical situations is either very limited or of uncertain efficiency. Thus, there is a sustained need for therapeutic devices combining efficiency and low adverse side effects.

IFNs

There are three major isoforms of IFN-α, -β, and -γ. Each of these unique isoforms has different biologic actions. There are multiple IFN-α subtypes, whereas there appear to be only single IFN-β and IFN-γ species. IFN-α and IFN-β are more closely related to each other, structurally and functionally, than they are to IFN-γ. Indeed, IFN-α and -β each bind to the same receptor. IFN-α is stimulated by viral infection, and IFN-γ is stimulated by mitogenic or antigenic stimulation of T lymphocytes or natural killer cells. IFN-α has much more potent antiviral effects than does IFN-γ, whereas IFN-γ is 100 to 10,000 fold more potent as an immunomodulator than IFN-α.[58] This observation has led to the concept that IFN-α and -β are primarily antiviral agents with some immunomodulatory effects, but IFN-γ is primarily an immunomodulatory agent with some antiviral effects.[59]

Several recent clinical reports have concluded that IFN-α treatment is effective for reducing and eliminating serum HCV RNA,[60,61] improving histological scores for hepatic fibrosis and necroinflammation,[59,60,62] and attenuating the hepatic iron deposition[23] in patients with chronic hepatitis C. It should be noted that HCV may promote hepatocyte damage by activating lipid peroxidation via iron-mediated mechanisms.[63] IFNs have also been known to reduce α-SMA expression in the liver,[64] to suppress collagen synthesis in vitro, and to decrease type I, III, and IV procollagen mRNA levels.[65] It was reported that hepatic mRNA levels of TGF-β1 in addition to type I procollagen are significantly decreased in responders to IFN-α therapy,[66] and that IFN-α can inhibit activation of human and rat HSCs in primary culture.[67,68] These findings

Table 1. Antifibrotic agents tested in vivo

Name	Inhibition of HSC Activation	Cytoprotection	Fibrolysis
IFN-α	+	+	+
IFN-γ	+	Unknown	Unknown
IL-10	Unknown	+	Unknown
Decorin	+	+	+
Soluble TGF-β type II receptor	+	+	+
HGF	Unknown	+	+
α-Tocopherol	+	+	Unknown
Silymarin	+	+	Unknown
Sho-saiko-to	+	+	Unknown
Arg-Gly-Asp peptides	+	Unknown	Unknown
HOE 77	+	–	+
Safironil	+	–	+
S 4682	+	–	+
Pentoxifylline	+	+	+
Polyenyl-phosphatidylcholine	+	Unknown	+

Note: The relative effectiveness of each of the compounds is arbitrarily designated by the following symbols: –: not effective; and +: effective.

suggest that the effect of IFN-α may be not only inhibiting HCV-mediated injury, but also directly inactivating HSC.

Ribavirin has a synergic effect when used together with IFN-α. However, ribavirin is not recommended as monodrug therapy because it has no effect on serum HCV RNA clearance. Administration of a combination of IFN-α and ribavirin produces a superior viral clearance response rate than IFN alone.[69,70] It was reported that IFN-α and ribavirin combination therapy significantly reduces the rate of fibrosis progression in patients with hepatitis C.[71]

In contrast to IFN-α, IFN-γ is ineffective at eradicating serum HCV RNA, but extensive experimental findings suggest that IFN-γ can inhibit HSC activation[67,72] and profoundly deduces hepatic fibrogenesis.[72,73] With regard to the use of IFN-γ in patients with HCV infection, there has been little clinical interest because overexpression of IFN-γ in the liver leads to chronic hepatitis[74] and because of the potential long-term side effects related to its immunomodulatory effects. In patients with pulmonary fibrosis, however, a study showing that IFN-γ treatment resulted in a reduction in fibrosis, suggests that it could be feasible to use IFN-γ in the setting of hepatic fibrosis.[75]

Anti-Inflammatory Agents

Hepatic inflammation precedes and promotes the progression of hepatic fibrosis. A strategy to inhibit hepatic inflammation is the neutralization of proinflammatory cytokines. In experimental fibrosis, interleukin (IL)-1 receptor antagonists[76] or soluble TNF-α receptors[77] are associated with diminished necrosis and inflammation in liver tissue. Recently, a promising study has evaluated the use of IL-10, a powerful anti-inflammatory and antifibrotic cytokine that is synthesized by activated HSCs in patients with chronic HCV infection.[78] The rationale is that IL-10 down-regulates proinflammatory Th1 responses and that IL-10 (-/-) mice develop more severe hepatic fibrosis than do wild-type mice.[79] Patients with chronic HCV infection treated with recombinant IL-10 showed not only an improvement in hepatic inflammation, but also resolution of the initial deposition of fibrous scarring. Multicenter trials are currently underway.

Anti-TGF-β Agents

TGF-β is a potent fibrogenic cytokine produced by Kupffer cells and HSCs. With the importance of TGF-β to ECM production firmly established, the TGF-β/HSC axis is a potentially important therapeutic avenue. Two signaling receptors, termed type I and type II, mediate the biologic actions of TGF-β. The extracellular domain of the type II receptor binds the ligand, causing formation of heteromeric complexes incorporating type I and type II receptors. The type II receptor then transphosphorylates the type I receptor, activating its kinase and initiating downstream signaling.[80] Thus, the type II receptor appears to be essential for the biological activity of TGF-β in vivo.[81,82] There are several approaches to neutralize TGF-β activity. These include use of proteoglycan decorin, soluble TGF-β type II receptors, and gene therapy approaches.

Decorin is the protein core component of proteoglycan that binds and sequesters TGF-β1 in the ECM. Intraveous injection of recombinant decorin as well as intramuscular injection of the decorin gene, reduced levels of TGF-β mRNA and TGF-β protein and prevented fibrosis induced by glomerulonephritis in a model of glomerular inflammation.[83,84]

Administration of a soluble receptor against the extracellular domain of the TGF-β type II receptor led to an inhibition of activation of cultured HSCs and hepatofibrogenesis in a model of hepatic fibrosis.[85] After intramuscular injection of an adenovirus expressing the soluble receptor (AdTβ-ExR), followed by induced experimental fibrosis, the soluble receptor protein was detectable in the blood, and hepatic fibrosis was attenuated.[86] The soluble receptor can bind active TGFβ, and thus prevent TGF-β from binding to cognate receptors. Our preliminary

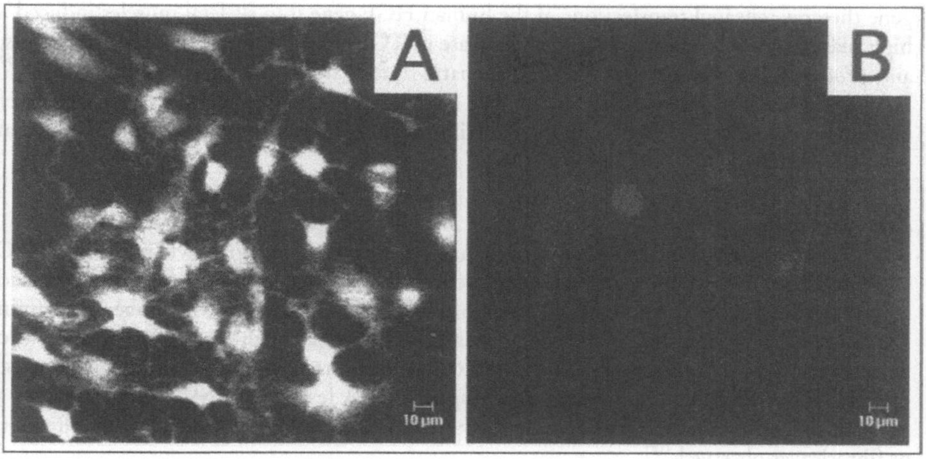

Figure 6. Effect of cultured rat HSCs, infected with AdTβ-ExR on intracellular ROS generation using the fluorescence probe dichlorofluorescein and a laser scanning confocal microscopic technique. Cells cultured in DMEM supplemented with 10% FBS for 5 days were infected with AdTβ-ExR (B) or AdLacZ (A) for 1 h, and then incubated in serum-free DMEM for an additional 3 days. AdTβ-ExR, adenovirus vector expressing a soluble receptor against an extracellular domain of the TGF-β type II receptor; AdLacZ, adenovirus vector expressing the bacterial β-galactosidase.

study showed that infection of HSCs with this adenovirus AdTβ-ExR attenuated intracellular TGF-β1 levels, ROS generation (Fig. 6), and prevented HSC activation, suggesting that the adenovirus-mediated soluble TGF-β receptor might lead to an interruption of the TGF-β autocrine loop in activated HSC (unpublished observations). Further, using adenovirus-mediated gene transfer to express a truncated TGF-β type II receptor in the liver, a single application of the adenovirus TGF-β receptor construct, intravenously or via the portal vein, induced prevention of fibrosis in a model of hepatic fibrosis.[87,88] The truncated receptor theoretically inhibits TGF-β activity, by competing with binding of the cytokine to endogenous TGF-β receptors.

In future studies it might be feasible to use a liver-specific gene-delivery system for in vivo transfection of liver cells with genes encoding the respective antagonists of soluble cytokine receptors,[89] thus providing a sustained supply of cytokine scavengers at the site of action. The use of TGF-β antisense oligomers also might be of value for the treatment of hepatic fibrosis in which TGF-β overexpression plays the most important pathogenetic role.[90]

An important consideration with the use of these compounds will be that, because TGF-β is important in all forms of wound healing, inhibition of TGF-β could conceivably impair beneficial wound healing.

Hepatocyte Growth Factor

Hepatocyte growth factor (HGF), originally identified and cloned as a potent mitogen for mature hepatocytes,[91,92] shows mitogenic, motogenic and morphogenic activities for a wide variety of cells. In the liver HGF binds to interstitial collagen types I, II, V, and VI at sites of excessive release.[93] Its binding to the ECM can protect this growth factor from proteolytic degradation and allow for rapid release when tissues are remodeled and the ECM is degraded during development or after injury, thus paving the way for wound healing, neovascularization, and organ growth.[93] It is now evident that HGF plays an essential part in the development and regeneration of the liver.[94] It has been reported that HGF reduces TGF-β expression, induces MMP activity, and inhibits fibrogenesis in animal models of hepatic fibrosis.[95,96] Based on gene therapy, repeated transfections of the human HGF gene into skeletal muscles induced a high plasma HGF level, suppressed the increase of TGF-β1, inhibited hepatocyte apoptosis, and produced the complete resolution of hepatic fibrosis in an animal model of cirrhosis.[97] Further studies assessing its efficacy and safety are needed to establish its utility in human liver disease, particularly in view of its potential stimulation of hepatocyte growth and theoretical risk of carcinogenesis.[30]

Antioxidants

Oxidative stress could represent a mechanism contributing to hepatocyte injury and HSC activation common to different pathways, such as inflammatory, ethanol, and iron overload stimuli, leading to hepatic fibrosis.[2,14] Therefore, a known antioxidant, α-tocopherol (vitamin E) has been examined in animals[98,99] and humans.[100] In patients with chronic hepatitis C, who were refractory to IFN therapy, treatment with α-tocopherol resulted in inhibition of parameters of HSC activation, including type I procollagen mRNA level and β-SMA expression, to a substantial degree. However, no change in the histological degree of hepatocellular inflammation or fibrosis was observed.[100]

In addition to α-tocopherol, silybinin, Sho-saiko-to,[101] and estradiol[102,103] have been reported to reduce lipid peroxidation and attenuate HSC activation in experimental models. Silybinin acts as an oxygen radical scavenger,[104] prevents the lipid profile of hepatocyte membranes,[105] inhibits the proliferation of HSCs in vitro, and retards collagen accumulation in an experimental model.[104] There have been some positive clinical trials involving silymarin, a standardized extract from milk thistle *Silybum marianum*, the principle active component of which is silybinin, comprising 60-70% of the silymarin.[106,107] However, the latest results of a randomized controlled trial suggested that silymarin had no beneficial effect in human liver diseases.[108]

The Chinese herbal medicine Sho-saiko-to is a mixture of seven herbal preparations, which is widely administered in Japan to patients with chronic hepatitis and cirrhosis. Sho-saiko-to functions as a fibrosuppressant, at least in part, by inhibition of lipid peroxidation in hepatocytes and HSCs in vivo.[101] Among the active components of Sho-saiko-to, two flavonoids, baicalin and baicalein were found to be mainly responsible for the antioxidative activity.[101] These flavonoids were reported to suppress the proliferation of vascular smooth muscle cells.[109,110] The chemical structures of the flavonoids are very similar to silybinin. Each molecule contains a 2-phenyl-1-benzophyrane-4-one (flavone) structure (Fig. 7).[111]

S-adenosyl-L-methionine (SAMc) is a substrate of glutathione synthesis that has hepatoprotective and antioxidant properties.[112,113] It attenuates hepatic fibrosis in experimental models. SAMc is currently used in several human liver diseases, such as alcoholic liver disease, and drug-induced liver disease. In alcoholic patients, it improves survival and delays the need for liver transplantation.[114]

Estradiol is a potent endogenous antioxidant that reduces lipid peroxide levels in the liver and serum.[115,116] It was reported that estradiol treatment results in a dose-dependent suppression of hepatic fibrosis in hepatic fibrosis models,[102] and that estradiol can inhibit the activation of AP-1 and NF-κB in cultured hepatocytes undergoing oxidative stress[15,16] and attenuate HSC activation in primary culture.[102,103] However, the administration of estrogen in women poses some potential risks, including breast cancer and endometrial abnormalities.[117-119]

Arg-Gly-Asp Peptides

Many integrins recognize and bind the amino acid sequence Arg-Gly-Asp (RGD) that presents in fibronectin, collagen type VI, and other ECM molecules.[120] The RGD tripeptide can inhibit fibrosis in a model of immune-mediated liver damage.[121] An interesting observation is that the extradomain III, which is present in a splice variant of fibronectin, can induce the

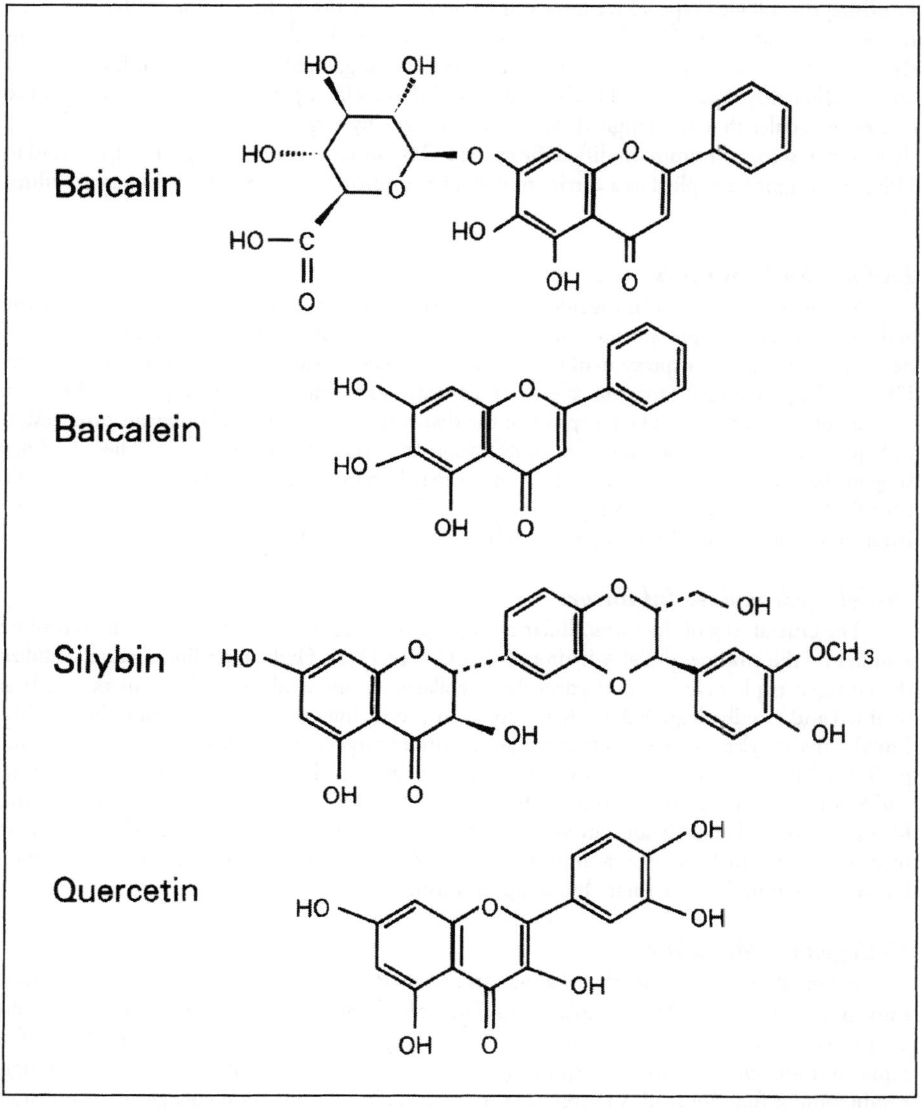

Figure 7. Comparison of the chemical structure of baicalin, baicalein, silybinin, and quercetin. Each of these molecules contain a 2-phenyl-1-benzopyrane-4-one (flavone) structure.

myofibroblastic phenotype in cultured HSCs and that a neutralizing antibody to this domain can reverse this step of HSC activation.[122] Collagen type VI is a major matrix protein involved in the adhesion of cells to the surrounding matrix. The main cellular sources of collagen type VI in normal and fibrotic livers are HSC.[123] Binding of cells to collagen type VI by integrins is RGD-dependent. Although several RDG sequences are found in collagen type VI, the cyclic RGD-containing peptide C*GRGDSPC*, in which C* denotes the cyclizing cysteine residues, specifically inhibited the attachment of collagen type VI to cells, while the RGD-dependent binding of fibronectin to these cells was not inhibited.[124] In addition, the linear GRGDSP peptide[125] failed to inhibit the cellular attachment of collagen type VI,[124] suggesting that a non-integrin collagen type VI receptor was antagonized by the cyclic peptide.[126] To date, drug carriers to hepatocytes, Kupffer, and sinusoidal endothelial cells have been described extensively.[120] In contrast to in vitro findings, in vivo studies suggested that most antifibrotic agents are not efficiently taken up by HSCs. To attain HSC-specific uptake, drugs can be coupled to carrier molecules that are designed for selective uptake by target cells. A recent study showed that human serum albumin modified the cyclic RGD-containing peptide specifically bound to HSC and might be applied as a carrier to deliver antifibrotic agents to HSCs in hepatic fibrosis.[126]

Endothelin Inhibitors

Although ET-1 is profibrogenic in other organs such as lung and kidney through the activation of ET_A receptors, its role in hepatic fibrosis remains controversial. In early phases of hepatofibrogenesis, the expression of ET_A receptors is predominant over ET_B receptors; therefore, ET-1 could play a profibrogenic role.[127] In contrast, as hepatic fibrosis progresses, there is a marked up-regulation of ET_B receptors in the diseased liver, which could prevent progression of hepatic fibrosis by inhibiting HSC proliferation and collagen synthesis.[128,129] These findings suggest that the use of selective ET_A antagonists may be a more rational strategy to treat hepatic fibrosis than nonselective ET receptor antagonists.[30] A recent study has shown that the administration of a selective ET_A antagonist inhibits hepatic fibrosis in a experimental model.[130]

Prolyl Hydroxylase Inhibitors

The crucial step of the intracellular collagen processing is the synthesis of hydroxyproline residues by the enzyme, prolyl 4-hydroxylase (EC 1.14.11.2). Hydroxyproline residues stabilize the collagen triple helix. Underhydroxylated collagen is not stable at body temperature. It is retained and rapidly degraded inside the collagen-producing cells, HSCs and endothelial cells. Prolyl 4-hydroxylase therefore represents an important target as an antifibrotic.[131] Several competitive inhibitors of prolyl 4-hydroxylase have been developed, and include HOE 77, Safironil, and S 4682. It has been shown that HOE 77,[132-134] Safironil,[134] and S 4682[131] inhibit hepatic fibrosis in animal models and attenuate HSC activation in primary culture. The molecular mechanism by which these compounds inhibit HSC activation remains unclear. Further studies will be required before their clinical application.

Collagenase Modulators

Potential targets include the modulation of MMP activity and its regulators such as plasminogen activator or TIMPs. uPA, an initiator of the matrix proteolysis cascade, has been examined in a rat model of cirrhosis.[135] A single intravenous administration of a replication-deficient adenovirus expressing a nonsecreted form of human uPA resulted in high production of functional uPA protein in the liver, leading to induction collagenase expression and reversal of fibrosis with subsequent hepatocyte regeneration and improved liver function.

Retinoids

Vitamin A includes a class of biologically active compounds also referred to as retinoids. This class of compounds in the diet is not only provided as retinyl esters, mainly retinyl palmitate, from animal sources, but also as carotenoids, mainly β-carotene, from plant sources. HSCs are the main storage sites for retinoids in intact liver lobules. The main storage type for retinoids in HSCs is retinyl palmitate. When activated in injured livers, HSCs transform into myofibroblast-like cells, during which process they gradually lose their intracellular retinoids.

Although a substantial body of evidence suggests indicates that retinoids play an important role in the regulation of HSC activation,[136,137] the findings from in vivo experimental models are mixed.

In cultured HSCs, retinyl palmitate supplementation produced many lipid droplets occupying the cytoplasmic space and attenuated HSC activation, leaving insufficient space for other organelles to develop.[138] Retinyl palmitate also inhibited hepatofibrogenesis in experimental models of hepatic fibrosis.[138,139] Further, a clear association in an animal model between collagen deposition and retinoid status existed.[140] In another study, however, 9-cis retinoic acid, a retinoic acid analog, enhanced hepatic fibrosis in an experimental model, which accompanied enhancement of cellular plasminogen activator and plasmin levels, thereby inducing plasmin-mediated activation of latent TGF-β.[141] It has been suggested that the divergent effects on fibrogenesis are related to differential effects of retinoic acid isomer; in cultured HSCs, all-trans retinoic acid exerts a significant inhibitory effect on the synthesis of certain ECM mRNAs, whereas 9-cis retinoic acid has the reverse effect.[142] Regardless of their effects in experimental models, it is unlikely that retinoids could be tested or used as antifibrotic agents in humans because long-term administration of retinol at high concentration appears to induce hepatic fibrosis in humans.[143]

Miscellaneous Compounds

Pentoxifylline is a xanthine derivative drug that displays vasodilating properties for peripheral blood vessels and reduces blood viscosity. It was reported that pentoxifylline inhibits myofibroblastic transformation of HSC,[144] being effective on TGF-β and PDGF.[145,146] In addition to its antifibrogenic properties pentoxifylline exerts hepatoprotective effects,[147] which are associated with a reduced production of TNF-α by Kupffer cells.[148] However, pentoxifylline has had mixed effects in animal models of hepatic fibrosis.[146] Further clinical studies have to be performed to establish its real therapeutic benefits in hepatic fibrosis.

Polyenylphosphatidylcholine is a mixture of polyunsaturated phosphatidylcholines extracted from soybeans. It has been reported that this compound protects against alcohol-induced fibrosis and cirrhosis. Alcoholic liver injury is often associated with oxidative stress results in lipid peroxidation, including phosphatidylcholines, the backbone of cell membranes. It has therefore been postulated that replacement of phosphatidylcholine in membranes might lead to reduced cellular injury and fibrogenesis.[149] Polyenylphosphatidylcholine reduces fibrogenesis in animal models of hepatic fibrosis,[150,151] and attenuates HSC activation and increases MMP activity in primary culture.[151] Where this compound therapy might be useful in patients with hepatic fibrosis is open.

Conclusions

The ultimate goal of research on hepatofibrogenesis is to develop a rational basis for effective antifibrotic therapy. The progress in understanding the cellular and molecular mechanisms underlying hepatofibrogenesis brings the development of effective therapies closer to reality. In the future, targeting of HSCs and fibrogenic mediators will be a mainstay of antifibrotic therapy. Points of therapeutic intervention may include:

1. removing the causative agent (i.e., eradicating HCV RNA);
2. suppressing inflammation and cell death;
3. inhibiting HSC activation;
4. promoting matrix degradation. The future prospects for effective antifibrotic treatment are more promising than ever for the millions of patients with chronic HCV infection worldwide.

References

1. Wake K. Cell-cell organization and functions of 'sinusoids' in liver microcirculation system. J Electron Microsc 1999; 48:89-98.
2. Shimizu I. Antifibrogenic therapies in chronic HCV infection. Current Drug Targets-Infectious Disorders 2001; 1:227-240.
3. Cossarizza A, Baccarani-Contri M, Kalashnikova G et al. A new method for the cytofluorimetric analysis of mitochondrial membrane potential using the J-aggregate forming lipophilic cation 5,5',6,6'-tetrachloro-1,1',3,3'-tetraethylbenzimidazolcarbocyanine iodide (JC-1). Biochem Biophys Res Commun 1993; 197:40-45.
4. Pinzani M. Novel insights into the biology and physiology of the Ito cell. Pharmac Ther 1995; 66:387-412.
5. Hagiwara H, Hayashi N, Mita E et al. Quantitation of hepatitis C virus RNA in serum of asymptomatic blood donors and patients with type C chronic liver disease. Hepatology 1993; 17:545-550.
6. Kita H, Hiroishi K, Moriyama T et al. A minimal and optimal cytotoxic T cell epitope within hepatitis C virus nucleoprotein. J Gen Virol 1995; 76 (Pt 12):3189-3193.
7. Koziel MJ, Dudley D, Wong JT et al. Intrahepatic cytotoxic T lymphocytes specific for hepatitis C virus in persons with chronic hepatitis. J Immunol 1992; 149:3339-3344.
8. Kagi D, Vignaux F, Ledermann B et al. Fas and perforin pathways as major mechanisms of T cell-mediated cytotoxicity. Science 1994; 265:528-530.
9. Koziel MJ, Dudley D, Afdhal N et al. HLA class I-restricted cytotoxic T lymphocytes specific for hepatitis C virus. Identification of multiple epitopes and characterization of patterns of cytokine release. J Clin Invest 1995; 96:2311-2321.
10. Braun MY, Lowin B, French L et al. Cytotoxic T cells deficient in both functional fas ligand and perforin show residual cytolytic activity yet lose their capacity to induce lethal acute graft-versus-host disease. J Exp Med 1996; 183:657-661.
11. Shimizu I, Yao D-F, Horie C et al. Mutations in a hydrophilic part of the core gene of hepatitis C virus in patients with hepatocellular carcinoma in China. J Gastroenterol 1997; 32:47-55.
12. Kato N, Yoshida H, Kioko Ono-Nita S et al. Activation of intracellular signaling by hepatitis B and C viruses: C- viral core is the most potent signal inducer. Hepatology 2000; 32:405-412.
13. Poli G. Pathogenesis of liver fibrosis: Role of oxidative stress. Mol Aspects Med 2000; 21:49-98.
14. Shimizu I. Impact of estrogens on the progression of liver disease. Liver Int 2003; 23:63-69.
15. Omoya T, Shimizu I, Zhou Y et al. Effects of idoxifene and estradiol on NF-κB activation in cultured rat hepatocytes undergoing oxidative stress. Liver 2001; 21:183-191.
16. Cui X, Shmizu I, Lu G et al. Idoxifene and estradiol enhance antiapoptotic activity through the estrogen receptor beta in cultured rat hepatocytes. Dig Dis Sci 2003; 48:570-580.
17. Fattovich G, Giustina G, Degos F et al. Morbidity and mortality in compensated cirrhosis type C: A retrospective follow-up study of 384 patients. Gastroenterology 1997; 112:463-472.
18. Lefkowitch JH, Schiff ER, Davis GL et al. Pathological diagnosis of chronic hepatitis C: A multicenter comparative study with chronic hepatitis B. The Hepatitis Interventional Therapy Group. Gastroenterology 1993; 104:595-603.
19. Gressner AM, Lotfi S, Gressner G et al. Identification and partial characterization of a hepatocyte-derived factor promoting proliferation of cultured fat-storing cells (parasinusoidal lipocytes). Hepatology 1992; 16:1250-1266.
20. Lee KS, Buck M, Houglum K et al. Activation of hepatic stellate cells by TGF alpha and collagen type I is mediated by oxidative stress through c-myb expression. J Clin Invest 1995; 96:2461-2468.
21. Hayashi H, Takikawa T, Nishimura N et al. Improvement of serum aminotransferase levels after phlebotomy in patients with chronic active hepatitis C and excess hepatic iron. Am J Gastroenterol 1994; 89:986-988.
22. Bonkovsky HL, Banner BF, Rothman AL. Iron and chronic viral hepatitis. Hepatology 1997; 25:759-768.
23. Shimizu I, Omoya T, Takaoka T et al. Serum amino-terminal propeptide of type III procollagen and 7S domain of type IV collagen correlate with hepatic iron concentration in patients with chronic hepatitis C following alpha-interferon therapy. J Gastroenterol Hepatol 2001; 16:196-201.

24. Stohs SJ, Bagchi D. Oxidative mechanisms in the toxicity of metal ions. Free Radic Biol Med 1995; 18:321-336.
25. Berger M, de Hazen M, Nejjari A et al. Radical oxidation reactions of the purine moiety of 2'-deoxyribonucleosides and DNA by iron-containing minerals. Carcinogenesis 1993; 14:41-46.
26. Letteron P, Duchatelle V, Berson A et al. Increased ethane exhalation, an in vivo index of lipid peroxidation, in alcohol-abusers. Gut 1993; 34:409-414.
27. Letteron P, Fromenty B, Terris B et al. Acute and chronic hepatic steatosis lead to in vivo lipid peroxidation in mice. J Hepatol 1996; 24:200-208.
28. Berson A, De BV, Letteron P et al. Steatohepatitis-inducing drugs cause mitochondrial dysfunction and lipid peroxidation in rat hepatocytes. Gastroenterology 1998; 114:764-774.
29. Maher JJ, McGuire RF. Extracellular matrix gene expression increases preferentially in rat lipocytes and sinusoidal endothelial cells during hepatic fibrosis in vivo. J Clin Invest 1990; 86:1641-1648.
30. Bataller R, Brenner DA. Hepatic stellate cells as a target for the treatment of liver fibrosis. Semin Liver Dis 2001; 21:437-451.
31. Knittel T, Mehde M, Grundmann A et al. Expression of matrix metalloproteinases and their inhibitors during hepatic tissue repair in the rat. Histochem Cell Biol 2000; 113:443-453.
32. Yoshiji H, Kuriyama S, Miyamoto Y et al. Tissue inhibitor of metalloproteinases-1 promotes liver fibrosis development in a transgenic mouse model. Hepatology 2000; 32:1248-1254.
33. Parola M, Pinzani M, Casini A et al. Stimulation of lipid peroxidation or 4-hydroxynonenal treatment increases procollagen (I) gene expression in human liver fat-storing cells. Biochem Biophys Res Commun 1993; 194:1044-1050.
34. Bedossa P, Houglum K, Trautwein C et al. Stimulation of collagen α1(I) gene expression is associated with lipid peroxidation in hepatocellular injury: A link to tissue fibrosis? Hepatology 1994; 19:1262-1271.
35. Baroni GS, D'Ambrosio L, Ferretti G et al. Fibrogenic effect of oxidative stress on rat hepatic stellate cells. Hepatology 1998; 27:720-726.
36. Inoue M. Protective mechanisms against reactive oxygen species. In: Arias IM, Bojer JL, Fausto N, eds. The Liver. Biology and Pathobiology. New York: Raven Press, 1994:443-459.
37. Kuiper J, Brouwer A, Knook DL. Kupffer and sinusoidal endothelial cells. In: Arias IM, Bojer JL, Fausto N, eds. The Liver. Biology and Pathobiology. New York: Raven Press, 1994:791-818.
38. Pietrangelo A. Metals, oxidative stress, and hepatic fibrogenesis. Semin Liver Dis 1996; 16:13-30.
39. Olynyk JK, Khan NA, Ramm GA et al. Aldehydic products of lipid peroxidation do not directly activate rat hepatic stellate cells. J Gastroenterol Hepatol 2002; 17:785-790.
40. Casini A, Pinzani M, Milani S et al. Regulation of extracellular matrix synthesis by transforming growth factor B1 in human fat-storing cells. Gastroenterology 1993; 105:245-253.
41. Sanchez A, Alvarez AM, Benito M et al. Apoptosis induced by transforming growth factor-beta in fetal hepatocyte primary cultures: involvement of reactive oxygen intermediates. J Biol Chem 1996; 271:7416-7422.
42. Enzan H, Himeno H, Iwamura S et al. Immunohistochemical identification of Ito cells and their myofibroblastic transformation in adult human liver. Virchows Arch 1994; 424:249-256.
43. Friedman SL, Arthur MJ. Activation of cultured rat hepatic lipocytes by Kupffer cell conditioned medium. Direct enhancement of matrix synthesis and stimulation of cell proliferation via induction of platelet-derived growth factor receptors. J Clin Invest 1989; 84:1780-1785.
44. Mullhaupt B, Feren A, Fodor E et al. Liver expression of epidermal growth factor RNA. Rapid increases in immediate-early phase of liver regeneration. J Biol Chem 1994; 269:19667-19670.
45. Rockey DC, Fouassier L, Chung JJ et al. Cellular localization of endothelin-1 and increased production in liver injury in the rat: potential for autocrine and paracrine effects on stellate cells. Hepatology 1998; 27:472-480.
46. Pinzani M, Failli P, Ruocco C et al. Fat-storing cells as liver-specific pericytes. Spatial dynamics of agonist-stimulated intracellular calcium transients. J Clin Invest 1992; 90:642-646.
47. Bataller R, Gines P, Nicolas JM et al. Angiotensin II induces contraction and proliferation of human hepatic stellate cells. Gastroenterology 2000; 118:1149-1156.
48. Rockey DC, Housset CN, Friedman SL. Activation-dependent contractility of rat hepatic lipocytes in culture and in vivo. J Clin Invest 1993; 92:1795-1804.

49. Sakamoto M, Ueno T, Kin M et al. Ito cell contraction in response to endothelin-1 and substance P. Hepatology 1993; 18:978-983.
50. Rockey D. The cellular pathogenesis of portal hypertension: Stellate cell contractility, endothelin, and nitric oxide. Hepatology 1997; 25:2-5.
51. Bauer M, Zhang JX, Bauer I et al. ET-1 induced alterations of hepatic microcirculation: sinusoidal and extrasinusoidal sites of action. Am J Physiol 1994; 267(1 Pt 1):G143-G149.
52. Adinolfi LE, Gambardella M, Andreana A et al. Steatosis accelerates the progression of liver damage of chronic hepatitis C patients and correlates with specific HCV genotype and visceral obesity. Hepatology 2001; 33:1358-1364.
53. Rockey DC, Chung JJ. Inducible nitric oxide synthase in rat hepatic lipocytes and the effect of nitric oxide on lipocyte contractility. J Clin Invest 1995; 95:1199-1206.
54. Helyar L, Bundschuh DS, Laskin JD et al. Induction of hepatic Ito cell nitric oxide production after acute endotoxemia. Hepatology 1994; 20:1509-1515.
55. Shah V, Haddad FG, Garcia-Cardena G et al. Liver sinusoidal endothelial cells are responsible for nitric oxide modulation of resistance in the hepatic sinusoids. J Clin Invest 1997; 100:2923-2930.
56. Nieto N, Greenwel P, Friedman SL et al. Ethanol and arachidonic acid increase alpha 2(I) collagen expression in rat hepatic stellate cells overexpressing cytochrome P450 2E1. Role of H2O2 and cyclooxygenase-2. J Biol Chem 2000; 275:20136-20145.
57. Pietrangelo A, Gualdi R, Casalgrandi G et al. Molecular and cellular aspects of iron-induced hepatic cirrhosis in rodents. J Clin Invest 1995; 95:1824-1831.
58. Reiter Z. Interferon-a major regulator of natural killer cell-mediated cytotoxicity. J Interferon Res 1993; 13:247-257.
59. Capra F, Casaril M, Gabrielli GB et al. Éø-Interferon in the treatment of chronic viral hepatitis: Effects on fibrogenesis serum markers. J Hepatol 1993; 18:112-118.
60. Hiramatsu N, Hayashi N, Kasahara A et al. Improvement of liver fibrosis in chronic hepatitis C patients treated with natural interferon alpha. J Hepatol 1995; 22:135-142.
61. Yamada G, Takatani M, Kishi F et al. Efficacy of interferon alfa therapy in chronic hepatitis C patients depends primarily on hepatitis C virus RNA level. Hepatology 1995; 22:1351-1354.
62. Manabe N, Chevallier M, Chossegros P et al. Interferon-Éø2b therapy reduces liver fibrosis in chronic non-A, non-B hepatitis: A quantitative histological evaluation. Hepatology 1993; 18:1344-1349.
63. Farinati F, Cardin R, De Maria N et al. Iron storage, lipid peroxidation and glutathione turnover in chronic anti-HCV positive hepatitis. J Hepatol 1995; 22:449-456.
64. Sakaida I, Nagatomi A, Hironaka K et al. Quantitative analysis of liver fibrosis and stellate cell changes in patients with chronic hepatitis C after interferon therapy. Am J Gastroenterol 1999; 94:489-496.
65. Granstein RD, Flotte TJ, Amento EP. Interferons and collagen production. J Invest Dermatol 1990; 95(6 Suppl):75S-80S.
66. Castilla A, Prieto J, Fausto N. Transforming growth factors beta 1 and alpha in chronic liver disease. Effects of interferon alfa therapy. N Engl J Med 1991; 324:933-940.
67. Mallat A, Preaux AM, Blazejewski S et al. Interferon alfa and gamma inhibit proliferation and collagen synthesis of human Ito cells in culture. Hepatology 1995; 21:1003-1010.
68. Lu G, Shimizu I, Cui X et al. Interferon-alpha enhances biological defense activities against oxidative stress in cultured rat hepatocytes and hepatic stellate cells. J Med Invest 2002; 49:172-181.
69. McHutchison JG, Gordon SC, Schiff ER et al. Interferon alfa-2b alone or in combination with ribavirin as initial treatment for chronic hepatitis C. Hepatitis Interventional Therapy Group. N Engl J Med 1998; 339:1485-1492.
70. Poynard T, Marcellin P, Lee SS et al. Randomised trial of interferon alpha2b plus ribavirin for 48 weeks or for 24 weeks versus interferon alpha2b plus placebo for 48 weeks for treatment of chronic infection with hepatitis C virus. International Hepatitis Interventional Therapy Group (IHIT). Lancet 1998; 352:1426-1432.
71. Poynard T, McHutchison J, Davis GL et al. Impact of interferon alfa-2b and ribavirin on progression of liver fibrosis in patients with chronic hepatitis C. Hepatology 2000; 32:1131-1137.
72. Baroni GS, D'Ambrosio L, Curto P et al. Interferon gamma decreases hepatic stellate cell activation and extracellular matrix deposition in rat liver fibrosis. Hepatology 1996; 23:1189-1199.

73. Shi Z, Wakil AE, Rockey DC. Strain-specific differences in mouse hepatic wound healing are mediated by divergent T helper cytokine responses. Proc Natl Acad Sci U S A 1997; 94:10663-10668.
74. Toyonaga T, Hino O, Sugai S et al. Chronic active hepatitis in transgenic mice expressing interferon-gamma in the liver. Proc Natl Acad Sci U S A 1994; 91:614-618.
75. Ziesche R, Hofbauer E, Wittmann K et al. A preliminary study of long-term treatment with interferon gamma-1b and low-dose prednisolone in patients with idiopathic pulmonary fibrosis. N Engl J Med 1999; 341:1264-1269.
76. Mancini R, Benedetti A, Jezequel AM. An interleukin-1 receptor antagonist decreases fibrosis induced by dimethylnitrosamine in rat liver. Virchows Arch 1994; 424:25-31.
77. Bruck R, Hershkoviz R, Lider O et al. The use of synthetic analogues of Arg-Gly-Asp (RGD) and soluble receptor of tumor necrosis factor to prevent acute and chronic experimental liver injury. Yale J Biol Med 1997; 70:391-402.
78. Nelson DR, Lauwers GY, Lau JY et al. Interleukin 10 treatment reduces fibrosis in patients with chronic hepatitis C: a pilot trial of interferon nonresponders. Gastroenterology 2000; 118:655-660.
79. Louis H, Van Laethem JL, Wu W et al. Interleukin-10 controls neutrophilic infiltration, hepatocyte proliferation, and liver fibrosis induced by carbon tetrachloride in mice. Hepatology 1998; 28:1607-1615.
80. Wrana JL, Attisano L, Carcamo J et al. TGF beta signals through a heteromeric protein kinase receptor complex. Cell 1992; 71:1003-1014.
81. ten Dijke P, Yamashita H, Ichijo H et al. Characterization of type I receptors for transforming growth factor-beta and activin. Science 1994; 264:101-104.
82. Bassing CH, Yingling JM, Howe DJ et al. A transforming growth factor beta type I receptor that signals to activate gene expression. Science 1994; 263:87-89.
83. Border WA, Noble NA, Yamamoto T et al. Natural inhibitor of transforming growth factor-beta protects against scarring in experimental kidney disease. Nature 1992; 360:361-364.
84. Isaka Y, Brees DK, Ikegaya K et al. Gene therapy by skeletal muscle expression of decorin prevents fibrotic disease in rat kidney. Nat Med 1996; 2:418-423.
85. George J, Roulot D, Koteliansky VE et al. In vivo inhibition of rat stellate cell activation by soluble transforming growth factor beta type II receptor: A potential new therapy for hepatic fibrosis. Proc Natl Acad Sci USA 1999; 96:12719-12724.
86. Ueno H, Sakamoto T, Nakamura T et al. A soluble transforming growth factor beta receptor expressed in muscle prevents liver fibrogenesis and dysfunction in rats. Hum Gene Ther 2000; 11:33-42.
87. Qi Z, Atsuchi N, Ooshima A et al. Blockade of type beta transforming growth factor signaling prevents liver fibrosis and dysfunction in the rat. Proc Natl Acad Sci U S A 1999; 96:2345-2349.
88 Nakamura T, Sakata R, Ueno T et al. Inhibition of transforming growth factor beta prevents progression of liver fibrosis and enhances hepatocyte regeneration in dimethylnitrosamine-treated rats. Hepatology 2000; 32:247-255.
89. Alt M, Caselmann WH. Liver-directed gene therapy: molecular tools and current preclinical and clinical studies. J Hepatol 1995; 23:746-758.
90. Khanna A, Li B, Li P et al. Transforming growth factor-beta 1: Regulation with a TGF-beta 1 antisense oligomer. Kidney Int Suppl 1996; 53:S2-S6.
91. Nakamura T, Nawa K, Ichihara A. Partial purification and characterization of hepatocyte growth factor from serum of hepatectomized rats. Biochem Biophys Res Commun 1984; 122:1450-1459.
92. Russell WE, McGowan JA, Bucher NL. Partial characterization of a hepatocyte growth factor from rat platelets. J Cell Physiol 1984; 119:183-192.
93. Schuppan D, Schmid M, Somasundaram R et al. Collagens in the liver extracellular matrix bind hepatocyte growth factor. Gastroenterology 1998; 114:139-152.
94. Jiang WG, Hallett MB, Puntis MC. Hepatocyte growth factor/scatter factor, liver regeneration and cancer metastasis. Br J Surg 1993; 80:1368-1373.
95. Yasuda H, Imai E, Shiota A et al.. Antifibrogenic effect of a deletion variant of hepatocyte growth factor on liver fibrosis in rats. Hepatology 1996; 24:636-642.
96. Matsuda Y, Matsumoto K, Yamada A et al. Preventive and therapeutic effects in rats of hepatocyte growth factor infusion on liver fibrosis/cirrhosis. Hepatology 1997; 26:81-89.

97. Ueki T, Kaneda Y, Tsutsui H et al. Hepatocyte growth factor gene therapy of liver cirrhosis in rats. Nat Med 1999; 5:226-230.
98. Parola M, Leonarduzzi G, Biasi F et al. Vitamin E dietary supplementation protects against carbon tetrachloride- induced chronic liver damage and cirrhosis. Hepatology 1992; 16:1014-1021.
99. Chojkier M, Houglum K, Lee KS et al. Long- and short-term D-alpha-tocopherol supplementation inhibits liver collagen alpha1(I) gene expression. Am J Physiol 1998; 275(6 Pt 1):G1480-G1485.
100. Houglum K, Venkataramani A, Lyche K et al. A pilot study of the effects of d-alpha-tocopherol on hepatic stellate cell activation in chronic hepatitis C. Gastroenterology 1997; 113:1069-1073.
101. Shimizu I, Ma Y-R, Mizobuchi Y et al. Effects of Sho-saiko-to, a Japanese herbal medicine, on hepatic fibrosis in rats. Hepatology 1999; 29:149-160.
102. Shimizu I, Mizobuchi Y, Shiba M et al. Inhibitory effect of estradiol on activation of rat hepatic stellate cells in vivo and in vitro. Gut 1999; 44:127-136.
103. Yasuda M, Shimizu I, Shiba M et al. Suppressive effects of estradiol on dimethylnitrosamine-induced fibrosis of the liver in rats. Hepatology 1999; 29:719-727.
104. Boigk G, Stroedter L, Herbst H et al. Silymarin retards collagen accumulation in early and advanced biliary fibrosis secondary to complete bile duct obliteration in rats. Hepatology 1997; 26:643-649.
105. Pietrangelo A, Borella F, Casalgrandi G et al. Antioxidant activity of silybin in vivo during long-term iron overload in rats. Gastroenterology 1995; 109:1941-1949.
106. Salmi HA, Sarna S. Effect of silymarin on chemical, functional, and morphological alterations of the liver. A double-blind controlled study. Scand J Gastroenterol 1982; 17:517-521.
107. Ferenci P, Dragosics B, Dittrich H et al. Randomized controlled trial of silymarin treatment in patients with cirrhosis of the liver. J Hepatol 1989; 9:105-113.
108. Pares A, Planas R, Torres M et al. Effects of silymarin in alcoholic patients with cirrhosis of the liver: results of a controlled, double-blind, randomized and multicenter trial. J Hepatol 1998; 28:615-621.
109. Huang HC, Wang HR, Hsieh LM. Antiproliferative effect of baicalein, a flavonoid from a Chinese herb, on vascular smooth muscle cell. Eur J Pharmacol 1994; 251:91-93.
110. Dethlefsen SM, Shepro D, D'Amore PA. Arachidonic acid metabolites in bFGF-, PDGF-, and serum-stimulated vascular cell growth. Exp Cell Res 1994; 212:262-273.
111. Shimizu I. Sho-saiko-to: Japanese herbal medicine for protection against hepatic fibrosis and carcinoma. J Gastroenterol Hepatol 2000; 15(s1):D84-D90.
112. Lieber CS. Role of S-adenosyl-L-methionine in the treatment of liver diseases. J Hepatol 1999; 30:1155-1159.
113. Cutrin C, Menino MJ, Otero X et al. Effect of nifedipine and S-adenosylmethionine in the liver of rats treated with CCl_4 and ethanol for one month. Life Sci 1992; 51:L113-L118.
114. Mato JM, Camara J, Fernandez dP, Caballeria L, Coll S, Caballero A et al. S-adenosylmethionine in alcoholic liver cirrhosis: a randomized, placebo-controlled, double-blind, multicenter clinical trial. J Hepatol 1999; 30:1081-1089.
115. Yoshino K, Komura S, Watanabe I et al. Effect of estrogens on serum and liver lipid peroxide levels in mice. J Clin Biochem Nutr 1987; 3:233-239.
116. Lacort M, Leal AM, Liza M et al. Protective effect of estrogens and catecholestrogens against peroxidative membrane damage in vitro. Lipids 1995; 30:141-146.
117. Grady D, Rubin SM, Petitti DB et al. Hormone therapy to prevent disease and prolong life in postmenopausal women. Ann Intern Med 1992; 117:1016-1037.
118. Colditz GA, Hankinson SE, Hunter DJ et al. The use of estrogens and progestins and the risk of breast cancer in postmenopausal women. N Engl J Med 1995; 332:1589-1593.
119. Collaborative Group on Hormonal Factors in Breast Cancer. Breast cancer and hormone replacement therapy: collaborative reanalysis of data from 51 epidemiological studies of 52705 women with breast cancer and 108411 women without breast cancer. Lancet 1997; 351:1047-1059.
120. Meijer DK, Molema G. Targeting of drugs to the liver. Semin Liver Dis 1995; 15:202-256.
121. Bruck R, Hershkoviz R, Lider O et al. Inhibition of experimentally-induced liver cirrhosis in rats by a nonpeptidic mimetic of the extracellular matrix-associated Arg-Gly-Asp epitope. J Hepatol 1996; 24:731-738.

122. Jarnagin WR, Rockey DC, Koteliansky VE et al. Expression of variant fibronectins in wound healing: cellular source and biological activity of the EIIIA segment in rat hepatic fibrogenesis. J Cell Biol 1994; 127(6 Pt 2):2037-2048.
123. Loreal O, Clement B, Schuppan D et al. Distribution and cellular origin of collagen VI during development and in cirrhosis. Gastroenterology 1992; 102:980-987.
124. Marcelino J, McDevitt CA. Attachment of articular cartilage chondrocytes to the tissue form of type VI collagen. Biochim Biophys Acta 1995; 1249:180-188.
125. Cardarelli PM, Yamagata S, Taguchi I et al. The collagen receptor alpha 2 beta 1, from MG-63 and HT1080 cells, interacts with a cyclic RGD peptide. J Biol Chem 1992; 267:23159-23164.
126. Beljaars L, Molema G, Schuppan D et al. Successful targeting to rat hepatic stellate cells using albumin modified with cyclic peptides that recognize the collagen type VI receptor. J Biol Chem 2000; 275:12743-12751.
127. Rockey DC. Vasoactive agents in intrahepatic portal hypertension and fibrogenesis: Implications for therapy. Gastroenterology 2000; 118:1261-1265.
128. Pinzani M, Milani S, De Franco R et al. Endothelin 1 is overexpressed in human cirrhotic liver and exerts multiple effects on activated hepatic stellate cells. Gastroenterology 1996; 110:534-548.
129. Gallois C, Habib A, Tao J et al. Role of NF-kappaB in the antiproliferative effect of endothelin-1 and tumor necrosis factor-alpha in human hepatic stellate cells. Involvement of cyclooxygenase-2. J Biol Chem 1998; 273:23183-23190.
130. Cho JJ, Hocher B, Herbst H et al. An oral endothelin-A receptor antagonist blocks collagen synthesis and deposition in advanced rat liver fibrosis. Gastroenterology 2000; 118:1169-1178.
131. Bickel M, Barinhause KH, Gerl M et al. Sective inhibition of hepatic collagen accumulation in experimental liver fibrosis in rats by a new prolyl 4-hydroxylase inhibitor. Hepatology 1998; 28:404-411.
132. Sakaida I, Matsumura Y, Kubota M et al. The prolyl 4-hydroxylase inhibitor HOE 077 prevents activation of Ito cells, reducing procollagen gene expression in rat liver fibrosis induced by choline-deficient L-amino acid-defined diet. Hepatology 1996; 23:755-763.
133. Matsumura Y, Sakaida I, Uchida K et al. Prolyl 4-hydroxylase inhibitor (HOE 077) inhibits pig serum-induced rat liver fibrosis by preventing stellate cell activation. J Hepatol 1997; 27:185-192.
134. Wang YJ, Wang SS, Bickel M et al. Two novel antifibrotics, HOE 077 and Safironil, modulate stellate cell activation in rat liver injury: differential effects in males and females. Am J Pathol 1998; 152:279-287.
135. Salgado S, Garcia J, Vera J et al. Liver cirrhosis is reverted by urokinase-type plasminogen activator gene therapy. Mol Ther 2000; 2:545-551.
136. Davis BH, Coll D, Beno DW. Retinoic acid suppresses the response to platelet-derived growth factor in human hepatic Ito-cell-like myofibroblasts: A post-receptor mechanism independent of raf/fos/jun/egr activation. Biochem J 1993; 294 (Pt 3):785-791.
137. Sato T, Kato R, Tyson CA. Regulation of differentiated phenotype of rat hepatic lipocytes by retinoids in primary culture. Exp Cell Res 1995; 217:72-83.
138. Mizobuchi Y, Shimizu I, Hori H et al. Retinyl palmitate reduces hepatic fibrosis in rats induced by dimethylnitrosamine or pig serum. J Hepatol 1998; 29:933-943.
139. Senoo H, Wake K. Suppression of experimental hepatic fibrosis by administration of vitamin A. Lab Invest 1985; 52:182-194.
140. Seifert WF, Bosma A, Brouwer A et al. Vitamin A deficiency potentiates carbon tetrachloride-induced liver fibrosis in rats. Hepatology 1994; 19:193-201.
141. Okuno M, Moriwaki H, Imai S et al. Retinoids exacerbate rat liver fibrosis by inducing the activation of latent TGF-beta in liver stellate cells. Hepatology 1997; 26:913-921.
142. Hellemans K, Grinko I, Rombouts K et al. All-trans and 9-cis retinoic acid alter rat hepatic stellate cell phenotype differentially. Gut 1999; 45:134-142.
143. Russell RM, Boyer JL, Bagheri SA et al. Hepatic injury from chronic hypervitaminosis a resulting in portal hypertension and ascites. N Engl J Med 1974; 291:435-440.
144. Preaux AM, Mallat A, Rosenbaum J et al. Pentoxifylline inhibits growth and collagen synthesis of cultured human hepatic myofibroblast-like cells. Hepatology 1997; 26:315-322.
145. Windmeier C, Gressner AM. Effect of pentoxifylline on the fibrogenic functions of cultured rat liver fat-storing cells and myofibroblasts. Biochem Pharmacol 1996; 51:577-584.

146. Windmeier C, Gressner AM. Pharmacological aspects of pentoxifylline with emphasis on its inhibitory actions on hepatic fibrogenesis. Gen Pharmacol 1997; 29:181-196.
147. Chazouilleres O, Ballet F, Chretien Y et al. Protective effect of vasodilators on liver function after long hypothermic preservation: A study in the isolated perfused rat liver. Hepatology 1989; 9:824-829.
148. Kozaki K, Egawa H, Bermudez L et al. Effects of pentoxifylline pretreatment on Kupffer cells in rat liver transplantation. Hepatology 1995; 21:1079-1082.
149. Lieber CS. Pathogenesis and treatment of liver fibrosis in alcoholics: 1996 update. Dig Dis 1997; 15:42-66.
150. Lieber CS, Robins SJ, Li J et al. Phosphatidylcholine protects against fibrosis and cirrhosis in the baboon. Gastroenterology 1994; 106:152-159.
151. Aleynik SI, Leo MA, Ma X et al. Polyenylphosphatidylcholine prevents carbon tetrachloride-induced lipid peroxidation while it attenuates liver fibrosis. J Hepatol 1997; 27:554-561.

Chapter 12

Reversibility of Liver Fibrosis:

Role of Matrix Metalloproteinases

Isao Okazaki, Tetsu Watanabe, Maki Niioka, Yoshihiko Sugioka and Yutaka Inagaki

Summary

Reversibility of liver fibrosis has been reported both experimentally and clinically, if the cause of liver damage is removed or adequately treated. We first reported the collagenase activity in the process of experimental liver fibrosis in 1974. The present review discusses the participation of matrix metalloproteinases (MMPs), especially MMP-1/MMP-13, in the spontaneous resolution of liver fibrosis, in association with tissue inhibitor of MMPs (TIMPs). A recent advance in molecular biology has revealed the important role of MMPs and TIMPs in the recovery from liver fibrosis and cirrhosis. In situ hybridization study showed that some stem/progenitor cells expressing MMP-13 mRNA may play an important role in the recovery from liver cirrhosis, and these cells were determined to be neural cells by us to be neural origin. Transfusion of stem cells derived from bone marrow will be applied to patients with liver cirrhosis in near future. Suitable conditions to make stem cells proliferate and differentiate for the expression of MMP-13 should be developed.

Introduction

Liver cirrhosis has been defined as an irreversible and fatal disease resulting from all types of chronic liver disease.[1] Although the etiology of liver cirrhosis is not uniform, the common pathophysiological feature is an abnormal and progressive "deposition of extracellular matrix" in the liver which causes a distorted structure of intrahepatic and extrahepatic circulation with portal hypertension, an impaired metabolic function with bleeding tendency, hepatic encephalopathy, and secondary renal failure. Ninety percent of cases with hepatocellular carcinoma are associated with liver cirrhosis. Hepatologists have had a dream to control the abnormal deposition of extracellular matrix.[2-4]

On the contrary, Cameron and Karunaratne[5] observed the reversibility of liver fibrosis of rats after cessation of chronic treatment with toxic reagents. Perez-Tamayo[6] introduced an interesting case of hemochromatosis with the evidence for the reversibility of liver cirrhosis. These studies stimulated us to study the mechanism of the degradation of extracellular matrix in the fibrotic liver. We reported the interstitial collagenase activity in the process of liver fibrosis[7] of rat induced by carbon tetrachloride and have investigated the role of matrix metalloproteinases in the spontaneous resolution of experimental liver fibrosis of rat.[2,4]

Fibrogenesis: Cellular and Molecular Basis, edited by Mohammed S. Razzaque.

Very recently we found that neural progenitor cells derived from hematopoietic stem cells express interstitial collagenase in the recovery phase of experimental cirrhosis.[4,8] The present review describes the reversibility of liver fibrosis in clinical cases and the reversibility mechanism in experimental models in relation to matrix metalloproteinases (MMPs) and their specific tissue inhibitors (TIMPs). Especially, we discuss the participation of stem/progenitor cells expressing MMPs and TIMPs in the degradation of newly formed fibrous tissue in the recovery from rat liver cirrhosis.

Reversibility of Human Liver Fibrosis

The pathology of liver cirrhosis is characterized by the abnormal deposition of large amount of extracellular matrix, and liver cirrhosis has been believed to be an irreversible disease which causes four major serious complications (the rupture of esophageal varices, uncontrolled ascites leading to renal failure, hemorrhagic tendency including gastrointestinal bleeding, brain bleeding and others, and hepatic encephalopathy) indicating poor prognosis of liver cirrhosis.

The reversibility of human liver cirrhosis, however, has also been observed after effective therapy among patients with alcoholic liver disease,[9] hemochromatosis,[6,10] primary biliary cirrhosis,[11,12] secondary biliary cirrhosis,[13,14] hepatitis B virus (HBV)-related chronic hepatitis and cirrhosis,[15,16] hepatitis C virus (HCV)-related hepatitis and cirrhosis,[17-26] hepatitis D virus (HDV)-related hepatitis,[27] autoimmune hepatitis[28] and other liver diseases[10] since the histological examination had prevailed in the clinical hepatology.

Maruyama et al[29] reported that the histologically documented reversibility of liver fibrosis in alcoholic liver diseases was closely correlated with a decrease in serum fibrosis marker of type IV collagen. Among out-clinic patients of alcoholic diseases, however, regression of fibrosis has not been clearly demonstrated due to continuous alcohol consumption.[30]

We demonstrated the disappearance of liver fibrosis during recovery from HBV-positive subacute hepatitis with massive fibrosis.[15] The improvement of fibrosis in patients with chronic hepatitis B after successful long-term lamivudine administration has also been reported in a larger cohort.[16]

Recent clinical experience with interferon therapy for HCV antibody positive-chronic hepatitis and cirrhosis has documented the regression of hepatic fibrosis in many cases[19-26] The close correlation between the histologically documented grade of liver fibrosis and the serum type IV collagen levels has been reported in patients who responded to interferon therapy.[17-19,31] We observed higher serum MMP-1 levels and the higher ratio of MMP-1 to TIMP-1 or TIMP-2 in patients who well responded to interferon therapy compared with nonresponders.[19] The effect of interferon on the recovery of liver fibrosis is not only due to the eradication of HCV but may also result from a direct and indirect action of this cytokine on fibrogenesis.[19] That is, interferon itself decreases the production of collagen[32,33] and attenuates the fibrogenic activity induced by TGF-β.[34-36] Interferon can also induce collagenase activity.[37,38]

Liver fibrosis assumed to be resulted from chronic pancreatitis and stenosis of the common bile duct, was resolved after biliary drainage.[14] Recovery was also reported after surgery in an infant with cirrhosis due to congenital biliary atresia.[13] Reversibility of hepatic fibrosis and cirrhosis has been reported in some patients with primary biliary cirrhosis who respond to treatment.[11,12]

It is now generally accepted that reversibility of liver cirrhosis occurs under some conditions where the cause of liver damage is removed or adequately treated. Further research focusing on the relationship between matrix metalloproteinases (MMPs), their specific inhibitors, tissue inhibitors for matrix-metalloproteinases (TIMPs) -1 to -4 and liver fibrosis, is expected to contribute to active lysis of remaining fibers after the removal of causes and/or conventional treatment.

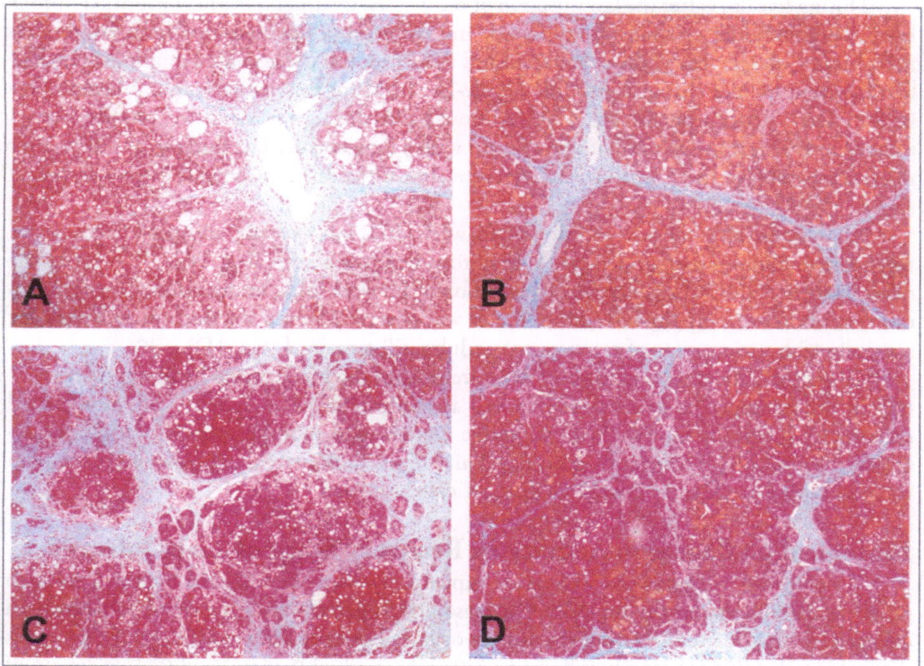

Figure 1. Azan staining of rat liver at day 2 (A) and at day 5 (B) after the last injection of CCl_4 for 8 weeks, and at day 2 (C) and at day 5 (D) after the last injection of 12 week CCl_4 treatment. Newly formed fibrous bands (A, C) are dramatically resolved at day 5 (B, D) after discontinuation of toxic reagents. Regenerative nodules are seen (C) but remarkable resolution of fibrous bands is observed (D).Original magnification: 25x.

Reversibility of Experimental Liver Fibrosis

Cameron and Karunaratne[5] reported the reversibility of liver fibrosis in the carbon tetrachloride (CCl_4)-induced cirrhosis model after removal of the toxic agent. The reversibility of liver fibrosis has been confirmed in this experimental model.[39-44] The reversibility has been observed in other several experimental models such as thioacetamide-,[41] α-naphthyl-isothiocyanate-,[42] ethionine,[40] choline-deficient-diet-induced liver cirrhosis[43] as well as by the ligation of bile duct[39,45] after removal of the causative agent.

We have investigated the mechanism of extracellular matrix degradation in the spontaneous resolution of CCl_4-induced rat liver fibrosis reported by Cameron and Karunaratne.[7,46,47] Wistar male rats were injected intraperitoneally with 1 mg/kg of a 30% solution of CCl_4 in olive oil twice a week for 8 or 12 weeks. These rats were sacrificed at 48 h, 5 days and 7 days after the last injection of CCl_4 treatment for 8 or 12 weeks.[46,47] Control rats were injected with the same amount of olive oil intraperitoneally.

At 48 hours after the last injection of 8-week CCl_4 intoxication, the liver showed moderate fibrosis, linking neighboring tracts completely and separating the lobules into small sublobules.[46] In some portions, the newly formed fibrous bands connecting the neighboring portal triads extended within lobules with multiple-branched bands, which were connected to the perihepatocellular fibrosis. Regenerative nodules were not observed[7,44,46-49] as shown in Figure 1A.

During five days after the last injection of the 8-week treatment, the newly formed fibrous bands became thiner or almost disappeared, and the hepatocytes recovered from fatty metamorphosis. Enlargement of the portal triad still remained, but the extended multiple branched bands within lobules changed into fine needle-like bands or disappeared around hepatocytes without fatty metamorphosis (Fig. 1B).[47] Fibrous bands during the resolution process showed a characteristic feature, somewhat like a pencil with a long needle. The findings at 7 days after the last injection of the 8-week treatment were almost the same as those at 5 days described above.[44,46-49]

At 48 hours after the last injection of 12-week CCl_4 intoxication, advanced fibrosis with regenerative nodules surrounded by thick fibrous bands was seen (Fig. 1C).[7,44,46-50] On the other hand, at 5 and 7 days after the last injection of 12-week CCl_4 intoxication, the spontaneous resolution of newly formed fibrous bands was observed even if the distortion of the liver structure with various sizes of regenerative nodules still remained (Fig. 1D).[7,44,46-49]

To estimate the degree of fibrosis, measurement of the biochemical hydroxyproline content in the homogenate of rat liver[49] as well as morphometric analysis[46,47] of Azan-stained sections were performed. The hydroxyproline content and morphological fibrotic rates decreased significantly in the rat liver on day 5 and day 7 in the recovery phase compared with those in both 8-week and 12-week CCl_4 intoxicated rat liver.

Biological Collagenase Activity in Recovery Stage of Liver Fibrosis

We observed collagenolysis around the explants of a slice of rat fibrotic liver on a collagen gel film, demonstrating the typical collagenase attack pattern against neutral salt-extracted collagen by disc electrophoresis of the sample collected from the reacted collagen gels.[4,7] This disc gel revealed β^A (3/4-length of β chain), α^A (3/4-length of α chain) and α^B (1/4-length of α chain) products, which are the typical products of limited collagen degradation by mammalian collagenase on polyacrylamide gels. This was the first publication to identify the presence of interstitial collagenase in the liver.

A semi-quantitative assay using this tissue culture technique revealed that the collagenase activity increased at the early stage of liver fibrosis and decreased at the advanced stage of liver fibrosis. Contrary to our expectations, the collagenase decreased at days 5 and 7 after 8- and 12-week CCl_4 intoxication.[51] Since we have believed the participation of collagenase activity in the recovery from liver fibrosis and cirrhosis, we wanted to develop a specific and sensitive assay method for interstitial collagenase activity in liver homogenate.

Liver homogenate contains lots of proteins and inhibitors, and the collagenase activity in liver homogenate is expected to be quite low. The reaction mixture containing a substrate of type I collagen extracted from rabbit skin in solution,[52] the homogenate of baboon liver as an enzyme and 3mM p-chloromercuribenzoate to inhibit thiol proteinase activity and to convert procollagenase into the active form, was incubated under neutral pH at 27°C, assayed by viscometer, and monitored by disc electrophoresis.[53] The levels of the reaction products of β^A and α^A and their ratio showed the predicted degradation pattern of type I collagen molecules by interstitial collagenase, and its activity was consistent with the degradation measured by viscometer (Fig. 2). Using the principle of this assay method but with a substrate of type I collagen in fibril form, we demonstrated an increased activity of interstitial collagenase in the early stage of liver fibrosis in baboons that had been fed alcohol chronically[53,54] and in patients with alcoholic liver fibrosis (Fig. 2).[54] Our original observation of an increased collagenase activity in the early stage of liver fibrosis and a reduced collagenase activity in advanced fibrosis were confirmed by other investigators.[55-57] The activity in the recovery phase from liver fibrosis and cirrhosis, however, was not accompanied by up-regulation of interstitial collagenase even by this sensitive and quantitative assay.[49] This question, "why collagenase activity doesn't increases for the destruction of extracellular matrix in the recovery from liver fibrosis?" remains unresolved.

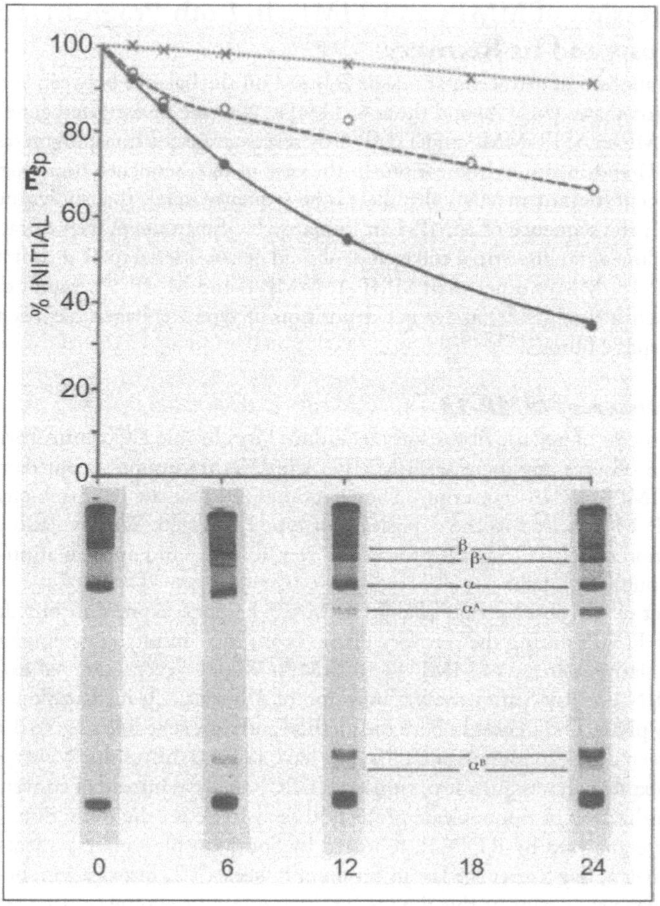

Figure 2. Viscometric assay of the collagenase from a human liver at autopsy (black circle), and from the biopsy of baboon liver (O) using collagen in solution at 25°C. The control reaction mixture did not contain liver homogenate (X). The results are expressed as a percentage of the initial specific viscosity of the relevant reaction mixture. (Black circle) $\eta sp=3.28$; (O) $=3.41$; (X) $=3.04$. Data are compiled from Maruyama et al[15] Disc gel electrophoresis on each reaction mixture revealed the typical collagenase action against type I collagen; that is, the gradual increase in the reaction products of β^A, α^A and α^B.

Arthur et al[58,59] reported that hepatic stellate cells secrete a neutral metalloproteinase that can degrade type IV collagen (a component of the basement membrane), which appears to be MMP-2. Hepatic stellate cells produce not only MMP-2[59-61] but also MMP-3,[62,63] membrane type (MT) 1-MMP,[64,65] MT2-MMP,[65] TIMP-1,[66-69] and TIMP-2.[70] The amount of MMP-2 mRNA and protein contents of MMP-2 increased in the process of experimental liver fibrosis and gradually decreased in the recovery stage of liver fibrosis.[71] Both MMP-2, a potent gelatinase, and MT1-MMP, an activator of MMP-2, can cleave native type I collagen,[72,73] but with less efficiency than does MMP-1.[72] Therefore biological activity against type I collagen contains activities of MMP-1, MMP-2 and MT1-MMP. In order to clarify the participation of MMP-1 in the recovery from liver fibrosis and cirrhosis, we moved to the study of gene expression of MMPs and TIMPs.

Gene Expression of MMPs and TIMPs in Both Progressive Stage of Liver Fibrosis and Its Recovery

The metabolism of extracellular matrix is based on the balance between activities of matrix metalloproteinases (MMPs) and those of TIMPs. We have investigated gene expression of MMP-13, MMP-2, MT1-MMP and TIMP-1 by reverse transcription-polymerase chain reaction (RT-PCR) and in situ hybridization. In the case of rats, sequence homology analysis revealed that except for human MMP-13, there is no sequence in rats that shows more than 90% similarity with the sequence of MMP-1 in humans.[74,75] Since the cDNA of rat MMP-1 has not yet been cloned, rat interstitial collagenase should be considered to be the rat homologue of human MMP-13. MMPs other than MMP-1, MMP-8 and MMP-13 cannot degrade type I collagen, which is very stable, and a net deposition of type I collagen has been observed in progressive hepatic fibrosis.[2,4,76-83]

Gene Expression of MMP-13

In the process of hepatic fibrosis in rats induced by chronic CCl_4 intoxication, Iredale et al[84,85] did not observe any increase in MMP-13 mRNA transcription but demonstrated an increase in TIMPs mRNA transcripts. They hypothesized that the relative increase of TIMPs against MMP-13 may lead to the deposition of type I collagen. We have also demonstrated down-regulation of MMP-13 in the progress of liver fibrosis[46] and up-regulation of TIMP-1by RT-PCR (unpublished data).

Iredale et al[84,85] observed no change in MMP-13 gene expression but decreased gene expression of TIMPs during the recovery phase. From their results, they emphasized the importance of relative increase of MMP-13 to TIMPs. We, however, observed up-regulation of MMP-13 mRNA at day 5 after the last injection of 8-week CCl_4 intoxication, that is, in the early recovery phase. Differences in both the method and time schedule used to detect MMP-13 mRNA between their experiment and ours may have affected the results.[46] First, since the level of MMP-13 expression was quite low, we used RT-PCR analysis instead of conventional Northern blot hybridization or ribonuclease protection assay to detect the transcripts. Transcripts of MMP-13 were analyzed by RT-PCR, followed by Southern blot analysis. Second, we examined the rat liver at day 5 after the last injection of 8-week CCl_4 intoxication, but they did not examine the gene expression at this time point. Intriguingly, the time course study of the gene expression of MMP-13 during the recovery phase showed quite similar findings to our previous histochemical demonstration of lysosomal enzyme activities, showing a dramatic up-regulation of enzyme activity at day 5 after discontinuing CCl_4 intoxication.[44,86] Our report is the first to demonstrate the up-regulation of MMP-13 in the very early recovery phase, that is, at day 5 after the last injection of CCl_4 for 8 or 12 weeks.[4,8]

Stronger MMP-13 gene expression was noted at day 5 after the last injection of 12-week CCl_4 intoxication compared with that at day 5 after the last injection of 8-week CCl_4 intoxication.[4] It is very interesting why the intense gene expression of MMP-13 appears in the recovery from liver cirrhosis. This enhanced expression may be related to the appearance of hepatic stem cells described later. These cells may appear in rat liver cirrhosis more frequently than in liver fibrosis.

By in situ hybridization no signal for MMP-13 mRNA was observed in normal rat liver. In the liver of rats treated with CCl_4 for 8 weeks, signals for MMP-13 mRNA were observed in a few cells lining the fibrous septa. Some of these cells were stained positive for α-smooth muscle actin (α-SMA).[46] On the other hand, in the cirrhotic liver of rats treated with CCl_4 for 12 weeks very weak signals of MMP-13 mRNAwere observed in stellate cells. No hepatocyte in

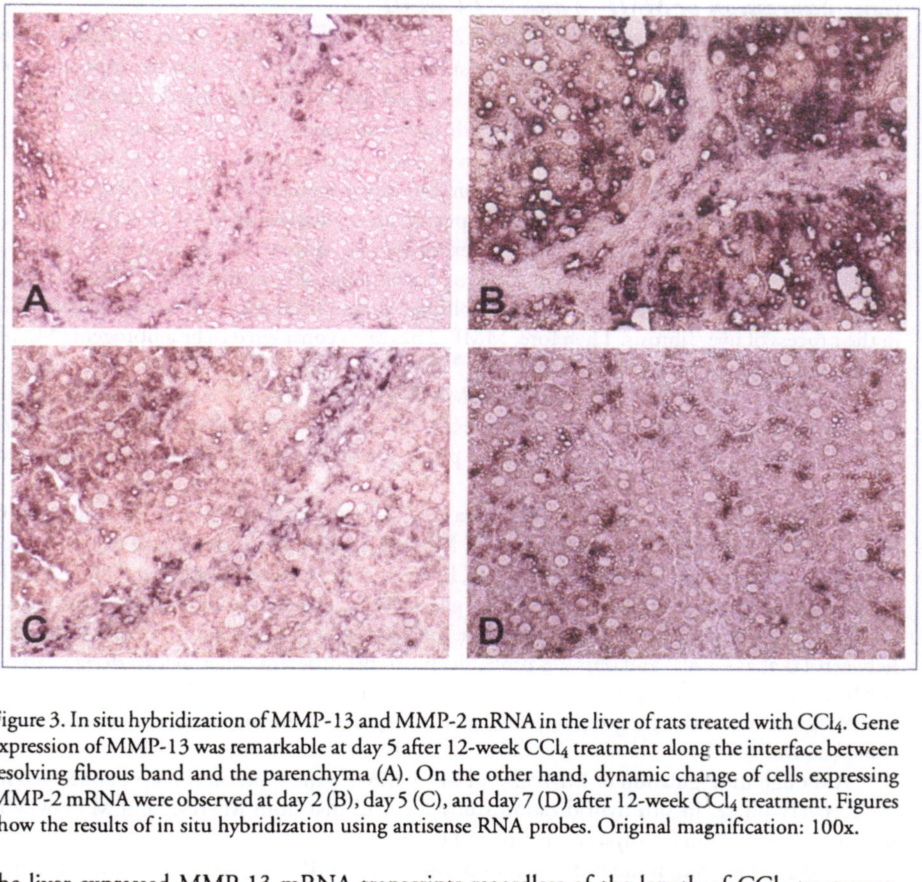

Figure 3. In situ hybridization of MMP-13 and MMP-2 mRNA in the liver of rats treated with CCl_4. Gene expression of MMP-13 was remarkable at day 5 after 12-week CCl_4 treatment along the interface between resolving fibrous band and the parenchyma (A). On the other hand, dynamic change of cells expressing MMP-2 mRNA were observed at day 2 (B), day 5 (C), and day 7 (D) after 12-week CCl_4 treatment. Figures show the results of in situ hybridization using antisense RNA probes. Original magnification: 100x.

the liver expressed MMP-13 mRNA transcripts regardless of the length of CCl_4 treatment. These data were consistent with the results of RT-PCR described above.

We demonstrated positive signals of MMP-13 mRNA during the recovery phase of liver fibrosis, and cells positive for MMP-13 mRNA were observed mainly at the interface between the resolving fibrous septa and the parenchyma by in situ hybridization. The very intense signals for gene expression of MMP-13 were seen at day 5 after the last injection of 12-week CCl_4 intoxication.[4] This was consistent with the results of RT-PCR described above. In particular, cells expressing MMP-13 at day 5 after 12-week CCl_4 treatment were seen in the margins of the nodules interfaced with the resolving fibrous bands (Fig. 3A).

Overlapping the images of in situ hybridization and immunohistochemical staining revealed that some, but not all, of the MMP-13-positive cells were stellate cells that were stained with anti-α-SMA antibody. This is the first report providing direct evidence for definite MMP-13 gene expression during the recovery phase, which is in contrast to the down-regulation of MMP-13 expression during the progression of fibrosis mentioned above.[4]

Most MMP-13-positive cells detected in the recovery phase were albumin-negative, α-fetoprotein-negative, α-SMA-negative, ED2-negative, and CK19-negative.[4] It is reasonable that some of MMP-13-positive cells in the recovery phase could be stem cells as described below.

Gene Expression of MMP-2 and MT1-MMP

The enzymatic characteristics of MMP-2 are quite different from those of MMP-1 and MMP-13. Not only are there differences in substrate specificity,[87] but also the expression of these MMPs is differentially regulated. That is, phorbol esters (TPA), interleukin-1β, tumor necrosis factor-α, and basic fibroblast-growth factor stimulate MMP-1 expression;[87] transforming growth factor (TGF) β1, interleukin-8, and concanavalin A induce MMP-2 expression.[88-91] p53 up-regulates MMP-2 expression, but down-regulates MMP-1 expression.[92,93]

Takahara et al[64,71] and Theret et al[94] reported up-regulation of MMP-2 mRNA during the remodeling of extracellular matrix in experimental animal and human liver fibrosis. MT1-MMP mRNA also increased in the progressive stage of liver fibrosis.[64,95] Ikeda et al[96] reported that MMP-2 is necessary for the proliferation and infiltration of hepatic stellate cells in the process of liver fibrosis. Therefore, MMP-2 has been considered to be a fibrogenic MMP, which acts in the formation of liver fibrosis.[2,82,83,97]

We demonstrated the up-regulation of MMP-2 mRNA and MT1-MMP mRNA in the progressive phase of liver fibrosis by RT-PCR.[47] The up-regulation of both MMP-2 mRNA and MT1-MMP mRNA in the progressive phase is consistent with the reports mentioned above.

In the recovery phase, however, Takahara et al[71] showed that after the discontinuation of chronic CCl_4 treatment, there was increased MMP-2 expression on day 3 and day 7 and reduced expression on day 14. We demonstrated, however, that increased MMP-2 and MT1-MMP gene expression in liver fibrosis or cirrhosis is gradually decreased to normal levels in the recovery.[47] These findings are not consistent with Takahara et al.[71] The reason for this difference between Takahara et al[71] and ours[47] is probably due to the different experimental design.[47] The MMP-2 and MT1-MMP gene expression is very different from that of MMP-13, which showed down-regulation in the progressive phase and up-regulation in the recovery phase as mentioned above.

Although up-regulation of MMP-2 and MT1-MMP gene expression was not observed in the recovery phase, in situ hybridization study revealed that both MMP-2 and MT1-MMP may play some role in fibrolysis as mentioned below.[47]

In situ hybridization showed that cells positive for MMP-2 mRNA were mainly observed around the fibrous bands and some positive cells were scattered in the lobules in the livers of rats treated with CCl_4 for 8 or 12 weeks (Fig. 3B). On day 5, MMP-2 mRNA-positive cells were observed exclusively along or within the resolving fibrous septa (Fig. 3C). That is, fewer MMP-2 mRNA-positive cells were seen in the lobules at day 5 compared with day 2. At day 7 after the last injection, the positive signals were still identifiable, scattered in the mesenchymal cells, including Kupffer cells, within the lobules (Fig. 3D). Surprisingly, these dynamic changes in the distribution of cells expressing MMP-2 mRNA occurred in the very early days of recovery. The same dynamic change was observed in cells expressing MT1-MMP, TIMP-1 and TIMP-2, but not in MMP-13 mRNA.

That is, the findings of MT1-MMP mRNA expression in both the 8-week- and 12-week-treated groups were essentially the same as those of MMP-2 mRNA expression, although the level of MT1-MMP mRNA was lower than that of MMP-2 mRNA.[47] Cells expressing MMP-2 mRNA were almost the same as those expressing MT1-MMP mRNA at day 5. Both enzymes seem to degrade the newly formed fibrous bands in the recovery phase. Therefore, MMP-2 and MT1-MMP may play some role in the recovery from liver fibrosis.

Colocalization of MMP-2-positive, MT1-MMP-positive and TIMP-2-positive cells was seen around the fibrous bands at day 5. Kinoshita et al[98] noted that TIMP-2, which is an inhibitor of MT1-MMP, paradoxically promotes activation of progelatinase A through

MT1-MMP. From these results it is suggested that colocalized MT1-MMP and TIMP-2 can activate pro-MMP-2 around fibrous bands to dissolve extracellular matrix.

At day 7 MMP-2 and/or MT1-MMP-positive cells were observed diffusely in the lobules,[47] implying their role in the degradation of the perihepatocellular fibrosis. Advanced liver fibrosis is associated with the appearance of perihepatocellular fibrosis, which contributes to the formation of sinusoidal capillarization.[50,99] Intralobular shunt vessels between portal vein tributaries and hepatic vein tributaries are formed and it has been believed that there is no metabolic exchange between the blood and hepatocytes, leading to the irreversibility of liver fibrosis. We previously confirmed the reversibility of sinusoidal capitalization that appeared in the experimental liver fibrosis of rats induced by chronic CCl_4 intoxication.[50,51] We also previously developed a method to assay type IV collagenase in the liver and demonstrated the reduced activity of type IV collagenase in human liver cirrhosis.[79,100,101] Although gene expression of both enzymes was down-regulated totally, the both enzymes may participate in the destruction of perihepatocellular fibrosis and result in the recovery from liver fibrosis. Biological type IV collagenase activity includes net activities of MMP-2, MMP-3 and MMP-9 under the influence of TIMPs. Although the reports[62,63,102] showed that MMP-3 and MMP-9 are involved in perihepatocellular fibrosis, no study has reported about gene expression of MT1-MMP and other MMPs in the context of fibrolysis in the liver.

During the recovery phase of liver fibrosis, most of the cells positive for MMP-2 mRNA were negative for desmin and α-SMA.[47] Most of the cells positive for MT1-MMP mRNA were also negative for desmin and α-SMA.[47] Some MMP-2 mRNA-positive cells seemed to be hepatocytes, although they did not express albumin. Both mesenchymal cell (mainly hepatic stellate cells) and Kupffer cells were the major sources of MMP-2 production during the recovery phase of liver fibrosis, although they did not express characteristic cell markers. A few hepatocytes within the fibrous bands also produced MMP-2.[47]

The size of cells expressing both enzymes seems to be larger than those positive for MMP-13, and we assume that both enzymes, in particular MMP-2, may be related with the proliferation of hepatocytes, stellate cells and Kupffer cells.

Gene Expression of TIMP-1 and TIMP-2

TIMP-1 mRNA increased during the early phase of CCl_4-induced liver fibrosis, and then decreased in the progressive stage of experimental liver cirrhosis,[69,84] and remained at a low level during the recovery from liver fibrosis.[85] The net activity of MMPs is determined by the balance between the activities of the MMPs and their inhibitors. Herbst et al[68] revealed that high levels of TIMP-1 and TIMP-2 transcripts were present in all fibrotic rat and human livers, predominantly in the stellate cells.

We also observed up-regulation of TIMP-1 and TIMP-2 mRNA in the progressive phase and down-regulation in the recovery phase by RT-PCR. In situ hybridization showed strong signals for TIMP-1 and TIMP-2 mRNA in the cells expressing MMP-2 and/or MT1-MMP at the interface between the newly formed fibrous bands and the parenchyma in the progressive phase (unpublished data).

In the recovery phase of liver fibrosis, increased levels of TIMPs gene expression gradually decreased, but gene expression is still high. In situ hybridization revealed that the localization of TIMP-1-and TIMP-2-positive cells drastically changed over 5 and 7 days after discontinuance of chronic CCl_4 treatment. That is, at day 5 mRNA for both TIMPs were observed mainly in the parenchymal cells along the resolving fibrous bands, while weak signals of TIMPs were observed mainly in the hepatocytes at day 7 (unpublished data). Gene expression of TIMPs in hepatocytes may promote the proliferation of hepatocytes. These results suggest that TIMPs may participate in the recovery from liver fibrosis.

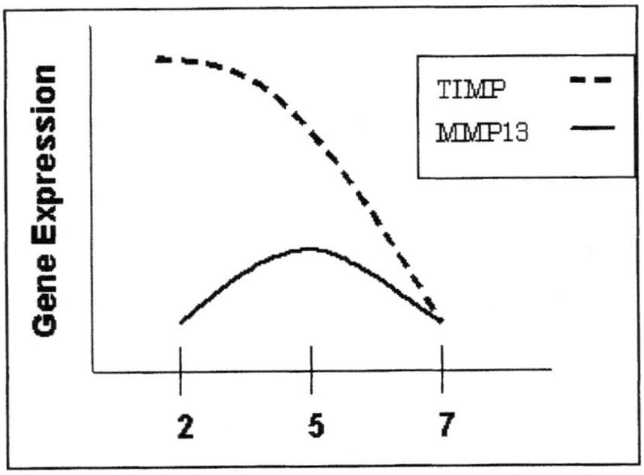

Figure 4. Schematic representation of MMPs and TIMPs mRNA expression observed in stem/progenitor cells appeared in the recovery phase of liver fibrosis. A solid line indicates MMP-13 gene expression and a dotted line indicates gene expression of MMP-2 and MT1-MMP, TIMP-1 and -2 (- - -- -).

Stem/Progenitor Cells Expressing MMP-13 mRNA Appear in Recovery from Liver Cirrhosis

Cells expressing MMP-13 mRNA were detected by in situ hybridization along the interface between the resolving fibrous bands and the parenchyma at day 5 after the last injection of CCl_4 for 8 or 12 weeks. In particular, strong gene expression was observed at day 5 after 12-week treatment, that is, in the recovery from liver cirrhosis. Some cells were confirmed to be hepatic stellate cells showing positive staining for α-SMA, but most cells did not show positive staining for α-SMA, albumin, α-fetoprotein, CK19 and ED2 immunohistochemically.[46] Based on these results we have investigated whether the cells expressing MMP-13 mRNA in the spontaneous resolution of liver fibrosis are stem cells, because recent evidence indicated that hematopoietic stem cells migrate in the liver and can differentiate into hepatic epithelium.[103]

It has been reported that some hepatic stellate cells are stained with GFAP, nestin and neural cell adhesion molecules.[104-107] Then, we used antibodies specific for Musashi-1, nestin and CD34. Musashi-1 protein, which is recognized as a neural RNA-binding protein, is involved in the development of neurons and glia by regulating gene expression at the post-transcriptional level.[108] Nestin is an intermediate filament, which appears transiently in neural progenitor cells.[109] Neuronal cells positive for Musashi-1 and nestin are considered to be progenitor cells in the central nervous system.[108-110] These cells have not previously been observed in normal liver.

Putative stem cells showing expression of the above three markers were observed within the resolving fibrous bands as well as at the interface between the resolving fibrous bands and the parenchyma at day 5 after the last injection of 12-week CCl_4 intoxication, in the early phase of the recovery process of liver fibrosis.[4,8] The Musashi-1-positive cell was also positive for CD34, and the nestin-positive cell was positive for both CD34 and MMP-13 in serial sections.[4,8] By in situ hybridization and immunohistochemical studies using mirror image section, some MMP-13 mRNA positive cells were found to express Musashi-1.[4]

Our study suggests that hematopoietic stem cells derived from bone marrow may differentiate into neural stem/progenitor cells. Neural stem/progenitor cells may migrate to the perilobular regions at the interface of the newly formed fibrous tissue from the perivenular regions where they first come from bone marrow. These cells changed phenotype to express MMP-13 mRNA (Fig. 4).

We are currently investigating the relationship between expression of stem cell markers (c-Kit, CD34, Thy-1.1, Lin, Sca-1) and that of MMP-13 mRNA, and how and when stem/progenitor cells do express MMP-13 mRNA in the process of cell growth and differentiation. Its clarification, as well as the development of the methods to make stem/progenitor cells proliferate and differentiate with expression of MMP-13, may lead to a new strategy for the treatment of liver cirrhosis.

Different Mechanisms in Recovery from Liver Fibrosis and That from Liver Cirrhosis

Strong gene expression of MMP-13 was observed in the recovery from liver cirrhosis, while few signals were seen in recovery phases of liver fibrosis after 8-week treatment as described above. We do not have further data to explain this difference directly.

Gelatinase activity of MMP-2 increased in the recovery phase of 8-week-treated rat liver by gelatin zymography.[47] Increased gelatinase activity in the recovery phase is consistent with the observation of Takahara et al[71] except for that in the recovery phase of 12-week-treated rats, which they did not examine. Decreased gelatinase activity in the recovery phase of 12-week-treated rats may reflect slow recovery from cirrhosis. We observed some discrepancy between MMP-2 mRNA expression and enzymatic activity in both progressive and recovery phases of liver fibrosis.[47] It seems that gelatinase activity depends on a balance between MMP-2 and their inhibitors.

Resolution of newly formed fibrous tissue in liver fibrosis of rats treated with CCl_4 for 8 weeks may be based on a positive balance of MMP-2 and MT1-MMP compared to TIMPs. On the other hand, resolution of fibrous tissue in liver cirrhosis of rats treated with CCl_4 for 12 weeks may need the enzyme of MMP-13, expressed by the stem cells derived from hematopoietic system, mainly bone marrow, in particular, neural progenitor cells.

The stem cells expressing MMP-13 mRNA proliferate and differentiate into hepatocytes, stellate cells and other cells. Some stem cells change their phenotype and express MMP-2 and MT1-MMP followed by TIMPs. MMP-2, MT1-MMP and TIMP participate in the destruction of the extracellular matrix.[73,98] MMP-2, MT1-MMP and TIMPs promote proliferation of liver cells, in particular hepatocytes and stellate cells. Hepatocyte growth factor (HGF) promotes the gene expression of MMP-13 as well as the proliferation of liver cells.

As one of the necessary cytokines, HGF is known to reduce experimental liver fibrosis.[111,112] Ueki, et al[112] showed that in dimethylnitrosamine-induced liver fibrosis, HGF down-regulated TGFβ expression, followed by the disappearance of predeposited fibrous tissue. Ozaki, et al[113] proposed that HGF induces MMP-1 expression in LI 90 (a cell line of human stellate cells) via Ets –1 in the promoter region of MMP-1. The recovery from liver fibrosis by external HGF treatment indicates that fibrolysis induced by increased collagenase activity could be related to regeneration of the liver.

Our results raise the possibility that hematopoietic stem cells could be used for the treatment of human liver fibrosis or cirrhosis. We are now investigating the possibility of gene therapy to expand stem cells in the liver and to enhance MMP-1 gene expression in stem cells with or without transfusion of stem cells derived from bone marrow in patients with liver cirrhosis.

The most important problems to be resolved are whether stem cells expressing MMP-13 mRNA transdifferentiate into cancer cells and whether HGF promotes this process. We previously reported that transient expression of interstitial collagenase was observed in differentiated, very early hepatocellular carcinomas (less than 2 cm), but not in moderately or undifferentiated hepatocellular carcinomas.[114]

Molecular mechanism of the transcriptional regulation of the MMP-1/MMP-13 gene has not been studied in hepatic stellate cells. Inflammatory cytokines such as tumor necrosis factor (TNF)-α and interleukin (IL)-1β are known to be involved in the activation of hepatic stellate cells. Most such cytokines activate the retrovirus-associated DNA-mitogen-activated protein kinase (Ras-MAPK) signaling pathway including c-Jun NH2-terminal kinase (JNK) and extracellular signal-regulated kinase (ERK), which in turn activate the transcription of early genes such as c-fos and c-jun. The Fos and Jun proteins contribute to the induction of MMP-1 gene transcription by their binding to the proximal AP-1 sites of the promoter in fibroblasts and immortalized cells.[115-117] In addition to the role of AP-1, Vincenti et al[118] recently reported the role of NF-κB in the induction of rabbit MMP-1 expression in IL1-treated synovial fibroblasts. TNF-α and IL-1β increase MAPK activity including JNK and ERK in rat hepatic stellate cells, by stimulating AP-1 activity in the hepatic stellate cells.[119] Activation of hepatic stellate cells is also closely related to NF-κB activity.[120] Activation of transcriptional factors such as AP-1 and NF-κB may also contribute to gene regulation of MMP-1 in hepatic stellate cells. The molecular mechanisms of the regulation of MMP-1 expression are still unknown. It should be determined how such transcriptional factors as NF-κB or AP-1 may be involved in MMP-1 gene expression in hepatic stellate cells. Hozawa et al[121] reported that Rac1 GTP may direct transcriptional induction of MMP-1 expression by TNF-α predominantly via NF-κB in human hepatic stellate cells.

Stem cells derived from bone marrow may be applied to patients with liver cirrhosis in near future. Stem cells should be programmed to proliferate, differentiate and express MMP-1 mRNA through the mechanisms mentioned above.

Conclusion

Reversibility of liver cirrhosis occurs under some conditions if the cause of liver damage is removed or adequately treated. The participation of MMPs, especially MMP-1/MMP-13, in the spontaneous resolution of liver fibrosis, has been clearly demonstrated by our group. Cells responsible for MMP-13 mRNA expression in the recovery phase may be neural stem/pregenitor cells. Transfusion of stem cells derived from bone marrow will be applied to patients with liver cirrhosis in near future. We will make efforts to find suitable conditions for stem cells to proliferate and differentiate for the expression of MMP-13.

References

1. Okuda K. Liver cirrhosis. Oxford: Blackwel; 2001.
2. Okazaki I, Watanabe T, Hozawa S et al. Reversibility of hepatic fibrosis: from the first report of collagenase in the liver to the possibility of gene therapy for recovery. Keio J Med 2001; 50(2):58-65.
3. Okazaki I, Yonezawa T, Watanabe T et al. New insights into the extracellular matrix. In: Okazaki I, Ninomiya Y, Friedman S, Tanikawa K, eds. Extracellular Matrix and the Liver -approach to gene therapy. New York: Academic Press, 2003:3-23.
4. Watanabe T, Niioka M, Hozawa S et al. Stem cells expressing matrix metalloproteinase-13 mRNA appear during regression reversal of hepatic cirrhosis. In: Okazaki I, Ninomiya Y, Friedman S, Tanikawa K, eds. Extracellular Matrix and the Liver—Approach to gene therapy. New York: Academic Press, 2003:362-390.
5. Cameron GR, Karunaratne WAE. Carbon tetrachloride cirrhosis in relation to liver regeneration. J Path Bact 1936; 42:1-21.
6. Perez-Tamayo R. Some aspects of connective tissue of the liver. Progress in Liver Diseases 1965; 2:192-210.
7. Okazaki I, Maruyama K. Collagenase activity in experimental hepatic fibrosis. Nature 1974; 252:49-50.
8. Watanabe T, Niioka M, Shimosakai K et al. Neural progenitor cells participate in spontaneous resolution of liver fibrosis by expressing MMP-13. Hepatology 2001; 34:359A.

9. Rubin E, Hutterer F. Hepatic fibrosis: Studies in the formation and resorption of hepatic collagen. In: Wagner B, Smith D, eds. The Connective Tissue. Baltimore: Williams & Wilkins, 1967:142-160.
10. Rojkind M, Dunn MA. Hepatic fibrosis. Gastroenterology 1979; 76:849-863.
11. Poupon RE, Balkau B, Eschwege E et al. A multicenter, controlled trial of ursodiol for the treatment of primary biliary cirrhosis. UDCA-PBC Study Group. N Engl J Med 1991; 324:1548-1554.
12. Kaplan MM, DeLellis RA, Wolfe HJ. Sustained biochemical and histologic remission of primary biliary cirrhosis in response to medical treatment. Ann Intern Med 1997; 126:682-688.
13. Bunton GL, Cameron R. Regeneration of liver after biliary cirrhosis. Ann N Y Acad Sci 1963; 111:412-421.
14. Hammel P, Couvelard A, O'Toole D et al. Regression of liver fibrosis after biliary drainage in patients with chronic pancreatitis and stenosis of the common bile duct. N Engl J Med 2001; 344(6):418-423.
15. Maruyama K, Okazaki I, Kashiwazaki K et al. A case of subacute hepatitis with reversible liver fibrosis. Gastroenterology 1981; 16:611-615.
16. Schiff ER, Heathcote J, Dienstag JL et al. Improvements in liver histology and cirrhosis with extended lamivudine therapy. Hepatology 2000; 32:296A.
17. Capra F, Casaril M, Gabrielli GB et al. alpha-Interferon in the treatment of chronic viral hepatitis: Effects on fibrogenesis serum markers. J Hepatol 1993; 18:112-118.
18. Teran JC, Mullen KD, Hoofnagle JH et al. Decrease in serum levels of markers of hepatic connective tissue turnover during and after treatment of chronic hepatitis B with interferon-alpha. Hepatology 1994; 19:849-856.
19. Arai M, Niioka M, Maruyama K et al. Changes in serum levels of metalloproteinases and their inhibitors by treatment of chronic hepatitis C with interferon. Dig Dis Sci 1996; 41:995-1000.
20. Bonis PA, Ioannidis JP, Cappelleri JC et al. Correlation of biochemical response to interferon alfa with histological improvement in hepatitis C: a meta-analysis of diagnostic test characteristics. Hepatology 1997; 26:1035-1044.
21. Dufour JF, DeLellis R, Kaplan MM. Regression of hepatic fibrosis in hepatitis C with long-term interferon treatment. Dig Dis Sci 1998; 43:2573-2576.
22. Lau DT, Kleiner DE, Ghany MG et al. 10-year follow-up after interferon-alpha therapy for chronic hepatitis C. Hepatology 1998; 28:1121-1127.
23. Shiffman ML, Hofmann CM, Contos MJ et al. A randomized, controlled trial of maintenance interferon therapy for patients with chronic hepatitis C virus and persistent viremia. Gastroenterology 1999; 117:1164-1172.
24. Sobesky R, Mathurin P, Charlotte F et al. Modeling the impact of interferon alfa treatment on liver fibrosis progression in chronic hepatitis C: A dynamic view. Gastroenterology 1999; 116:378-386.
25. Poynard T, McHutchinson J, Davis GL et al. Impact of interferon alfa-2b and ribavirin on progression of liver fibrosis in patients with chronic hepatitis C. Hepatology 2000; 32:1131-1137.
26. Shiratori Y, Imazeki F, Moriyama M et al. Histologic improvement of fibrosis in patients with hepatitis C who have sustained response to interferon therapy. Ann Intern Med 2000; 132:517-524.
27. Lau DT, Kleiner DE, Park Y et al. Resolution of chronic delta hepatitis after 12 years of interferon alfa therapy. Gastroenterology 1999; 117:1229-1233.
28. Dufour JF, DeLellis R, Kaplan MM. Reversibility of hepatic fibrosis in autoimmune hepatitis. Ann Intern Med 1997; 127:981-985.
29. Maruyama K, Okazaki I, Sato S et al. Type IV collagen (IV-C) is a useful marker of hepatic fibrosis in patients with alcoholic liver diseases. Alcohol & Alcoholism 1992; 27(Supple 1):176.
30. Lieber CS. Prevention and treatment of liver fibrosis based on pathogenesis. Alcohol Clin Exp Res 1999; 23:944-949.
31. Ishibashi K, Kashiwagi T, Ito A et al. Changes in serum fibrogenesis markers during interferon therapy for chronic hepatitis type C. Hepatology 1996; 24:27-31.
32. Jimenez SA, Freundlich B, Rosenbloom J. Selective inhibition of human diploid fibroblast collagen synthesis by interferons. J Clin Invest 1984; 74:1112-1116.
33. Czaja MJ, Weiner FR, Takahashi S et al. Gamma-interferon treatment inhibits collagen deposition in murine schistosomiasis. Hepatology 1989; 10:795-800.

34. Varga J, Olsen A, Herhal J et al. Interferon-gamma reverses the stimulation of collagen but not fibronectin gene expression by transforming growth factor-beta in normal human fibroblasts. Eur J Clin Invest 1990; 20:487-493.
35. Castilla A, Prieto J, Fausto N. Transforming growth factors beta 1 and alpha in chronic liver disease. Effects of interferon alfa therapy. N Engl J Med 1991; 324:933-940.
36. Nemoto T, Inagaki Y. Molecular basis for anti-fibrotic effects of interferon alpha. Connective Tissue 2001; 33:104.
37. Berman B, Duncan MR. Short-term keloid treatment in vivo with human interferon alfa-2b results in a selective and persistent normalization of keloidal fibroblast collagen, glycosaminoglycan, and collagenase production in vitro. J Am Acad Dermatol 1989; 21(4 Pt 1):694-702.
38. Duncan MR, Berman B. Differential regulation of glycosaminoglycan, fibronectin, and collagenase production in cultured human dermal fibroblasts by interferon-alpha, -beta, and -gamma. Arch Dermatol Res 1989; 281:11-18.
39. Jacques WE, McAdams AI. Reversible biliary cirrhosis in rat after partial ligation of common bile duct. Arch Path 1957; 63:149-153.
40. Hutterer F, Rubin E, Popper H. Mechanism of collagen resorption in reversible hepatic fibrosis. Exp Mol Path 1964; 3:215-223.
41. Quinn PS, Higginson J. Reversible and irreversible changes in experimental cirrhosis. Am J Pathol 1965; 47:353-369.
42. Morrione TG, Levine J. Collagenolytic activity and collagen resorption in experimental cirrhosis. Arch Path 1967; 84:59-63.
43. Takada A, Porta EA, Hartroft WS. The recovery of experimental dietary cirrhosis. II. Turnover changes of hepatic cells and collagen. Am J Pathol 1967; 51:959-976.
44. Okazaki I, Oda M, Maruyama K et al. Mechanism of collagen resorption in experimental hepatic fibrosis with special reference to the activity of lysosomal enzymes. Biochem Exp Biol 1974; 11:15-28.
45. Zimmerman H, Reichen J, Zimmerman A et al. Reversibility of secondary biliary fibrosis by biliodigestive anastomosis in the rat. Gastroenterology 1992; 103:579-589.
46. Watanabe T, Niioka M, Hozawa S et al. Gene expression of interstitial collagenase in both progressive and recovery phase of rat liver fibrosis induced by carbon tetrachloride. J Hepatol 2000; 33:224-235.
47. Watanabe T, Niioka M, Ishikawa A et al. Dynamic change of cells expressing MMP-2 mRNA and MT1-MMP mRNA in the recovery from liver fibrosis in the rat. J Hepatol 2001; 35:465-473.
48. Okazaki I, Maruyama K. Mammalian collagenase in the process of hepatic fibrosis. J UOEH 1980; 2:401-424.
49. Okazaki I, Maruyama K, Ono A et al. Reversibility of hepatic fibrosis in toxic liver injury. J UOEH 1982; 4:169-181.
50. Okazaki I, Tsuchiya M, Kamegaya K et al. Capillarization of hepatic sinusoids in carbon tetrachloride-induced hepatic fibrosis. Bibl Anat 1973; 12:476-483.
51. Maruyama K, Tsuchiya M, Oda M et al. Role of collagenase in capillarization of hepatic sinusoid. Microcirculation 1976; 1:333-334.
52. Okazaki I, Brinckerhoff CE, Sinclair JR et al. Iron increases collagenase production by rabbit synovial fibroblasts. J Lab Clin Med 1981; 97:396-402.
53. Maruyama K, Feinman L, Okazaki I et al. Direct measurement of neutral collagenase activity in homogenates from baboon and human liver. Biochim Biophy Acta 1981; 658:124-131.
54. Maruyama K, Feinman L, Fainsilber Z et al. Mammalian collagenase increases in early alcoholic liver disease and decreases with cirrhosis. Life Sciences 1982; 30:1379-1384.
55. Carter E, Mccarron M, Alpert E et al. Lysyl oxidase and collagenase in experimental acute and chronic liver injury. Gastroenterology 1982; 82:526-534.
56. Lindblad W, Fuller G. Hepatic collagenase activity during carbon tetrachloride-induced fibrosis. Fund. Toxicol 1983; 3:34-40.
57. Montfort I, Perez-Tamayo R, Alvizouri A et al. Collagenase of hepatocytes and sinusoidal liver cells in the reversibility of experimental cirrhosis of the liver. Virchows Arch B Cell Pathol 1990; 59:281-289.
58. Arthur MJ, Friedman SL, Roll FJ et al. Lipocytes from normal rat liver release a neutral metalloproteinase that degrades basement membrane (type IV) collagen. J Clin Invest 1989; 84:1076-1085.

59. Arthur MJ, Stanley A, Iredale JP et al. Secretion of 72 kDa type IV collagenase/gelatinase by cultured human lipocytes Analysis of gene expression, protein synthesis and proteinase activity. Biochem J 1992; 287:701-707.
60. Casini A, Ceni E, Salzano R et al. Acetaldehyde regulates the gene expression of matrix-metalloproteinase-1 and -2 in human fat-storing cells. Life Sci 1994; 55:1311-1316.
61. Milani S, Herbst H, Schuppan D et al. Differential expression of matrix-metalloproteinase-1 and -2 genes in normal and fibrotic human liver. Am J Pathol 1994; 144:528-537.
62. Herbst H, Heinrichs O, Schuppan D et al. Temporal and spatial patterns of transin/stromelysin RNA expression following toxic injury in rat liver. Virchows Arch B Cell Pathol Incl Mol Pathol 1991; 60:295-300.
63. Vyas SK, Leyland H, Gentry J et al. Rat hepatic lipocytes synthesize and secrete transin (stromelysin) in early primary culture. Gastroenterology 1995; 109:889-898.
64. Takahara T, Furui K, Yata Y et al. Dual expression of matrix metalloproteinase-2 and membrane-type 1-matrix metalloproteinase in fibrotic human livers. Hepatology 1997; 26:1521-1529.
65. Theret N, Musso O, L'Helgoualc'h A. Clement B. Differential expression and origin of membrane-type 1 and 2 matrix metalloproteinases (MT-MMPs) in association with MMP2 activation in injured human livers. Am J Pathol 1998; 153:945-954.
66. Iredale JP, Murphy G, Hembry RM et al. Human hepatic lipocytes synthesize tissue inhibitor of metalloproteinases-1. Implications for regulation of matrix degradation in liver. J Clin Invest 1992; 90:282-287.
67. Iredale JP, Goddard S, Murphy G et al. Tissue inhibitor of metalloproteinase-I and interstitial collagenase expression in autoimmune chronic active hepatitis and activated human hepatic lipocytes. Clin Sci (Lond) 1995; 89(1):75-81.
68. Herbst H, Wege T, Milani S et al. Tissue inhibitor of metalloproteinase-1 and -2 RNA expression in rat and human liver fibrosis. Am J Pathol 1997; 150:1647-1659.
69. Roeb E, Purucker E, Breuer B et al. TIMP expression in toxic and cholestatic liver injury in rat. J Hepatol 1997; 27:535-544.
70. Benyon RC, Iredale JP, Goddard S et al. Expression of tissue inhibitor of metalloproteinases 1 and 2 is increased in fibrotic human liver. Gastroenterology 1996; 110:821-831.
71. Takahara T, Furui K, Funaki J et al. Increased expression of matrix metalloproteinase-II in experimental liver fibrosis in rats. Hepatology 1995; 21:787-795.
72. Aimes RT, Quigley JP. Matrix metalloproteinase-2 is an interstitial collagenase. Inhibitor-free enzyme catalyzes the cleavage of collagen fibrils and soluble native type I collagen generating the specific 3/4- and 1/4-length fragments. J Biol Chem 1995; 270:5872-5876.
73. Ohuchi E, Imai K, Fujii Y et al. Membrane type 1 matrix metalloproteinase digests interstitial collagens and other extracellular matrix macromolecules. J Biol Chem 1997; 272:2446-2451.
74. Freije JM, Diez-Itza I, Balbin M et al. Molecular cloning and expression of collagenase-3, a novel human matrix metalloproteinase produced by breast carcinomas. J Biol Chem 1994; 269:16766-16773.
75. Knauper V, Lopez-Otin C, Smith B et al. Biochemical characterization of human collagenase-3. J Biol Chem 1996; 271:1544-1550.
76. Friedman SL. Seminars in medicine of the Beth Israel Hospital, Boston. The cellular basis of hepatic fibrosis. Mechanisms and treatment strategies. N Engl J Med 1993; 328:1828-1835.
77. Arthur MJ. Role of Ito cells in the degradation of matrix in liver. J Gastroenterol Hepatol 1995; 10 Suppl 1:S57-62.
78. Friedman SL. Molecular mechanisms of hepatic fibrosis and principles of therapy. J Gastroenterol 1997; 32:424-430.
79. Okazaki I, Watanabe T, Hozawa S et al. Molecular Pathology of Liver Fibrosis. In: Yamanaka M, Toda G, eds. Liver Cirrhosis Update. Amsterdam: Excerpta Medica, 1998:4 23-33.
80. Okazaki I, Watanabe T, Hozawa S et al. Hepatic fibrolysis and hepatic sinusoidal cells. In: Tanikawa K, Ueno T, eds. Liver Diseases and Hepatic Sinusoidal Cells. Tokyo: Springer, 1999:242-251.
81. Rojkind M. Role of metalloproteinases in liver fibrosis. Alcohol Clin Exp Res 1999; 23:934-939.
82. Arthur MJ. Fibrogenesis II. Metalloproteinases and their inhibitors in liver fibrosis. Am J Physiol Gastrointest Liver Physiol 2000; 279:G245-249.
83. Okazaki I, Watanabe T, Hozawa S et al. Molecular mechanism of the reversibility of hepatic fibrosis: with special reference to the role of matrix metalloproteinases. J Gastroenterol Hepatol 2000; 15 Suppl:D26-32.

84. Iredale JP, Benyon RC, Arthur MJ et al. Tissue inhibitor of metalloproteinase-1 messenger RNA expression is enhanced relative to interstitial collagenase messenger RNA in experimental liver injury and fibrosis. Hepatology 1996; 24:176-184.
85. Iredale JP, Benyon RC, Pickering J et al. Mechanisms of spontaneous resolution of rat liver fibrosis. Hepatic stellate cell apoptosis and reduced hepatic expression of metalloproteinase inhibitors. J Clin Invest 1998; 102:538-549.
86. Maruyama K, Okazaki I, Kashiwazaki K et al. Different appearance of collagenase and lysosomal enzymes in recovery of experimental hepatic fibrosis. Biochem Exp Biol 1978; 14:191-201.
87. Yu A, Murphy A, Stetler-Stevenson W. 72-kDa gelatinase (gelatinase A): Structure, activation, regulation, and substrate specificity. Matrix metalloproteinases 1995: 85-113.
88. Hanemaaijer R, Koolwijk P, le Clercq L et al. Regulation of matrix metalloproteinase expression in human vein and microvascular endothelial cells. Effects of tumour necrosis factor alpha, interleukin 1 and phorbol ester. Biochem J 1993; 296 (Pt 3):803-809.
89. Gervasi DC, Raz A, Dehem M et al. Carbohydrate-mediated regulation of matrix metalloproteinase-2 activation in normal human fibroblasts and fibrosarcoma cells. Biochem Biophys Res Commun 1996; 228:530-538.
90. Bar-Eli M. Role of interleukin-8 in tumor growth and metastasis of human melanoma. Pathobiology 1999; 67:12-18.
91. Stearns ME, Garcia FU, Fudge K et al. Role of interleukin 10 and transforming growth factor beta1 in the angiogenesis and metastasis of human prostate primary tumor lines from orthotopic implants in severe combined immunodeficiency mice. Clin Cancer Res 1999; 5:711-720.
92. Bian J, Sun Y. Transcriptional activation by p53 of the human type IV collagenase (gelatinase A or matrix metalloproteinase 2) promoter. Mol Cell Biol 1997; 17:6330-6338.
93. Sun Y, Wenger L, Rutter JL et al. Cheung HS. p53 down-regulates human matrix metalloproteinase-1 (Collagenase-1) gene expression. J Biol Chem 1999; 274:11535-11540.
94. Theret N, Lehti K, Musso O et al. MMP2 activation by collagen I and concanavalin A in cultured human hepatic stellate cells. Hepatology 1999; 30:462-468.
95. Preaux AM, Mallat A, Nhieu JT et al. Matrix metalloproteinase-2 activation in human hepatic fibrosis regulation by cell-matrix interactions. Hepatology 1999; 30:944-950.
96. Ikeda K, Wakahara T, Wang YQ et al. In vitro migratory potential of rat quiescent hepatic stellate cells and its augumentation by cell activation. Hepatology 1999; 29:1760-1767.
97. Li D, Friedman SL. Liver fibrogenesis and the role of hepatic stellate cells: new insights and prospects for therapy. J Gastroenterol Hepatol 1999; 14:618-633.
98. Kinoshita T, Sato H, Okada A et al. TIMP-2 promotes activation of progelatinase A by membrane-type 1 matrix metalloproteinase immobilized on agarose beads. J Biol Chem 1998; 273:16098-16103.
99. Popper H, Uenfriend S. Hepatic fibrosis. Correlation of biochemical and morphologic investigations. Am J Med 1970; 49:707-721.
100. Maruyama K, Okazaki I, Kashiwazaki K et al. Biosynthesis of Type IV collagenase by hepatic sinusoidal cells in rats. Acta Hep Jpn 1987; 28:973-974.
101. Maruyama K, Okazaki I, Takagi T et al. Formation and degradation of basement membrane collagen. Alcohol Alcohol Suppl 1991; 1:369-374.
102. Winwood PJ, Schuppan D, Iredale JP et al. Kupffer cell-derived 95-kd typeIV collagenase/gelatinase B: Characterization and expression in cultured cells. Hepatology 1995; 22:304-315.
103. Lagasse E, Connors H, Al-Dhalimy M et al. Purified hematopoietic stem cells can differentiate into hepatocytes in vivo. Nat Med 2000; 6:1229-1234.
104. Neubauer K, Knittel T, Aurisch S et al. Glial fibrillary acidic protein—A cell type specific marker for Ito cells in vivo and in vitro. J Hepatol 1996; 24:719-730.
105. Knittel T, Aurisch S, Neubauer K et al. Cell-type-specific expression of neural cell adhesion molecule (N-CAM) in Ito cells of rat liver. Up-regulation during in vitro activation and in hepatic tissue repair. Am J Pathol 1996; 149:449-462.
106. Knittel T, Kobold D, Saile B et al. Rat liver myofibroblasts and hepatic stellate cells: Different cell populations of the fibroblast lineage with fibrogenic potential. Gastroenterology 1999; 117:1205-1221.

107. Niki T, Pekny M, Hellemans K et al. Class VI intermediate filament protein nestin is induced during activation of rat hepatic stellate cells. Hepatology 1999; 29:520-527.
108. Sakakibara S, Imai T, Hamaguchi K et al. Mouse-Musashi-1, a neural RNA-binding protein highly enriched in the mammalian CNS stem cell. Dev Biol 1996; 176:230-242.
109. Kawaguchi A, Miyata T, Sawamoto K et al. Nestin-EGFP transgenic mice: visualization of the self-renewal and multipotency of CNS stem cells. Mol Cell Neurosci 2001; 17:259-273.
110. Roy NS, Wang S, Jiang L et al. In vitro neurogenesis by progenitor cells isolated from the adult human hippocampus. Nat Med 2000; 6:271-277.
111. Yasuda H, Imai E, Shiota A et al. Antifibrogenic effect of a deletion variant of hepatocyte growth factor on liver fibrosis in rats. Hepatology 1996; 24:636-642.
112. Ueki T, Kaneda Y, Tsutsui H et al. Hepatocyte growth factor gene therapy of liver cirrhosis in rats. Nat Med 1999; 5:226-230.
113. Ozaki I, Zhao G, Mizuta T et al. Induction of collagenase (matrix metalloproteinase-1) by hepatocytes growth factor in human Ito cell LI90. Hepatology 1999; 30:491A.
114. Okazaki I, Wada N, Nakano M et al. Difference in gene expression for matrix metalloproteinase-1 between early and advanced hepatocellular carcinomas. Hepatology 1997; 25:580-584.
115. Vincenti MP, White LA, Schroen DJ et al. Regulating expression of the gene for matrix metalloproteinase-1 (collagenase): Mechanisms that control enzyme activity, transcription, and mRNA stability. Crit Rev Eukaryot Gene Expr 1996; 6:391-411.
116. Grumbles RM, Shao L, Jeffrey JJ et al. Regulation of the rat interstitial collagenase promoter by IL-1 beta, c-Jun, and Ras-dependent signaling in growth plate chondrocytes. J Cell Biochem 1997; 67:92-102.
117. Newberry EP, Willis D, Latifi T et al. Fibroblast growth factor receptor signaling activates the human interstitial collagenase promoter via the bipartite Ets-AP1 element. Mol Endocrinol 1997; 11:1129-1144.
118. Vincenti MP, Coon CI, Brinckerhoff CE. Nuclear factor kappaB/p50 activates an element in the distal matrix metalloproteinase 1 promoter in interleukin-1beta-stimulated synovial fibroblasts. Arthritis Rheum 1998; 41:1987-1994.
119. Poulos JE, Weber JD, Bellezzo JM et al. Fibronectin and cytokines increase JNK, ERK, AP-1 activity, and transin gene expression in rat hepatic stellate cells. Am J Physiol 1997; 273(4 Pt 1):G804-811.
120. Elsharkawy AM, Wright MC, Hay RT et al. Persistent activation of nuclear factor-kappaB in cultured rat hepatic stellate cells involves the induction of potentially novel Rel-like factors and prolonged changes in the expression of IkappaB family proteins. Hepatology 1999; 30:761-769.
121. Hozawa S, Suzuki H, Watanabe T et al. Rho family GTPases regulate transcriptional induction of matrix metalloproteinase-1 by tumor necrosis factor A in human hepatic stellate cells. Gastroenterology 2000; 118:A1013.

CHAPTER 13

Can Manipulation of Apoptotic Cell Death Benefit Tissue Scarring?

Wesam Ahmed, Mohammed S. Razzaque and Takashi Taguchi

Abstract

Cell death by apoptosis is an active process of cell removal that is initiated and regulated by activation of specific enzymes and signaling molecules. In contrast to necrotic cell death, apoptotic cell death holds the potential for therapeutic manipulation. Recent studies document important roles for apoptosis in both normal and pathological processes, ranging from embryonic development to tissue scarring. Apoptosis is thought to help reduce inflammation by the selective removal of inflammatory cells and to help change fibroproliferative mass into an acellular scar tissue by deleting cellular components. Our understanding of the mechanisms of cell-specific apoptosis during various pathophysiological processes show the opportunity to modulate the rate of apoptosis in a cell type-specific manner, to delay or suppress the progression of such immunoinflammatory diseases as pulmonary fibrosis and formations of pathologic scar tissue. We will briefly present the possible mechanisms of apoptotic cell death and their impact on the formation of fibrotic mass.

Introduction

Tissue scarring is an active process which is closely involved in end-stage organ failure. Although advanced tissue scarring is believed to be mostly irreversible, identification of important molecules that regulate crucial steps of fibrogenesis has provided an opportunity to delay or reverse the disease process. The nature of initiating events and subsequent healing responses after an injury partly determine whether the outcome will be controlled wound healing, or uncontrolled tissue scarring. It is the persistent activation of fibrogenic molecules that differentiates uncontrolled tissue scarring from controlled wound healing. Therefore, regulating the healing response by exogenous manipulation of fibrogenic molecules in patients with scarring diseases offers the tantalizing prospect of modulating uncontrolled tissue scarring to a controlled healing.

Advanced-stage tissue scar is mostly composed of matrix proteins with no (or very few) cellular components. Recent studies have documented the role of cell death in the pathogenesis of tissue scarring. The presence of dead cells during the early inflammatory phase of scarring diseases triggers the body's defensive responses, such as the release of additional cytokines, chemokines, and growth factors, resulting in a vicious cycle of inflammation, cell death, and scarring. During the late-stage of wound healing, a decrease in matrix-producing cells is seen along with an increased rate of apoptosis. This mechanism has been implicated

Fibrogenesis: Cellular and Molecular Basis, edited by Mohammed S. Razzaque.

in evolution of granulation tissue into an acellular scar tissue.[1] However, when removal of cellular components from granulation tissue is not adequate, keloid and hypertrophic scars form, both of which are characterized by an increased degree of cellular components.[2]

Apoptosis has a determinant role in the clinical outcome of such human scarring diseases as atherosclerosis. Atherosclerotic plaques are delicate tissue, composed of smooth muscle cells, lymphocytes, macrophages, and matrix proteins. Apoptotic deletion of smooth muscle cells from atherosclerotic lesions has been linked to destabilizing plaque, which is then prone to rupture and form emboli, but apoptotic deletion of macrophages has been linked to plaque stabilization, and thus reducing the risk of plaque rupture.[3] In the following chapter, Kuwano and colleagues describe the role of apoptosis in pulmonary fibrosis, and present beneficial effects of modulation of apoptosis by targeting proapoptotic molecules in the experimental models. To provide readers with background information on different modes of cell death, we will review molecular mechanisms of cell death, emphasizing apoptosis.

Apoptosis

Apoptosis is a major form of cell death. This term was coined because the release of apoptotic bodies by dying cells resembled the picture of falling leaves from deciduous trees, called in Greek "apoptosis".[4] During apoptosis, cells induce their own death by chopping their DNA into small fragments after receiving either internal or external death signals. Apoptosis is a multistep, and multifactorial complex process, which occurs in a well-choreographed sequence. These morphological events begin with nuclear and cytoplasmic condensation and blebbing of the plasma membrane. Consequently, the cell breaks up into apoptotic bodies, which are membrane-bound fragments containing structurally intact organelles as well as portions of the nucleus. The apoptotic bodies are rapidly recognized by neighboring cells, then ingested and cleared by degradation.

Mechanisms of Apoptosis

The molecular basis for classical apoptosis is related to the activation of a family of intracellular cysteine proteases, termed caspases. Caspases are present in the cell in a latent state (pro-caspases), but are activated in response to a wide variety of cell death stimuli.[5] Caspases are organized in a cascade which contains the initiator (upstream) caspases (caspase 8 and 9) responsible for activating the effector (downstream) caspases to perform proteolytic cleavage. Once activated, the downstream caspases cleave specific protein substrates to induce apoptosis.[6-8] At present, at least two major pathways of caspase activation have been identified: (i) The mitochondria-mediated apoptosis pathway, in which cytochrome c is released and activates upstream caspase 9 in the presence of cytosolic protein Apaf-1[9] and, (ii) The receptor-mediated apoptosis pathway, in which the stimulation of death receptors activate upstream caspase 8.[10] Both pathways activate a major downstream effector caspase 3.[11] A third, less common, pathway of caspase activation involving granzyme B has also been demonstrated. It bypasses both the mitochondria-mediated and receptor-mediated pathways of caspase activation.[12] Granzyme B is a serine protease synthesized in cytotoxic T lymphocytes and stored in secretory granules. Once activated, the cytotoxic T cells secrete perforin, which induces pores in the membrane of the target cells. These pores will be used by the cytotoxic T cells to inject granzyme B into target cells. The enzymatic activity of granzyme B is crucial to its ability to induce apoptosis through the direct activation of caspases like the downstream caspase 3. In addition, it has been reported that granzyme B can induce apoptosis in a mitochondria-dependent mechanism via Bid-dependent[13] or independent[14] pathways. Bid is a proapoptotic member of the BH3-only family.

Although the molecular basis of classical apoptosis is related to caspase activation, cell death with apoptotic features can also occur independent of caspase[15] through the release of apoptosis-inducing factor (AIF).[16] The two major apoptotic pathways, i.e., the intrinsic (mitochondrial) and the extrinsic (receptor-mediated) are described in the following sections.

The Intrinsic (Mitochondrial) Death Pathway

The mitochondria, which are known collectively as the powerhouse of the cell because it is the place where the oxidative phosphorylation and ATP synthesis take place, play a critical role in apoptosis. Various cell death pathways involve permeabilisation of mitochondrial membranes and release of mitochondrial proteins.[17] Two types of stimuli can induce the mitochondria to release proteins which activate apoptosis. These stimuli are internal, such as DNA damage, or external, e.g., cytotoxic drugs, heat-shock, hypoxia, growth-factor withdrawal, and irradiation. Several chemotherapeutic agents and anticancer drugs also act on mitochondria, although their exact mechanism of action is unclear.[18]

The intrinsic (mitochondrial) apoptotic pathway begins with the release of some mitochondrial proteins to the cytoplasm (Fig. 1). One protein is cytochrome c. The discovery that cytochrome c is released from mitochondria during cell death, as well as its role in triggering apoptosis, has had a dramatic impact on our understanding of the mechanism by which apoptosis is activated and regulated.[19] Cytochrome c is normally located in the space between the outer and inner membranes of mitochondria where it participates in the process of oxidative phosphorylation. Cytochrome c is released from the mitochondria into the cytosol in response to apoptotic stimuli. The released cytochrome c forms a complex with the cytosolic protein Apaf-1 and caspase 9 in the presence of either ATP or ADP. This complex induces autocatalytic activation of caspase 9, the first step in the proteolytic activation of caspase 3 and the other effector caspases 2, 6, 7 and 10, which are instrumental in causing cell death.[20,21] Although it is widely accepted that cytochrome c release is secondary to the loss of the mitochondrial membrane potential, cytochrome c release can also precede this loss.[22]

The detailed mechanisms of the mitochondrial membrane's permeabilization during apoptosis are not clear yet. Many competing models are trying to explain this mechanism.[23] In one model, mitochondrial swelling and rupture is induced by the opening of megachannel permeability transition pores which are formed by the adenine nucleotide translocator (ANT), located in the inner mitochondrial membrane, and the voltage-dependent anion channel (VDAC), located in the outer mitochondrial membrane. These pores span both the inner and the outer mitochondrial membranes where the two membranes are opposed. Bax, the pro-apoptotic Bcl-2 family member, binds to the ATN. This binding induces mitochondrial depolarization and increases the permeability of the inner membrane, which opens the permeability transition pores.[24] Consequently, water and solutes enter into the matrix, causing mitochondrial swelling. Another model proposes that mitochondrial swelling occurs as a consequence of defects in the mitochondrial ATP/ADP exchange induced by the closure of the VDACs, which causes hyperpolarization of the inner mitochondrial membrane and subsequent matrix swelling. On the other hand, at least in some cell types, the decline in membrane potential follows the release of cytochrome c, which contradicts the previous scenario.

Other models predict the formation of pores in the outer mitochondrial membrane, allowing the passage of cytochrome c and other mitochondrial proteins into the cytosol without damaging the membrane. Bax, in addition to its ability to bind to the ANT in the inner mitochondrial membrane, as mentioned above, forms large conducting channels in lipid planar bilayers.[23] Addition of Bax directly to isolated mitochondria induces the release of cytochrome c through a mechanism that is not blocked by permeability transition pore blockers and does not involve mitochondrial swelling. In addition, Bax cooperates with VDAC to form a cytochrome c-conducting channel.[25]

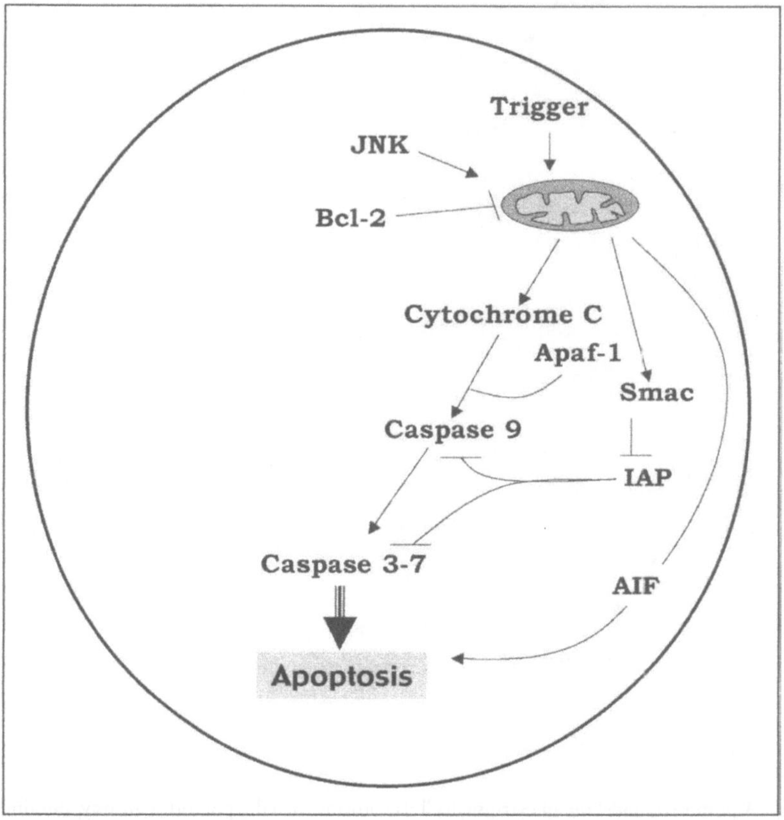

Figure 1. The intrinsic (mitochondrial) apoptotic pathway. Triggering the mitochondria induce the release of certain mitochondrial proteins. Cytochrome C release, with the interaction of Apaf-1, activates caspase 9, which activates caspase 3, among others, inducing apoptosis. Smac, another mitochondrial protein, is released in response to mitochondrial triggering to relieve the inhibitory effect of IAP on caspase activation. AIF release from the mitochondria induces apoptosis in a caspase-independent mechanism. Bcl-2 blocks apoptosis by inhibiting the release of mitochondrial proteins.

In addition to cytochrome c, several proteins are released from mitochondria in cells undergoing apoptosis. The recently identified Smac/Diablo molecule that binds to, and inactivates, the inhibitors of apoptosis proteins (IAPs), is an example of these proteins.[26,27] IAPs inhibit cell death by binding to procaspases and activated caspases, blocking their processing and activity. Smac/Diablo is released from the mitochondria along with cytochrome c during apoptosis and inactivates the IAPs, thus removing inhibition of caspase 9 activation.[28] Because Smac/Diablo is required for the inactivation of IAP by preventing the caspase 8-dependent cleavage and activation of caspase-3,[29] in some cells, Smac/Diablo is more crucial in induction of apoptosis than cytochrome c. Other proteins, in addition to cytochrome c and Smac/Diablo, which regulate caspase activation and apoptosis, are localized between the outer and the inner mitochondrial membrane. Caspase 9, caspase 3, caspase 2, and AIF, which induce caspase-independent apoptosis,[16] are examples of these proteins. These proteins can be released from the mitochondria to the cytosol during the induction of apoptosis.[30]

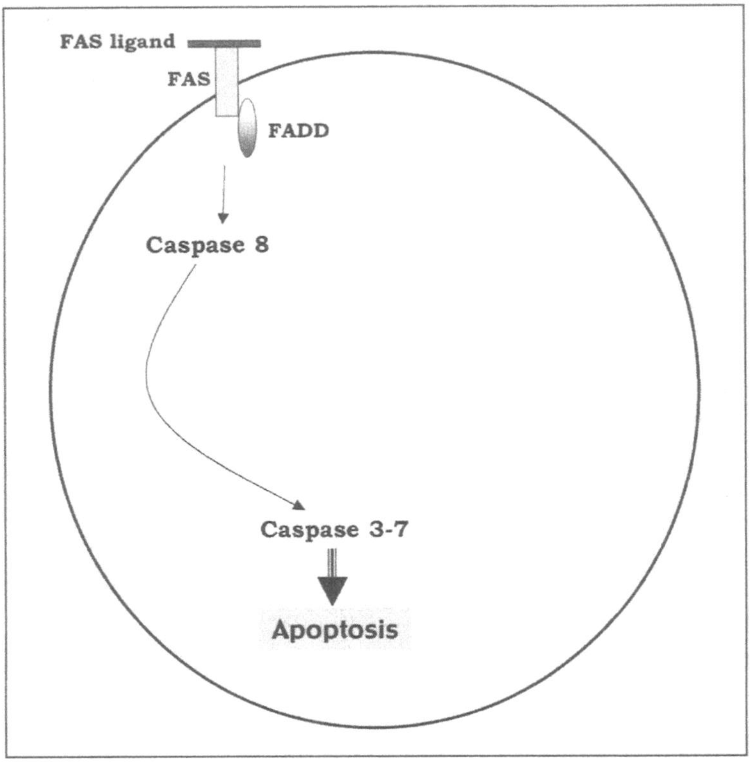

Figure 2. A proposed model for an extrinsic (cell receptor-mediated) apoptotic pathway. Binding of death ligand to a cell membrane receptor activates caspase 8. Activated caspase 8 stimulates downstream caspases like caspase 3, which induces apoptosis.

The Extrinsic (Receptor Mediated) Death Pathway

The other major mechanism of apoptosis induction is the receptor-mediated pathway which involves the activation of death cell receptors. Ligand binding stimulates downstream effectors (Fig. 2). The tumor necrosis factor (TNF) family of cytokine receptors, which differ in their ligand specificity, is an example of the death receptors. Upon ligand binding, the receptor is activated then it recruits and binds the death effector cell protein Fadd/Mort-1.[31] Fadd/Mort-1 binding recruits and cleaves procaspase 8.[32] Cleavage of procaspase 8 induces the active caspase 8, which is released to the cytosol, where it cleaves and activates the downstream effector caspase 3.[11]

Caspase 8 is not the only link between death receptors and induction of apoptosis. Activation of Fas can induce apoptosis by alternative pathways. For example, Fas receptors, via adapter proteins, activate the mitogen-activated protein 3 (MAP3) kinase, ASK1. When triggered, the proapoptotic ASK1 activates a phosphorylation cascade that culminates in the stimulation of c-Jun N terminal kinase (JNK).[31] Activated JNK phosphorylates c-Jun and p53, among other substrates, and induces apoptosis through modifying and transcriptionally regulating proteins in the Bcl-2 family. Activation of Fas can induce apoptosis through multiple pathways.[33]

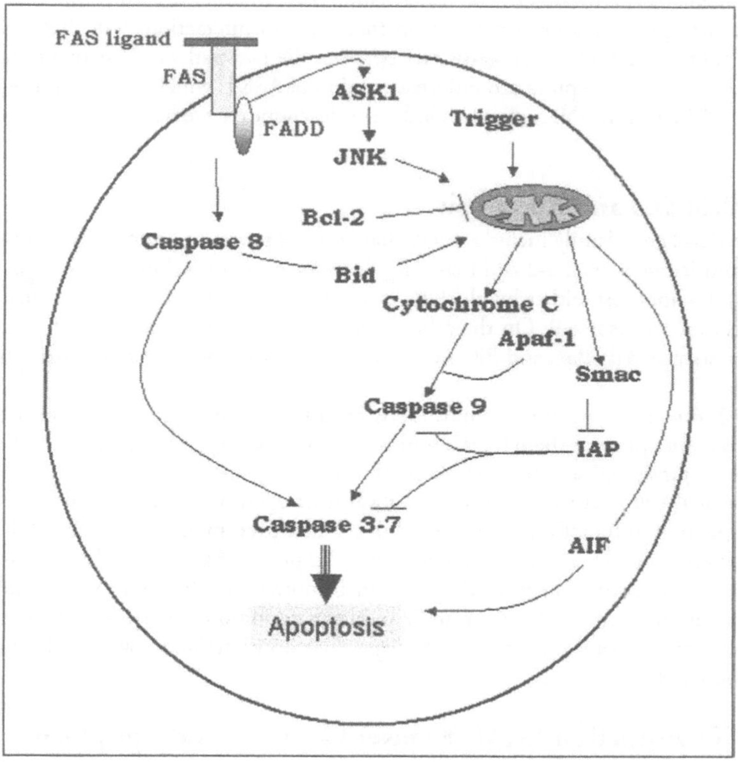

Figure 3. The cross talk between the intrinsic (mitochondrial) and the extrinsic (cell receptor-mediated) apoptotic pathways. One level of cross talk occurs at the level of Bid, where Bid cleavage by caspase 8 triggers the mitochondria to induce apoptosis. Another level of cross talk occurs when receptor mediated apoptotic pathway activates ASK 1. Active ASK 1 activates JNK, which stimulates the mitochondria to induce apoptosis. This figures also shows a different mechanism of inducing apoptosis, by the ability of granzyme B to directly activate caspase 3.

Cross Talk between Pathways

Despite the fact that mitochondrial-mediated and receptor-mediated pathways are separate, cross talk between them occurs (Fig. 3). Bid is an example of the cross talk between the two pathways. Bid is cleaved and activated by caspase 8. When cleaved, Bid activates the mitochondria-mediated apoptotic pathways via the interaction with Bax as well as ANT.[34] In addition, cleaved Bid forms ion channels in liposomes in vitro suggesting that it can directly disrupt the mitochondria and induce cytochrome c release.[35]

The mechanisms by which Bid rapidly and selectively targets the mitochondria are not known. Granzyme B, the cytotoxic T-cell specific serine protease, cleaves Bid and activates the mitochondrial pathway.[12] Granzyme B also affects mitochondria in a Bid-independent way resulting in mitochondrial depolarization and cell death, even though these mitochondria fail to release cytochrome c.[14]

Another level of cross talk is BAR, which inhibits apoptosis in both pathways. BAR contains a protein-interacting domain that binds and inhibits procaspase 8 and thereby blocks the receptor-mediated pathway. On the other hand, BAR inhibits the mitochondrial-mediated pathway through an unknown mechanism that involves interactions with Bcl-2 and Bax.[36] ASK1 is another example of the cross talk between the two pathways. As mentioned before, stimulation of death receptors can induce activation of ASK1, which phosphorylates and activates the JNK pathway. JNK activation induces mitochondria-mediated cell death in some cell lines.

Bcl-2 Proteins and Apoptosis

The Bcl-2 gene family includes genes that either suppress or promote apoptosis.[37,38] The anti-apoptotic members Bcl-2 and Bcl-x_L reside in the outer mitochondrial membrane, where they suppress apoptosis either by blocking cytochrome c release or binding to Apaf-1 to prevent activation of caspase 9. On the other hand, the mammalian pro-apoptotic Bcl-2 family members, such as Bak, Bax, and Bik, may promote apoptosis by displacing Apaf-1 from Bcl-2 and Bcl-x_L.

Bcl-2, the prototype of the family, exerts an anti-apoptotic action. Bcl-2 over-expression in some cell lines protects them from apoptosis.[39,40] Conversely, inhibition of Bcl-2 expression increases apoptosis and sensitivity to chemotherapeutic agents.[41]

The final cell death decision either towards or against apoptosis could be a result of protein-protein interactions between Bcl-2 and other members of the Bcl-2 family. Heterodimerization of Bcl-2 and the proapoptotic Bax prevent Bax-mediated apoptosis.[42] Similar results have been reported between Bcl-2 and Bak, another member of the same family.[43] Bcl-2 phosphorylation at serine-70 and serine-87 results in inhibition of Bcl-2 dimidiation with Bax. Many microtubule-damaging agents can trigger this phosphorylation, which releases Bax from the inhibitory effect of Bcl-2.[44,45]

Signal Transduction (MAP Kinase) Pathways and Apoptosis

During the last decade, considerable research has been focused on the role of signaling transduction pathways in cell proliferation, differentiation, and survival. mitogen-activated protein kinase (MAPK) represents one of the most extensively studied pathways (Fig. 4).[46]

The MAPK family comprises c-Jun N-terminal kinase (JNK), p38 MAPK, and extracellular signal-regulated kinase (ERK). The ERK pathway is activated by such stimuli as growth factors,[47] lipopolysaccharide,[48] and chemotherapeutic agents.[49] ERK activation exerts in most cases an antiapoptotic,[50,51] and less commonly proapoptotic[49,52] influence, depending upon the cellular context and by yet unclarified regulatory mechanisms. The JNK and p38 MAPK signaling pathways are also activated by various and overlapping stimuli, such as heat or osmotic shock, radiation, and growth factors.[47,53] The JNK and p38 MAPK signaling pathways are associated, with few exceptions, with enhanced activation of the apoptotic pathway.[54]

The JNK family was first described as stress kinases, and their response to cellular stress has been studied extensively.[55] The JNK family consists of JNK1 and JNK2, which are widely distributed, and of JNK3, which is primarily expressed in the heart, brain and testis. Differential splicing yields a total of 10 different isoforms.[56] After activation, JNKs phosphorylate various substrates in the nucleus, e.g., c-Jun or activating transcription factor-2 (ATF-2) and in the cytoplasm, e.g., Bcl-2.[57]

PI3K/AKT Pathway and Apoptosis

An important pathway downstream of the receptor tyrosine kinase (RTK), in many cases via RAS, involves the phosphatidylinositol 3-kinase (PI3K) cascade. This pathway is a major regulator of mammalian cell proliferation and survival became apparent when it was described that PI3Ks activity was physically and functionally associated with the transforming activity of viral oncogenes.[58]

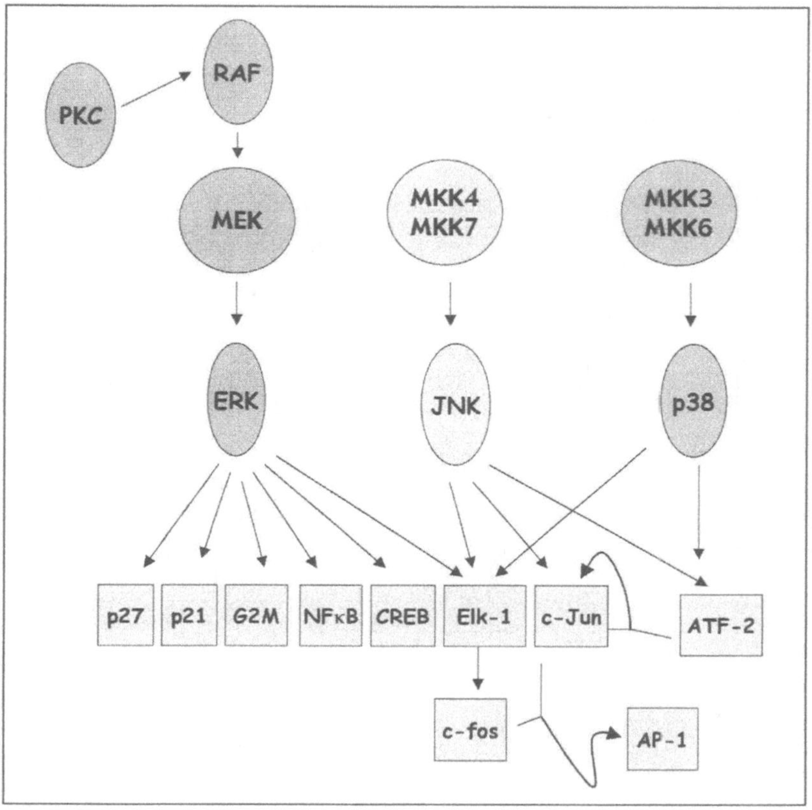

Figure 4. Different MAP kinases pathways: ERK, JNK, and p38 pathways, with the possible cross talk between them at the level of downstream targets.

In addition to their roles in signaling pathways, PI3Ks also control many important cell responses.[59,60] Members of the PI3K family can also be considered oncogenes because they control cell cycle progression, differentiation, survival, invasion and metastasis, and angiogenesis.[61] Several downstream targets for PI3Ks have been detected, and many biological effects of PI3Ks are mediated through the activation of the downstream target AKT/protein kinase B (PKB).[62]

At present, three members of the AKT family have been identified and are termed AKT1, AKT2, and AKT3. They are closely related, exhibiting more than 80% sequence homology, although they are products of different genes.[63] After exposure to hormones, growth factors, or cytokines, inactive (cytosolic) AKT is translocated to the plasma membrane by the products of PI3K, PIP_2, and PIP_3, where it is phosphorylated and activated by phosphoinositide-dependent kinase 1 (PDK1).

Many AKT substrates have been identified such as BAD, CREB, procaspase 9, $p21^{WAF1}$, members of the forkhead family of transcriptions factor, Iκ-B kinase, GSK-3- kinase, and mTOR/FRAP.[64] The large number of downstream targets of AKT explains why this kinase is rapidly emerging as a key mediator of cell proliferation, differentiation, and survival, and an inhibitor of apoptosis (Fig. 5). AKT stimulates many anti-apoptotic proteins like NFκB, and c-FLIP. In addition, it inhibits many proapoptotic proteins such as BAD and ASKI. AKT also

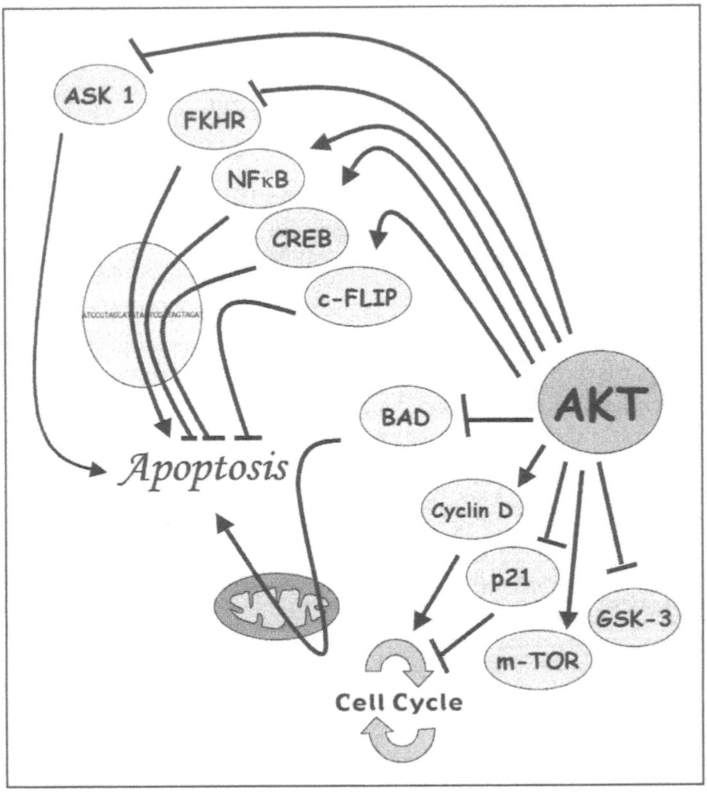

Figure 5. The schematic illustration of the large number of downstream targets of AKT and the roles that they play in cell proliferation, differentiation, and survival. AKT stimulates many anti-apoptotic proteins like NFκB, and c-FLIP. In addition, it inhibits many pro-apoptotic proteins such as BAD and ASKI. AKT also stimulates cyclin D, which is needed for the cell cycle and inhibits p21, the cyclin-dependent kinase inhibitor that blocks the cell cycle. The overall effects of AKT are to antagonize apoptosis and stimulate cell survival.

stimulates cyclin D, which is needed for the cell cycle and inhibits p21, the cyclin-dependent kinase inhibitor that blocks the cell cycle. The overall effects of AKT are to antagonize apoptosis and stimulate cell survival. Because both the MAPK (ERK) pathway and the PI3/AKT pathway play important roles in cell survival, cross talks occur between these two pathways at different levels (Fig. 6).

Other Modes of Cell Death

Necrosis

Necrosis represents another form of cell death, which occurs as a consequence of a cell exposure to toxic stimuli such as hyperthermia, hypoxia, ischemia, complement attack, metabolic poisons, and direct cell trauma. Necrosis is characterized by irreversible swelling of the cytoplasm and organelles, including the mitochondria, followed by loss of membrane integrity, resulting in cell lysing and release of noxious cellular constituents.[65] In contrast to apoptosis that can occur in a single cell surrounded by a group of viable cells, necrosis simultaneously involves a group of cells

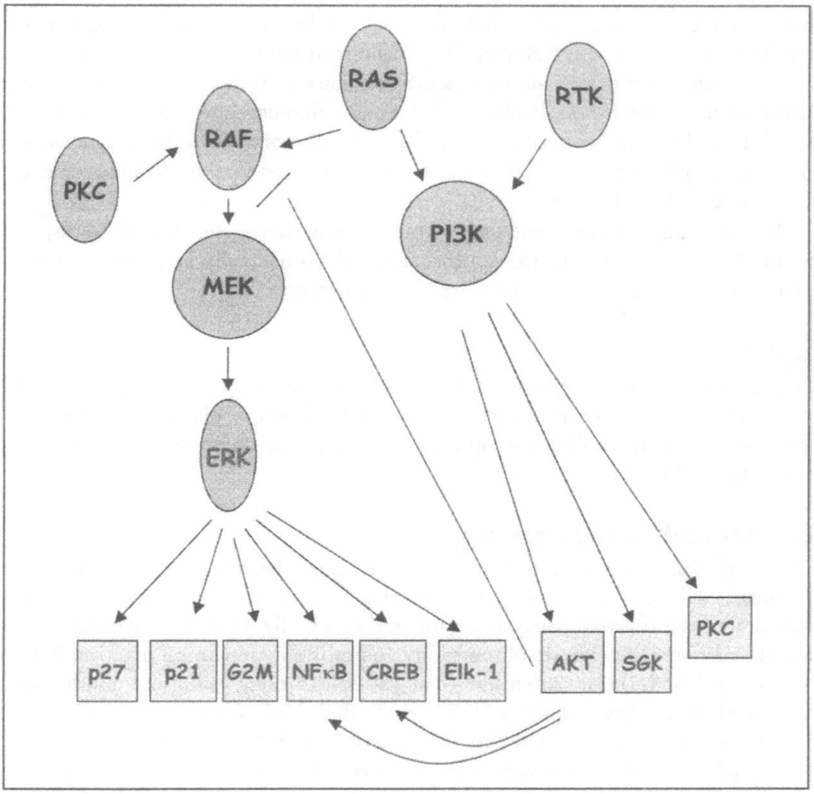

Figure 6. The possible cross talk between the MAPK (ERK) pathway and the PI3/AKT pathway at different levels. RAS, the upstream activator of the RAF/MEK/ERK pathway, activates PI3K, which stimulates AKT. AKT can inhibit RAF activity. CREB and NFκB are downstream targets for both pathways.

in which the necrotic cell ruptures and releases chemical mediators that induce inflammatory reactions. However, under some pathological conditions, both necrosis and apoptosis may occur concurrently. For example, ischemic injury is frequently characterized by damaged necrotic cells surrounded by cells that undergo delayed apoptotic death.[66] The severity and nature of the stimulus may determine if the cell will die by necrosis or apoptosis.[67]

Senescense

Senescense is another mechanism of cell death. Normal human cells in culture, after dividing for about 60 populations, undergo growth arrest known as replicative senescence in which the cells remain metabolically active, but lose permanently their proliferating activities.[68] Senescense is characterized by a cell enlargement and elevated expression of a pH-dependent galactosidase activity, the cyclin-dependent kinases $p16^{Ink4a}$, $p21^{Cip1/Waf1}$, and hypophosphorylated Rb.[69] The exact underlying mechanisms of senescence are not known yet. Progressive telomere shortening, which occurs at every cell division, is a proposed mechanism for the induction of replicative senescence. When telomeres reach the critical length of less than 5 kb, the Rb and p53 pathways become activated by a mechanism that is not yet well understood. This activation triggers irreversible growth arrest.[70]

Sublethal stress may induce a state that closely resembles replicative senescence.[71] Stress-induced senescence may occur following H_2O_2 or hyperoxia-induced oxidative stress,[72,73] UV- or irradiation-induced DNA damage,[71,74] and treatment with histone deacetylase inhibitors.[75] The mechanisms of stress-induced senescence are very similar to replicative senescence,[76] but neither is fully understood. In chronic hyperoxia, telomere shortening occurs 5 to 10 times faster.[77,78] In addition, it has been reported that induction of stress-induced senescence is triggered when the p53-dependent cell cycle is arrested via generation of nonspecific as well as telomerespecific DNA damage.[71]

It has been suggested that senescence represents a barrier against tumorgenesis,[79] and that an essential step in the malignant transformation of normal cells, comes when they acquire ability to proliferate an unlimited number of times (immortalization).

Paraptosis

Paraptosis, another mechanism of cell death, is a nonapoptotic programmed cell death with somewhat different cellular features, characterized by cytoplasmic vacuolation without nuclear fragmentation. Similar to apoptosis, it appears to be an active process requiring transcription and protein synthesis.[80]

Apoptosis and Tissue Scarring

Although scar formation is part of the normal wound-healing process, excessive scarring is the hallmark of fibrotic disease. Extensive scarring from chronic injury severely compromises the function of the affected tissues or organs, and eventually becomes irreversible. Irreversible fibrotic diseases are major causes of morbidity and mortality in clinical practice. For instance, cirrhosis of the liver is the seventh leading cause of death among young and middle-aged adults in the United States. Approximately 10,000 to 24,000 deaths from cirrhosis may be attributable to alcohol consumption each year. This number goes higher when other causes of liver cirrhosis, such as hepatitis, are taken into consideration. Alcohol-induced cell death and inflammation can initiate the scarring process that distorts structural integrity of the liver and impairs its functional abilities.

Similar to other fibrotic diseases, the complex interaction of certain cytokines, chemokines, and growth factors with resident hepatic cells (i.e., stellate cells) could initiate and propagate the disease process, which eventually lead to advanced stages of cirrhosis. Upon activation by cytokines and other factors, stellate cells start to proliferate, and begin to produce excessive amounts of matrix proteins, which are the main components of scar tissue. In addition to hepatic stellate cell activation, proliferation, differentiation and matrix synthesis,[81] increased rates of apoptotic cell deletion have been documented during hepatic fibrogenesis.[82,83] Hepatic stellate cell-secreted TGFβ1 in alcoholic liver disease might have a dual impact on the hepatic injury, possibly by promoting fibrogenesis and eliminating hepatocytes by apoptosis.[84]

In a similar study, experimental ligation of the common bile duct initiated inflammatory events, proliferation of ductal epithelial cells and periportal fibrosis but release of such obstruction resulted in reversal of the fibrotic process, presumably due to apoptotic deletion of matrix-producing cells and matrix degradation.[85] Theoretically, the inhibition or inactivation of hepatic stellate cells in the early fibroproliferative phase could serve as a basis for developing new therapeutic strategies to control the progression of fibrotic liver diseases; such a therapeutic approach might have clinical impact in the general treatment of fibrotic diseases. Although this hypothetical concept is provocative, it still needs to be proven.

Recent studies have documented that selective induction of an increased rate of apoptosis of lung cells, by activating the proapoptotic system in mice, could develop pulmonary fibrosis.[86] Suppressing such apoptosis by caspase inhibitors attenuated the fibrotic process in the lung,[87] suggesting that selective modulation of apoptosis has therapeutic implications in fibrotic diseases. Similarly, a recent study with a rat glomerulonephritis model showed that inducing apoptosis of the mesangial cells could attenuate the matrix accumulation in the glomeruli.[88]

The pathogenesis of fibrotic disorders is mostly similar regardless of the tissues or organs involved.[89-95] Early phases of fibrosis show inflammatory changes such as inflammatory infiltrates and proliferation of matrix-producing cells. In later stages, most of these infiltrating and proliferating cells are cleared (possibly by apoptosis), leaving mostly an acellular fibrotic mass. The microenvironment influences the intracellular signaling cascade. The altered microenvironment during fibrosis might facilitate the apoptotic removal of cellular components, by either influencing apoptotic stimuli or by immobilizing the survival state of the surrounding and neighboring cells.[96] Such cytokines and growth factors as platelet-derived growth factor (PDGF), transforming growth factor (TGF-β) and fibroblast growth factor (FGF) have both profibrogenic and proapoptotic effects.[97-99] PDGF, TGF-β_1 and bFGF not only have proliferating and migrating effects on both fibroblasts and myofibroblasts, but also induce excessive matrix production.[100-103]

For instance, PDGF is a potent mitogen for various types of cells including vascular smooth muscle cells. Interestingly, PDGF-BB has the ability to induce apoptosis in the vascular smooth muscle cells, while PDGF-AA could prevent such apoptosis.[103] Similarly, a cyclic peptide analog of PDGF-BB could induce apoptotic cell death in growing cardiac fibroblasts.[104] In contrast to the vascular smooth muscle cells, normal rat kidney fibroblasts undergo apoptosis, when stimulated by either PDGF-AA or PDGF-BB homodimers.[105] Similarly, bFGF-induced apoptosis was much higher in fibroblasts isolated from the rat palatal scar tissue than normal palatal fibroblasts.[106] TGF-β1 is able to induce apoptosis in a wide range of cells, such as microglia.[107]

Although knowledge about the contribution of apoptosis to fibrotic diseases is expanding, the exact pathways leading to the development of fibrosis remain to be elucidated. The slow and progressive nature of the disease process in humans may explain why the direct role of apoptosis and its initiating events during fibrogenesis is not always conclusive. In some cases, it takes years to form a fibroproliferative mass into an acellular scar tissue. In experimental models, where fibrotic lesions can be induced in relatively short period of time, it has been demonstrated that an increased rate of apoptosis is closely associated with fibrotic changes in the affected organs. For instance, in an experimental model of Thy 1.1 nephritis, glomerular hypercellularity and expansion of mesangial matrix is followed by resolution of glomerular hypercellularity, which is mediated by apoptotic deletion of mesangial cells.[108]

Although, apoptotic removal of proliferating mesangial cells and infiltrating inflammatory cells has a beneficial effect on the resolution of glomerular hypercellularity in anti-Thy1.1 nephritis, an injurious role for apoptosis has been documented in several human and experimental renal diseases. In nephrotoxic nephritis, an increased rate of apoptosis is related to tubular atrophy and subsequent interstitial fibrosis.[109] The degree of apoptosis was not only associated with renal scarring, but also positively correlated with levels of serum creatinine and proteinuria in the remnant kidney model.[110,111] The types of cells undergoing apoptosis during renal scarring varied with different phases of the disease process. For instance, apoptosis of such infiltrating inflammatory cells as neutrophils and macrophages contributed to the resolution of relatively early inflammatory events,[112 113] but apoptotic deletion of resident and transforming cells was noted in the later stages of the disease process, and led to interstitial fibrosis. Using a nephrotoxic nephritis model, El Nahas and colleagues have shown two distinct peaks of apoptosis on day 7 and at 4 to 6 weeks.[109] The first peak of apoptosis was mostly inflammatory infiltrates of glomeruli, while the second peak was due to removal of tubulointerstitial cells, resulting in tubular atrophy and interstitial fibrosis. Association between apoptotic cell death and accumulation of matrix proteins was found in age-associated renal scarring in Fischer

344 rats.[114,115] Similar induction of apoptosis in the kidney has been documented in hypertensive nephrosclerosis of Dahl rat, which was associated with augmented expression of the pro-apoptotic molecules Fas, Bax, and Bcl-X_S.[116]

Conclusions

Fibrosis is a complex pathological process that involves prolonged inflammation, proliferation of matrix-producing cells, increased deposition of matrix proteins, and eventual tissue remodeling.[117-119] As the fibrotic tissue matures, a striking decrease in cellularity occurs, due to apoptotic removal of cellular components, forming mostly an acellular fibrotic mass.[120-122] Given the dire consequences of scarring diseases, to find more effective therapies to reverse or prevent the damage related to widespread tissue scarring is an active field of study. The search continues to develop novel treatments that can modulate the scarring response to control healing, but the complexity of the scarring process makes it difficult to develop such magical therapy. Will the advances in molecular biology open doors to new magical therapies? The molecular knowledge of nonmammalian systems allows us to selectively transfect human cells to target a specific cell population to modulate their undesirable functions. Targeted removal of the unwanted cells, or rescuing essential cells from dying, might move us closer to develop site- and phase-specific therapeutic strategies to treat mostly irreversible scarring diseases. Finally, further research is needed to understand how normal cells may switch to fibrotic phenotypes in response to stimuli received from the extracellular environment. For example diverse stimuli that induce inflammatory and fibrotic responses generate such complex intracellular molecular events as synthesis of second messengers, initiation of phosphorylation cascades, and protein-protein interactions. Understanding the precise nature of such intracellular events is pivotal for the development of effective strategies of artificial manipulation to alter, amplify, or prevent cellular responses, so as to control the progression of fibrosis.

Acknowledgements

Our apology goes to the authors whose primary works could not be cited due to limitation of space. Part of this chapter is modified from the review article entitled "Role of apoptosis in fibrogenesis", by Razzaque et al and published in Nephron 2002; 90:365-372. We appreciate Jeena Ahmed and Joan Merriman's help in language editing.

References

1. Desmouliere A, Redard M, Darby I et al. Apoptosis mediates the decrease in cellularity during the transition between granulation tissue and scar. Am J Pathol 1995; 146:56-66.
2. Rockwell WB, Cohen IK, Ehrlich HP. Keloids and hypertrophic scars: A comprehensive review. Plast Reconstr Surg 1989; 84:827-837.
3. Kockx MM, Herman AG: Apoptosis in atherosclerosis: Beneficial or detrimental? Cardiovasc Res 2000; 45:736-746.
4. Kerr JF, Wyllie AH, Currie AR. Apoptosis: A basic biological phenomenon with wide-ranging implications in tissue kinetics. Br J Cancer 1972; 26:239-257.
5. Nunez G, Benedict MA, Hu Y et al. Caspases: The proteases of the apoptotic pathway. Oncogene 1998; 17:3237-3245.
6. Hengartner MO. The biochemistry of apoptosis. Nature 2000 ; 407:770-776.
7. Nagata S. Apoptotic DNA fragmentation. Exp Cell Res 2000; 256:12-18.
8. Stegh AH, Herrmann H, Lampel S et al. Identification of the cytolinker plectin as a major early in vivo substrate for caspase 8 during. Mol Cell Biol 2000; 20:5665-5679.
9. Saleh A, Srinivasula SM, Acharya S et al. Cytochrome c and dATP-mediated oligomerization of Apaf-1 is a prerequisite for procaspase-9 activation. J Biol Chem 1999; 274:17941-17945.
10. Fadeel B, Orrenius S, Zhivotovsky B. The most unkindest cut of all: On the multiple roles of mammalian caspases. Leukemia 2000; 14:1514-1525.

11. Stennicke HR, Jurgensmeier JM, Shin H et al. Pro-caspase-3 is a major physiologic target of caspase-8. J Biol Chem 1998; 273:27084-27090.
12. Barry M, Heibein JA, Pinkoski MJ et al. Granzyme B short-circuits the need for caspase 8 activity during granule-mediated cytotoxic T-lymphocyte killing by directly cleaving Bid. Mol Cell Biol 2000; 20:3781-3794.
13. Heibein JA, Goping IS, Barry M et al. Granzyme B-mediated cytochrome c release is regulated by the Bcl-2 family members bid and Bax. J Exp Med 2000; 192:1391-1402.
14. Alimonti JB, Shi L, Baijal PK et al. Granzyme B induces BID-mediated cytochrome c release and mitochondrial permeability transition. J Biol Chem 2001; 276:6974-6982.
15. Joza N, Susin SA, Daugas E et al. Essential role of the mitochondrial apoptosis-inducing factor in programmed cell death. Nature 2001; 410:549-554.
16. Susin SA, Lorenzo HK, Zamzami N et al. Molecular characterization of mitochondrial apoptosis-inducing factor. Nature 1999; 397:441-446.
17. Ferri KF, Kroemer G. Mitochondria-the suicide organelles. Bioessays 2001; 23:111-115.
18. Costantini P, Jacotot E, Decaudin D et al. Mitochondrion as a novel target of anticancer chemotherapy. J Natl Cancer Inst 2000; 92:1042-1053.
19. Liu X, Kim CN, Yang J et al. Induction of apoptotic program in cell-free extracts: Requirement for dATP and cytochrome c. Cell 1996; 86:147-157.
20. Slee EA, Harte MT, Kluck RM et al. Ordering the cytochrome c-initiated caspase cascade: Hierarchical activation of caspases-2, -3, -6, -7, -8, and -10 in a caspase-9-dependent manner. J Cell Biol 1999; 144:281-292.
21. Zou H, Henzel WJ, Liu X et al. Apaf-1, a human protein homologous to C. elegans CED-4, participates in cytochrome c-dependent activation of caspase-3. Cell 1997; 90:405-413.
22. Von Ahsen O, Waterhouse NJ, Kuwana T et al. The 'harmless' release of cytochrome c. Cell Death Differ 2000; 7:1192-1199.
23. Martinou JC, Desagher S, Antonsson B. Cytochrome c release from mitochondria: All or nothing. Nature Cell Biol 2000; 2:E41-E43.
24. Marzo I, Brenner C, Zamzami N et al. Bax and adenine nucleotide translocator cooperate in the mitochondrial control of apoptosis. Science 1998; 281:2027-2031.
25. Shimizu S, Narita M, Tsujimoto Y. Bcl-2 family proteins regulate the release of apoptogenic cytochrome c by the mitochondrial channel VDAC. Nature 1999; 399:483-487.
26. Verhagen AM, Ekert PG, Pakusch M et al. Identification of DIABLO, a mammalian protein that promotes apoptosis by binding to and antagonizing IAP proteins. Cell 2000; 102:43-53.
27. Du C, Fang M, Li Y et al. Smac, a mitochondrial protein that promotes cytochrome c-dependent caspase activation by eliminating IAP inhibition. Cell 2000; 102:33-42.
28. Goldstein JC, Waterhouse NJ, Juin P et al. The coordinate release of cytochrome c during apoptosis is rapid, complete and kinetically invariant. Nature Cell Biol 2000; 2:156-162.
29. Silke J, Verhagen AM, Ekert PG et al. Sequence as well as functional similarity for DIABLO/Smac and Grim, Reaper and Hid? Cell Death Differ 2000; 7:1275.
30. Vander Heiden MG, Thompson CB. Bcl-2 proteins: Regulators of apoptosis or of mitochondrial homeostasis? Nature Cell Biol 1999; 1:E209-E216.
31. Chang HY, Yang X, Baltimore D. Dissecting Fas signaling with an altered-specificity death-domain mutant: Requirement of FADD binding for apoptosis but not Jun N-terminal kinase activation. Proc Natl Acad Sci USA 1999; 96:1252-1256.
32. Berglund H, Olerenshaw D, Sankar A et al. The three-dimensional solution structure and dynamic properties of the human FADD death domain. J Mol Biol 2000; 302:171-188.
33. Buschmann T, Potapova O, Bar-Shira A et al. Jun NH2-terminal kinase phosphorylation of p53 on Thr-81 is important for p53 stabilization and transcriptional activities in response to stress. Mol Cell Biol 2001; 21:2743-2754.
34. Zamzami N, El Hamel C, Maisse C et al. Bid acts on the permeability transition pore complex to induce apoptosis. Oncogene 2000; 19:6342-6350.
35. Tang D, Lahti JM, Kidd VJ. Caspase-8 activation and bid cleavage contribute to MCF7 cellular execution in a caspase-3-dependent manner during staurosporine-mediated apoptosis. J Biol Chem 2000; 275:9303-9307.

36. Zhang H, Xu Q, Krajewski S et al. BAR: An apoptosis regulator at the intersection of caspases and Bcl-2 family proteins. Proc Natl Acad Sci USA. 2000; 97:2597-2602.
37. Boise LH, Gonzalez-Garcia M, Postema CE et al. bcl-x, a bcl-2-related gene that functions as a dominant regulator of apoptotic cell death. Cell 1993; 74:597-608.
38. Craig RW. The bcl-2 gene family. Semin Cancer Biol 1995; 6:35-43.
39. Molica S, Dattilo A, Giulino C et al. Increased bcl-2/bax ratio in B-cell chronic lymphocytic leukemia is associated with a progressive pattern of disease. Haematologica 1998; 83:1122-1124.
40. Karakas T, Maurer U, Weidmann E et al. High expression of bcl-2 mRNA as a determinant of poor prognosis in acute myeloid leukemia. Ann Oncol 1998; 9:159-165.
41. Konopleva M, Tari AM, Estrov Z et al. Liposomal Bcl-2 antisense oligonucleotides enhance proliferation, sensitize acute myeloid leukemia to cytosine-arabinoside, and induce apoptosis independent of other antiapoptotic proteins. Blood 2000; 95:3929-3938.
42. Oltvai ZN, Milliman CL, Korsmeyer SJ. Bcl-2 heterodimerizes in vivo with a conserved homolog, Bax, that accelerates programmed cell death. Cell 1993; 74:609-619.
43. Chittenden T, Flemington C, Houghton AB et al. A conserved domain in Bak, distinct from BH1 and BH2, mediates cell death and protein binding functions. EMBO J 1995; 14:5589-5596.
44. Shitashige M, Toi M, Yano T et al. Dissociation of Bax from a Bcl-2/Bax heterodimer triggered by phosphorylation of serine 70 of Bcl-2. J Biochem (Tokyo) 2001; 130:741-748.
45. Wang LG, Liu XM, Kreis W et al. The effect of antimicrotubule agents on signal transduction pathways of apoptosis: A review. Cancer Chemother Pharmacol 1999; 44:355-361.
46. Cobb MH. MAP kinase pathways. Prog Biophys Mol Biol 1999; 71:479-500.
47. McCawley LJ, Li S, Wattenberg EV et al. Sustained activation of the mitogen-activated protein kinase pathway. A mechanism underlying receptor tyrosine kinase specificity for matrix metalloproteinase-9 induction and cell migration. J Biol Chem 1999; 274:4347-4353.
48. Foey AD, Parry SL, Williams LM et al. Regulation of monocyte IL-10 synthesis by endogenous IL-1 and TNF-alpha: Role of the p38 and p42/44 mitogen-activated protein kinases. J Immunol 1998; 160:920-928.
49. Persons DL, Yazlovitskaya EM, Pelling JC. Effect of extracellular signal-regulated kinase on p53 accumulation in response to cisplatin. J Biol Chem 2000; 275:35778-35785.
50. Bueno OF, De Windt LJ, Tymitz KM et al. The MEK1-ERK1/2 signaling pathway promotes compensated cardiac hypertrophy in transgenic mice. EMBO J 2000; 19:6341-6350.
51. Remacle-Bonnet MM, Garrouste FL, Heller S et al. Insulin-like growth factor-I protects colon cancer cells from death factor-induced apoptosis by potentiating tumor necrosis factor alpha-induced mitogen-activated protein kinase and nuclear factor kappaB signaling pathways. Cancer Res 2000; 60:2007-2017.
52. Bhat NR, Zhang P. Hydrogen peroxide activation of multiple mitogen-activated protein kinases in an oligodendrocyte cell line: Role of extracellular signal-regulated kinase in hydrogen peroxide-induced cell death. J Neurochem 1999; 72:112-119.
53. Paumelle R, Tulasne D, Leroy C et al. Sequential activation of ERK and repression of JNK by scatter factor/hepatocyte growth factor in madin-darby canine kidney epithelial cells. Mol Biol Cell 2000; 11:3751-3763.
54. Ichijo H, Nishida E, Irie K et al. Induction of apoptosis by ASK1, a mammalian MAPKKK that activates SAPK/JNK and p38 signaling pathways. Science 1997; 275:90-94.
55. Kyriakis JM, Avruch J. Mammalian mitogen-activated protein kinase signal transduction pathways activated by stress and inflammation. Physiol Rev 2001; 81:807-869.
56. Ip YT, Davis RJ. Signal transduction by the c-Jun N-terminal kinase (JNK)—from inflammation to development. Curr Opin Cell Biol 1998; 10:205-219.
57. Widmann C, Gibson S, Jarpe MB et al. Mitogen-activated protein kinase: Conservation of a three-kinase module from yeast to human. Physiol Rev 1999; 79:143-180.
58. Penuel E, Martin GS. Transformation by v-Src: Ras-MAPK and PI3K-mTOR mediate parallel pathways. Mol Biol Cell 1999; 10:1693-1703.
59. Vanhaesebroeck B, Waterfield MD. Signaling by distinct classes of phosphoinositide 3-kinases. Exp Cell Res 1999; 253:239-254.
60. Cantrell DA. Phosphoinositide 3-kinase signalling pathways. J Cell Sci 2001; 114:1439-1445.

61. Roymans D, Slegers H. Phosphatidylinositol 3-kinases in tumor progression. Eur J Biochem 2001; 268:487-498.
62. Brazil DP, Hemmings BA. Ten years of protein kinase B signalling: A hard Akt to follow. Trends Biochem Sci 2001; 26:657-664.
63. Lawlor MA, Alessi DR. PKB/Akt: A key mediator of cell proliferation, survival and insulin responses? J Cell Sci 2001; 114:2903-2910.
64. Nicholson KM, Anderson NG. The protein kinase B/Akt signalling pathway in human malignancy. Cell Signal 2002; 14:381-395.
65. Wyllie AH, Kerr JF, Currie AR. Cell death: The significance of apoptosis. Int Rev Cytol 1980; 68:251-306.
66. Schulz JB, Weller M, Moskowitz MA. Caspases as treatment targets in stroke and neurodegenerative diseases. Ann Neurol 1999; 45:421-429.
67. Bonfoco E, Krainc D, Ankarcrona M et al. Apoptosis and necrosis: Two distinct events induced, respectively, by mild and intense insults with N-methyl-D-aspartate or nitric oxide/superoxide in cortical cell cultures. Proc Natl Acad Sci USA 1995; 92:7162-7166.
68. Campisi J. Replicative senescence: An old lives' tale? Cell 1996; 84:497-500.
69. Dimri GP, Lee X, Basile G et al. A biomarker that identifies senescent human cells in culture and in aging skin in vivo. Proc Natl Acad Sci USA 1995; 92:9363-9367.
70. von Zglinicki T, Burkle A, Kirkwood TB. Stress, DNA damage and ageing-an integrative approach. Exp Gerontol 2001; 36:1049-1062.
71. Toussaint O, Medrano EE, von Zglinicki T. Cellular and molecular mechanisms of stress-induced premature senescence (SIPS) of human diploid fibroblasts and melanocytes. Exp Gerontol 2000; 35:927-945.
72. Chen QM, Bartholomew JC, Campisi J et al. Molecular analysis of H2O2-induced senescent-like growth arrest in normal human fibroblasts: p53 and Rb control G1 arrest but not cell replication. Biochem J 1998; 332(Pt 1):43-50.
73. Dumont P, Burton M, Chen QM et al. Induction of replicative senescence biomarkers by sublethal oxidative stresses in normal human fibroblast. Free Radic Biol Med 2000; 28:361-373.
74. Oh CW, Bump EA, Kim JS et al. Induction of a senescence-like phenotype in bovine aortic endothelial cells by ionizing radiation. Radiat Res 2001; 156:232-240.
75. Ogryzko VV, Hirai TH, Russanova VR et al. Human fibroblast commitment to a senescence-like state in response to histone deacetylase inhibitors is cell cycle dependent. Mol Cell Biol 1996; 16:5210-5218.
76. Saretzki G, Feng J, von Zglinicki T et al. Similar gene expression pattern in senescent and hyperoxic-treated fibroblasts. J Gerontol. A Biol Sci Med Sci 1998; 53:B438-B442.
77. von Zglinicki T, Saretzki G, Docke W et al. Mild hyperoxia shortens telomeres and inhibits proliferation of fibroblasts: A model for senescence? Exp Cell Res 1995; 220:186-193.
78. Vaziri H, West MD, Allsopp RC et al. ATM-dependent telomere loss in aging human diploid fibroblasts and DNA damage lead to the post-translational activation of p53 protein involving poly(ADP-ribose) polymerase. EMBO J 1997; 16:6018-6033.
79. Sager R. Senescence as a mode of tumor suppression. Environ Health Perspect 1991; 93:59-62.
80. Sperandio S, de BI, Bredesen DE. An alternative, nonapoptotic form of programmed cell death. Proc Natl Acad Sci USA 2000; 97:14376-14381.
81. Gressner AM, Bachem MG. Cellular communications and cell-matrix interactions in the pathogenesis of fibroproliferative diseases: Liver fibrosis as a paradigm. Ann Biol Clin (Paris) 1994; 52:205-26.
82. Desmouliere A, Badid C, Bochaton-Piallat ML et al. Apoptosis during wound healing, fibrocontractive diseases and vascular wall injury. Int J Biochem Cell Biol 1997; 29:19-30.
83. Gressner AM. The cell biology of liver fibrogenesis - an imbalance of proliferation, growth arrest and apoptosis of myofibroblasts. Cell Tissue Res 1998; 292:447-452.
84. Rust C, Gores GJ. Apoptosis and liver disease. Am J Med 2000; 108:567-74.
85. Abdel-Aziz G, Lebeau G, Rescan PY et al. Reversibility of hepatic fibrosis in experimentally induced cholestasis in rat. Am J Pathol 1990; 137:1333-1342.
86. Hagimoto N, Kuwano K, Miyazaki H et al. Induction of apoptosis and pulmonary fibrosis in mice in response to ligation of Fas antigen. Am J Respir Cell Mol Biol 1997; 17:272-8.

87. Wang R, Ibarra-Sunga O, Verlinski L et al. Abrogation of bleomycin-induced epithelial apoptosis and lung fibrosis by captopril or by a caspase inhibitor. Am J Physiol Lung Cell Mol Physiol 2000; 279:L143-51.
88. Daniel C, Duffield J, Brunner T et al. Matrix metalloproteinase inhibitors cause cell cycle arrest and apoptosis in glomerular mesangial cells. J Pharmacol Exp Ther 2001; 297:57-68.
89. Razzaque MS, Foster CS, Ahmed AR. Role of collagen-binding heat shock protein 47 and transforming growth factor beta 1 in conjunctival scarring in ocular cicatritial pemphigoid. Invest Ophthalmol Vis Sci 2003; 44:1616-1621.
90. Marshall RP, McAnulty RJ, Laurent GJ. The pathogenesis of pulmonary fibrosis: Is there a fibrosis gene? Int J Biochem Cell Biol 1997; 129:107-120.
91. Razzaque MS, Nazneen A, Taguchi T. Immunolocalization of collagen and collagen-binding heat shock protein 47 in fibrotic lung diseases. Modern Pathol 1998; 11:1183-1188.
92. Mutsaers SE, Bishop JE, McGrouther G et al. Mechanisms of tissue repair: From wound healing to fibrosis. Int J Biochem Cell Biol 1997; 29:5-17.
93. Desmouliere A, Gabbiani G. Myofibroblast differentiation during fibrosis. Exp Nephrol 1995; 3:134-139.
94. Razzaque MS, Taguchi T. Factors that influence and contribute to the regulation of fibrosis. Contrib Nephrol 2003; 139:1-11.
95. Phan SH, Zhang K, Zhang HY et al. The myofibroblast as an inflammatory cell in pulmonary fibrosis. Curr Top Pathol 1999; 93:173-182.
96. Ruoslahti E, Reed JC. Anchorage dependence, integrins, and apoptosis. Cell 1994; 177:477-478.
97. Betsholtz C, Raines EW. Platelet-derived growth factor: A key regulator of connective tissue cells in embryogenesis and pathogenesis. Kidney Int 1997; 51:1361-1369.
98. Gauldie J, Sime PJ, Xing Z et al. Transforming growth factor-beta gene transfer to the lung induces myofibroblast presence and pulmonary fibrosis. Curr Top Pathol 1999; 93:35-45.
99. Zhang K, Phan SH. Cytokines and pulmonary fibrosis. Biol Signals 1996; 5:232-239.
100. Hishikawa K, Nakaki T, Fujii T. Transforming growth factor-beta (1) induces apoptosis via connective tissue growth factor in human aortic smooth muscle cells. Eur J Pharmacol 1999; 385:287-290.
101. Coleman AB, Momand J, Kane SE. Basic fibroblast growth factor sensitizes NIH 3T3 cells to apoptosis induced by cisplatin. Mol Pharmacol 2000; 57:324-333.
102. Inman GJ, Allday MJ. Apoptosis induced by TGF-beta1 in Burkitt's lymphoma cells is caspase 8 dependent but is death receptor independent. J Immunol 2000; 165:2500-2510.
103. Okura T, Igase M, Kitami Y et al. Platelet-derived growth factor induces apoptosis in vascular smooth muscle cells: Roles of the Bcl-2 family. Biochim Biophys Acta 1998; 1403:245-253.
104. Brennand DM, Scully MF, Kakkar VV et al. A cyclic peptide analogue of loop III of PDGF-BB causes apoptosis in human fibroblasts. FEBS Lett 1997; 419:166-170.
105. Kim HR, Upadhyay S, Li G et al. Platelet-derived growth factor induces apoptosis in growth-arrested murine fibroblasts. Proc Natl Acad Sci USA 1995; 92:9500-9504.
106. Funato N, Moriyama K, Shimokawa H et al. Basic fibroblast growth factor induces apoptosis in myofibroblastic cells isolated from rat palatal mucosa. Biochem Biophys Res Commun 1997; 240:21-26.
107. Xiao BG, Bai XF, Zhang GX et al. Transforming growth factor-beta1 induces apoptosis of rat microglia without relation to bcl-2 oncoprotein expression. Neurosci Lett 1997; 226:71-74.
108. Shimizu A, Kitamura H, Masuda Y et al. Glomerular capillary regeneration and endothelial cell apoptosis in both reversible and progressive models of glomerulonephritis. Contrib Nephrol 1996; 118:29-40.
109. Yang B, Johnson TS, Thomas GL et al. Apoptosis and caspase-3 in experimental anti-glomerular basement membrane nephritis. J Am Soc Nephrol 2001; 12:485-495.
110. Kitamura H, Shimizu A, Masuda Y et al. Apoptosis in glomerular endothelial cells during the development of glomerulosclerosis in the remnant-kidney model. Exp Nephrol 1998; 6:328-336.
111. Yang B, Johnson TS, Thomas GL et al. Expression of apoptosis-related genes and proteins in experimental chronic renal scarring. J Am Soc Nephrol 2001; 12:275-288.
112. Lan HY, Mitsuhashi H, Ng YY et al. Macrophage apoptosis in rat crescentic glomerulonephritis. Am J Pathol 1997; 151:531-538.

113. Savill J, Smith J, Sarraf C et al. Glomerular mesangial cells and inflammatory macrophages ingest neutrophils undergoing apoptosis. Kidney Int 1992; 42:924-936.
114. Razzaque MS, Shimokawa I, Nazneen A et al. Age-related nephropathy in the Fischer 344 rat is associated with overexpression of collagens and collagen-binding heat shock protein 47. Cell Tissue Res 1998; 293:471-478.
115. Razzaque MS, Shimokawa I, Koji T et al. Life-long caloric restriction suppresses age-associated Fas expression in the Fischer 344 rat kidney. Mol Cell Biol Res Commun 1999; 1:82-85.
116. Ying WZ, Wang PX, Sanders PW. Induction of apoptosis during development of hypertensive nephrosclerosis. Kidney Int 2001; 58:2007-2017.
117. Razzaque MS, Taguchi T. Pulmonary fibrosis: Cellular and molecular events. Pathol Int 2003; 53:133-145.
118. Razzaque MS, Taguchi T. Cellular and molecular events leading to renal tubulointerstitial fibrosis. Med Electron Microsc 2002; 35:68-80
119. Razzaque MS, Taguchi T. The possible role of colligin/HSP47, a collagen-binding protein, in the pathogenesis of human and experimental fibrotic diseases. Histol Histopathol 1999; 14:1199-1212.
120. Uhal BD, Gidea C, Bargout R et al. Captopril inhibits apoptosis in human lung epithelial cells: A potential antifibrotic mechanism. Am J Physiol 1998; 275:L1013-1017.
121. Thomas GL, Yang B, Wagner BE et al. Cellular apoptosis and proliferation in experimental renal fibrosis. Nephrol Dial Transplant 1998; 13:2216-2226.
122. Thomas SE, Andoh TF, Pichler RH et al. Accelerated apoptosis characterizes cyclosporine-associated interstitial fibrosis. Kidney Int 1998; 53:897-908.

CHAPTER 14

Pulmonary Fibrogenesis:

The Role of Apoptosis and Its Clinical Potentials

Kazuyoshi Kuwano, Naoki Hagimata and Nobuyuki Hara

Abstract

Pulmonary fibrosis is a common response to various insults or injuries to the lung. Although there are various initiating factors or causes, the terminal stages are characterized by proliferation and progressive accumulation of connective tissue replacing normal functional parenchyma. Conventional therapy consisting of glucocorticoids or immunosuppressive drugs is usually ineffective in preventing progression of the disease. Further understanding of the molecular mechanisms of endothelial and epithelial cell injury, inflammatory reaction, fibroblast proliferation, collagen deposition and tissue remodeling, should lead to the development of effective treatments against pulmonary fibrosis. An overview is presented regarding each of pathogenic events in pulmonary fibrosis, especially the role of apoptosis and its clinical potentials, which have emerged from the animal models and human tissue studies.

Introduction

Lung injury is believed to be due to inhalation of injurious agents or to blood- borne agents. Both acute and chronic inflammation can lead to an irreversible process characterized by pulmonary fibrosis. The term "idiopathic pulmonary fibrosis (IPF)" is used for those cases where there are no causative agents. The incidence of this devastating disease is estimated at 7 to 10 cases per 100,000 people per year, and its mortality is 50 to 70% at 5 years after the diagnosis.[1] Familial occurrence of IPF is well known. There have been studies into the association between genetic factors, such as HLA typing and gene polymorphism,[2,3] immunological abnormalities, viral infection, mineral dust, smoking and the development of IPF. The mechanisms by which these factors initiate or affect the development of IPF still need to be determined.

Alveolar epithelial damage is an important initial event in pulmonary fibrosis. Epithelial cell damage and cell death during alveolitis induce the formation of gaps in the epithelial basement membranes. The migration of fibroblasts through these gaps into the alveolar space facilitates intra-alveolar fibrosis.[4] Interstitial fibrosis and the subsequent relining of intra-alveolar fibrosis by alveolar and bronchiolar epithelial cells results in structural remodeling after lung injury. The fibrosing process is common to all interstitial lung diseases, including IPF, interstitial pneumonia associated with collagen vascular diseases, drug-induced pneumonitis, and sarcoidosis, as well as radiation pneumonitis, pneumoconiosis, asbestosis, and chronic hypersensitivity pneumonitis (Fig. 1).

Fibrogenesis: Cellular and Molecular Basis, edited by Mohammed S. Razzaque.

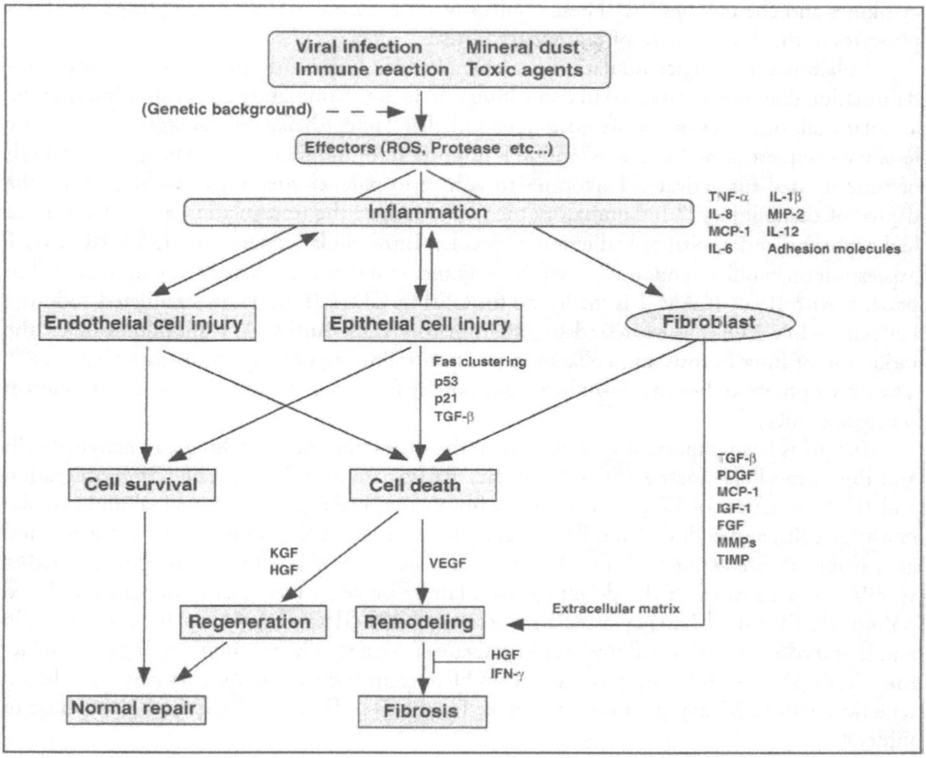

Figure 1. Molecular mechanisms of pulmonary fibrosis. Recent advances in our understanding of the molecular mechanisms of pulmonary fibrosis concerning parenchymal cell injury, inflammation, fibroblast proliferation and remodeling, result in the increasing recognition of its complexity. Various inflammatory mediators and death factors induce tissue damage, it is unlikely that a single treatment is effective in severe lung injury. Once parenchymal cells are damaged, to accelerate repair and regeneration of healing of tissue damage could be effective treatment to prevent irreversible fibrosis. Combination of effective treatments may overcome this devastating illness.

Inflammatory Mediators

Alveolar epithelial cells are damaged or destroyed in association with the accumulation of inflammatory cells and the presence of their mediators. Inflammatory cytokines, chemokines and growth factors produced by inflammatory and parenchymal cells have been involved not only in the recruitment of inflammatory cells into the alveolar walls and spaces, but also in the remodeling process through the stimulation of fibroblast proliferation and collagen synthesis.

Neutrophils are the earliest inflammatory cells to sequestrate to lung tissue in inflammatory lung diseases. Eosinophils produce toxic agents and proteases as well as neutrophils. The number of eosinophils is correlated with the severity of clinical symptoms, and with radiological and functional abnormalities in IPF.[5,6] Mast cell-deficient or neutrophil-depleted mice develop bleomycin-induced pulmonary fibrosis,[7,8] whereas nude mice or CD4 and CD8 T lymphocyte-depleted mice actually attenuate bleomycin-induced pulmonary fibrosis.[9,10] The CD28-deficient mice showed markedly impaired lung fibrosis after injection of bleomycin.[11] Alveolar macrophages isolated from patients with IPF and from animal models of pulmonary fibrosis produce arachidonic acid metabolites, reactive oxygen species (ROS), and inflammatory cytokines and chemokines.[12,13] These results suggest a central role for macrophages and lymphocytes in the development of pulmonary fibrosis.

Inflammatory cell accumulation is a characteristic feature of lung injury. However, inflammation does not always lead to tissue injury. In fact, inflammatory cell infiltration into the alveolar walls or spaces can occur without endothelial[14] or epithelial cell damage.[15,16] Not only leukocyte sequestration but also retention is important for inducing tissue damage. The length of time needed for activated leukocytes to adhere to parenchymal cells may determine the degree of tissue injury.[16] Inflammatory mediators induce the upregulation and activation of leukocyte and endothelial cell adhesion molecules. Intracellular adhesion molecule (ICAM)-1 expression on capillary endothelial cells in lung tissues has been reported to be upregulated in patients with IPF.[17] ICAM-1 is the ligand for CD11a/CD18 (LFA-1) on stimulated endothelial cells, while LFA-1 is expressed by inflammatory cells. Anti-CD11 antibodies block the induction of lung fibrosis and collagen deposition following bleomycin or silica challenge.[18] The development of bleomycin-induced pulmonary fibrosis is abrogated in soluble E-selectin transgenic mice.[19]

Integrins have important roles to play in the interaction between lung parenchymal cells and the extracellular matrix (ECM) and they participate in cellular proliferation, migration and the production of ECM. Leukocytes, fibroblasts, macrophages, bronchial and alveolar epithelial cells, and endothelial cells, express cell surface receptors, in particular integrins, and are capable of interacting with ECM.[20] Pretreatment of animals with anti-integrin antibodies specific for a member of the β_2 integrin subfamily prevents bleomycin- and silica-induced pulmonary fibrosis.[21] Latency-associated peptide (LAP)-β1 is a ligand for the integrin αvβ6 and that αvβ6-expressing cell lines could activate endogenous latent transforming growth factor (TGF)-β1. $\beta6^{-/-}$ mice are protected from bleomycin-induced pulmonary fibrosis. The interaction with αvβ6 appears to be important for the TGF-β1 activity required in response to injury.[22]

The chemokine superfamily is divided into two groups according to the position of the first two-cysteine residues. The C-X-C chemokines include interleukin-8 (IL-8) and macrophage inflammatory protein-2 (MIP-2), which are generally chemotactic for neutrophils but not for monocytes. The C-C chemokines include monocyte chemotactic protein-1 (MCP-1), -2, -3, MIP-1α, -1β, and RANTES (Regulated upon Activation in Normal T cells Expressed and Secreted), which are generally chemotactic for leukocytes other than neutrophils.[23] Increased levels of MCP-1 and MIP-1α protein in BALF from patients with IPF have been demonstrated compared to healthy volunteers.[24] Measuring MCP-1 concentrations in both BALF and serum may help to differentiate IPF from other interstitial lung diseases, and to predict the clinical outcome of IPF.[25] Immediately following intratracheal bleomycin instillation, early proinflammatory mediators such as IL-1, IL-6, and tumor necrosis factor (TNF)-α are present within the lung, which induces subsequent increases in MCP-1 and MIP-1α expression by alveolar macrophages and epithelial cells. Treatment of bleomycin-challenged mice with soluble TNF receptor (sTNFR) or anti-IL-6 antibodies significantly decreased MCP-1 and MIP-1α protein expression in the lung.[26] MIP-1α also modulates ICAM-1 expression, while MCP-1 stimulates dose-dependent expression of CD11b and CD11c on the surface of monocytes.[27,28] CCR2, MCP-1 receptor, knockout mice are protected from bleomycin-induced pulmonary fibrosis.[29] Neutralization of MCP-1 or MIP-1α protein with antibodies significantly attenuates pulmonary fibrosis in mice.[30]

IL-8 is a potent chemoattractant and activator of neutrophils. Although alveolar macrophages are the major source of IL-8 in lung tissues, many other cell types, also produce IL-8. IL-8 recruits and activates neutrophils in vitro and in vivo to release proteases and LTB_4. IL-8 levels are increased in the serum and BALF of patients with IPF, and are significantly correlated with the percentage of neutrophils in BALF.[31,32] Serum IL-8 levels are correlated significantly with impairment of lung function parameters and PaO_2.[31]

RANTES is expressed in macrophages, eosinophils and human airway epithelial cells. RANTES has been shown to attract monocytes, and T-lymphocytes, particularly CD45RO$^+$ T-cells, to activate eosinophils, and to contribute towards the progression of inflammatory lung diseases. BALF RANTES levels are elevated in patients with interstitial lung diseases compared with normal controls.[33] The expression of CCR1 mRNA is increased, parallel to the expression of its ligands, MIP-1α and RANTES, in bleomycin-induced pulmonary fibrosis. Furthermore, treatment with anti-CCR1 antibody significantly reduced the accumulation of inflammatory cells and collagen deposition, resulting in dramatic improvement of survival. These results suggest that CCR1 could be a novel target of therapy against pulmonary fibrosis.[34]

TNF-α has pleiotropic effects that overlap with IL-1β as regards inflammatory and immunologic responses. TNF-α affects leukocyte activation, adhesion molecule expression and endothelial cell activation. TNF-α is mitogenic and chemotactic for fibroblasts, whereas it inhibits collagen gene expression. TNF-α induces fibroblasts to produce PDGF, prostaglandin E_2 (PGE_2), collagenase, gelatinase, chemokines, granulocyte-macrophage colony-stimulating factor (GM-CSF), IL-1β and IL-6. Although TNF-α and its mRNA have been identified in alveolar macrophages in normal lung tissue, they are predominantly detected in type II epithelial cells lining the thickened alveolar septa, as well as in macrophages in fibrotic lung tissue.[35] Upregulation of TNF-α in vivo does not always result in pulmonary fibrosis. Administration of recombinant TNF-α to the rat lung induces neutrophilia, which resolves without any evidence of tissue damage. In contrast, transgenic mice, in which TNF-α gene expression has been driven from the lung-specific SP-C promoter, go on to develop an early lymphocytic alveolitis, followed by a fibrogenic response.[36] Mice exposed to silica upregulate their expression of TNF-α mRNA and protein in lung tissue and in BAL cells. Enhanced TNF-α expression precedes the onset of fibroblast proliferation and collagen deposition in the lung.[37,38] Anti-TNF-α antibodies or soluble TNFR can prevent the development of pulmonary fibrosis induced by silica or bleomycin exposure in mice.[37] Mice deficient in two different cell surface receptors for TNF (TNFR1 and TNFR2), p55 and p75, are protected from the fibroproliferative effects of inhaled asbestos fibers[39] and silica, or bleomycin instillation.[40]

The IL-1 family of cytokines consists of IL-1α IL-1β and IL-1 receptor antagonist (IL-1ra). IL-1ra can attenuate a variety of IL-1 actions in both in vitro and in vivo models. IL-1, both directly and indirectly through the induction of PDGF, induces fibroblast proliferation and affects collagen synthesis.[41] Although IL-1 enhances type I and type III collagen synthesis,[42] it also induces PGE_2, which inhibits fibroblast proliferation.[43] Immunohistochemistry demonstrates that IL-1ra is localized to hyperplastic type II pneumocytes, macrophages and reactive fibroblasts, while IL-1β is weakly detected in the lung tissue of IPF patients.[44] This imbalance between IL-1ra and IL-1β may result in disease progression through the failure of IL-1β-dependent successful tissue remodeling.[44] In contrast, the intratracheal administration of adenovirus expressing IL-1β in rat lungs induces acute inflammation followed by progressive tissue fibrosis.[45] The role of IL-1 in the pathophysiology of pulmonary fibrosis would appear to be complex.

T helper subsets are classified on the basis of cytokine profiles. The Th1 cytokines, IFN-γ and IL-12, and the Th2 cytokines, IL-4, IL-5 and IL-10, cross-regulate with each other. These cytokines regulate the balance between Th1 and Th2 cell-mediated allergic inflammation and also appear to be involved in the pathophysiology of pulmonary fibrosis. IL-4 could result in the development of fibrosis by inducing fibroblast proliferation and increased production of extracellular collagen. IL-4 transgenic mice are susceptible to bleomycin-induced pulmonary fibrosis.[46] There is upregulation of IL-5 mRNA and protein by T-lymphocytes and eosinophils at the sites of active fibrosis in mice with bleomycin-induced pulmonary fibrosis.[47]

The cytokine profile in human pulmonary fibrosis appears to be Th2 predominant, since Th2 cytokines such as IL-4, IL-5 and IL-10 are predominantly present in IPF and since IFN-γ is decreased.[48] However, another study has shown that not only IL-5 but also IL-2 and IFN-γ are significantly increased in IPF.[49] IFN-γ inhibits the proliferation of fibroblasts and suppresses collagen synthesis both in vitro and in vivo.[50,51] IFN-γ administration down regulates the transcription of the TGF-β gene in mice with bleomycin-induced pulmonary fibrosis.[52] Ziesche et.al demonstrated that 12 months of treatment with IFN-γ plus prednisolone was associated with substantial improvement in the condition of those patients with IPF who had shown no response to prednisolone alone.[53] Further studies are needed to deermine the definitive roles of Th1 and Th2 cytokines in the pathophysiology of pulmonary fibrosis.

IL-6 has various functions including the regulation of immunologic and inflammatory responses. IL-6 is synthesized by a variety of cells, and may stimulate proliferation of lung fibroblasts and promote accumulation of the ECM component. However, the contribution of IL-6 to fibrogenesis is controversial. TNF-α and IL-6 stimulate MIP-1α expression and that these mediators participate in profibrotic inflammatory reaction in bleomycin-induced pulmonary fibrosis.[54] On the other hand, a transgenic mouse with both human IL-6 and the IL-6 receptor showed marked lung infiltration of lymphocytes, but no remarkable pulmonary fibrosis.[55]

IL-10 is secreted by Th2 cells, B cells, monocytes, macrophages, and keratinocytes, and suppresses various inflammatory reactions. IL-10 inhibits the synthesis of inflammatory cytokines, nitric oxide and ROS by monocytes/macrophages. Human IL-10 gene transfer significantly decreases myeloperoxidase activity in BALF as well as TNF-α mRNA expression in BAL cells in mice with bleomycin-induced pulmonary fibrosis.[56] Although persistent expression of human IL-10 by gene transfer appears to be an effective therapy for IPF, however, in IL-10 knockout mice exposured to silica, the amount of total protein and the number of total cells in BALF are increased, while the hydroxyproline content of the lung is decreased compared with those in normal littermates.[57] Therefore, although IL-10 suppresses pulmonary inflammation, it may promote the process of pulmonary fibrosis.

IL-12 is an inducible, antigen presenting cell-derived cytokines composed of 35-kDa and 40-kDa subunits. IL-12 plays an important role in regulating Th1-dependent immune responses by priming T cells for high INF-γ production and inducing them to differentiate toward the Th1 subset. IL-12 also potentiates T and natural killer (NK) cell-mediated cytotoxicity and stimulates their proliferation. IL-12 attenuates bleomycin-induced pulmonary fibrosis via modulation of IFN-γ production.[58] In contrast, the subunit of IL-12 p40 can antagonize the biological activity of p70 IL-12 via competition for its receptor. The production of the p40 subunit and Th2 polarization may play important roles in fibrotic responses.[59]

Endothelin-1 was originally identified as a potent vasoconstrictor, and is also recognized to be a chemoattractant and stimulant of fibroblast collagen synthesis. In patients with IPF, there is an increase in endothelin-converting enzyme-1 (ECE-1) and ET-1 production in lung tissue.[60] ECE-1 immunoreactivity is seen predominantly in the macrophages and neutrophils in the lung tissue of IPF.[61] In bleomycin-induced pulmonary fibrosis, ET-1 is detected in macrophages, bronchiolar and alveolar epithelial cells and endothelial cells in fibrotic regions, but not in unaffected areas using immunohistochemistry.[61] These results suggest a potential role for ET-1 in the development of pulmonary fibrosis. The endothelin receptor antagonist significantly abrogates bleomycin-induced pulmonary fibrosis in mice.[62]

Growth Factors

TGF-βs are multifunctional cytokines that exist in three isoforms designed as TGF-β1, TGF-β2 and TGF-β3. Although the biological activity of these isoforms overlaps, TGF-β1 appears to be predominant among them in being expressed in pulmonary fibrosis.[63] There are three TGF-β receptors, type I (TGFR1), type II (TGFRII) and type III (TGFRIII). All three receptors bind to all three TGF-βs with a high affinity. TGF-β is the most potent promoter of ECM production, and also a strong chemotactic factor for monocytes and macrophages. TGF-β activates them to release a number of cytokines such as PDGF, IL-1β, basic fibroblast growth factor (bFGF), TNF-α, and TGF-β autoregulates its own expression. There is a consistent increase in TGF-β production in epithelial cells and macrophages in lung tissue from patients with IPF[64] and in bleomycin-induced pulmonary fibrosis in rodents.[65] Transient overexpression of active TGF-β through the transfection of porcine TGF-β cDNA to the rat lung, results in prolonged and severe interstitial and pleural fibrosis.[66] The increase in lung collagen accumulation in bleomycin-induced lung fibrosis is reduced by treatment with either TGF-β2, TGF-β1 antibody, or the recombinant TGFRII.[67,68] Decorin, a naturally occurring biological molecule that antagonizes TGF bioactivity, may ameliorate excessive TGF signaling in injured lungs. Adenovirus-mediated decorin gene transfer reduces fibrotic response to bleomycin.[69] Smad proteins regulate intracellular signals from the membrane to the nucleus of TGF-β.[70] The activated TGF-β receptors induce phosphorylation of Smad2 and Smad3 by association with activated TGF-β receptors. The complexes translocate to the nucleus and regulate transcriptional responses. Smad 3 deficiency attenuates bleomycin-induced pulmonary fibrosis in mice.[71] Smad7 prevents the phosphorylation of Smad2 and Smad3 by association with activated TGF-β receptors.[72,73] Transient gene transfer and the expression of exogenous Smad7 into the lung by adenoviral vectors prevent bleomycin-induced lung fibrosis.[74]

PDGF is one of the most potent mitogens and chemoattractants for mesenchymal cells and induces the proliferation of fibroblasts and the synthesis of extracellular matrix. Active PDGF consists of two homologous subunits, A and B, which can form three dimeric PDGF isoforms. Alveolar macrophages in the lung tissue of patients with IPF produce increased amounts of PDGF-B mRNA and protein.[75,76] Alveolar macrophages, type II epithelial cells and mesenchymal cells abnormally express PDGF in animal models.[77] The B chain of PDGF transgenic mice shows the development of lung disease with diffusely distributed emphysematous lesions, as well as inflammation and fibrosis within the focal area.[78] Intratracheal instillation of recombinant human PDGF-BB into rats produces fibrosis concentrated around the large airways and blood vessels.[79] In vivo gene transfer of an extracellular domain of the PDGF receptor ameliorates bleomycin-induced pulmonary fibrosis in mice.[80] IGF-1 acts synergistically with PGDF to promote fibroblast proliferation.[81] IGF-1 and PDGF mRNA expression are significantly upregulated in BAL cells in mice with bleomycin-induced pulmonary fibrosis.[82] Alveolar macrophages obtained from patients with IPF express more IGF-1 mRNA and protein compared with normal alveolar macrophages.[81,83]

Basic FGF is a potent stimulator of both fibroblast and endothelial cell proliferation and is associated with the fibroproliferative response, similar to PDGF. Basic FGF expression has been found to be upregulated in healing wounds, recombinant bFGF has been shown to accelerate wound healing, and anti-bFGF antibody inhibits granulation tissue formation and normal wound repair. Alveolar macrophages are predominant source of bFGF in intra-alveolar fibrotic lesions following acute lung injury.[84] Mast cells are predominant bFGF-producing cells in IPF, and bFGF levels are correlated with bronchoalveolar lavage cellularity and with the severity of gas exchange abnormality.[85]

TGF-α induces the proliferation of endothelial and epithelial cells, and fibroblasts. TGF-α is detected in fibrotic areas.[86] TGF-α expression is upregulated in alveolar epithelial cells and macrophages in fibroproliferative lesions in rats with asbestos- or bleomycin-induced pulmonary fibrosis.[87] Transgenic mice, in which human TGF-α is expressed in the lung in an epithelial cell-specific manner, develop a fibroproliferative response in the interstitium and pleural surface.[88] These results suggest that TGF-α may participate in fibroproliferative response following lung injury.

Keratinocyte growth factor (KGF) has been demonstrated to enhance the functional differentiation of rat alveolar type II cells, to increase DNA synthesis in these cells in vitro, and to stimulate the proliferation of these cells in vivo. KGF is produced in the mesenchymal cells, and the KGF receptor is expressed in the epithelium of the developing lung. Intratracheal instillation of KGF significantly attenuates bleomycin-induced pulmonary fibrosis in rats.[89] KGF may participate in maintaining and repairing the alveolar epithelium and has the potential to become an effective agent against lung injury and pulmonary fibrosis.

Hepatocyte Growth Factor (HGF), composed of β-subunit (35 kDa) and α-subunit (64 kDa), is produced by mesenchymal cells and has been identified as a potent mitogen for mature hepatocytes. HGF is known to act not only as a mitogen but also as a motogen or a morphogen for many kinds of epithelial cells. The receptor for HGF is the c-Met proto-oncogene product, which is predominantly expressed in various types of epithelial cells. HGF levels in BALF and the sera of patients with IPF are higher than those of healthy control subjects.[90,91] Hyperplastic alveolar epithelial cells as well as macrophages strongly express HGF in lung tissues from patients with IPF.[90] As well as other epithelial cells, HGF promotes DNA synthesis in alveolar type II cells in vitro.[92] A simultaneous or delayed administration of HGF equally represses the fibrotic changes in murine lung injury induced by bleomycin.[93] The combination of HGF and IFN-γ enhances the migratory activity of A549 cells through the upregulation of the c-Met/HGF receptor.[94] HGF administration may be a novel strategy in the effort to promote repairing and healing of inflammatory lung damage in cases of pulmonary fibrosis.

Epithelial Cell Apoptosis and Its Clinical Potentials

Lung epithelium is not only the primary site of lung damage but it also participates in inflammatory reaction through a number of mechanisms, including the release of inflammatory mediators. Alterations in the structure and function of lung epithelial cells may affect the expression of these molecules. Epithelial cells in IPF can secrete a number of molecules, such as growth factors and their receptors, proteases, surfactant proteins, adhesion molecules and matrix component, which may regulate the inflammatory and fibrotic response within the lung. Prominent alveolar epithelial cell injury is the characteristic feature of IPF. Although type I pneumocytes comprise 40% of the alveolar epithelial cell population and over 90% of the alveolar surface in the normal lung,[95] they are markedly decreased in the area of severe inflammation following extensive injury and cell death in the lung tissue from patients with IPF. The alveolar type II cell is a reparative cell and rapidly proliferates following epithelial cell injury. In areas most severely damaged, the basement membrane is covered by proliferating type II pneumocytes which are cuboidal, and death of both type I and type II pneumocytes are replaced by abundant fibroblasts and smooth muscle cells.[96] Apoptotic type II pneumocyte death occurs also in normal alveoli of IPF[97] (Fig. 2).

Apoptosis plays a major role in homeostasis to maintain a balance between cell proliferation and cell death. There are two principle signaling pathways of apoptosis, one is the death receptor-mediated pathway while the other pathway, which is triggered by many stimuli such as reactive oxygen radicals, anti-cancer drugs, radiation, and growth factor deprivation, is initiated within the mitochondria. The vulnerability to apoptosis induced by death receptors or other apoptosis stimulators, and the ability to survive by inhibitors of apoptosis varies according to the cell type. Apoptosis may play important roles in lung diseases in two different ways. First, failure to clear unwanted cells by apoptosis will prolong the inflammation because of the release of their toxic contents. Repair after an acute lung injury requires the elimination of

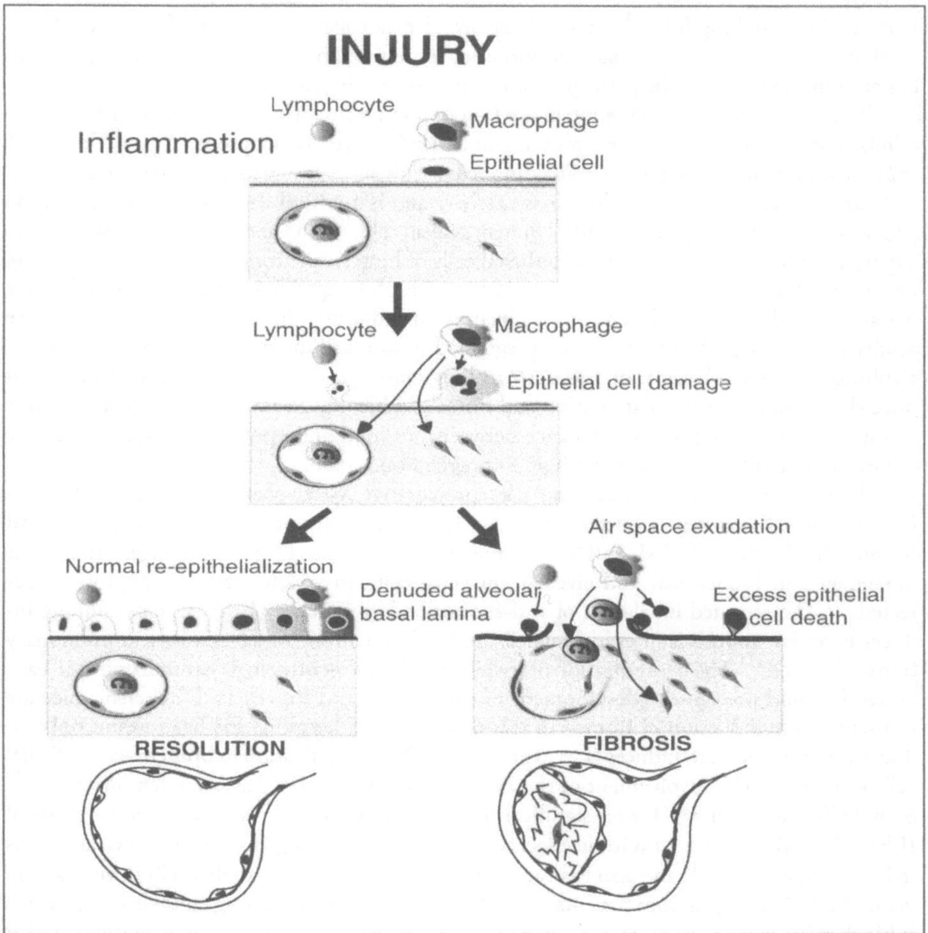

Figure 2. The alveolitis as an initial event in the pathogenesis of pulmonary fibrosis. The alveolitis is an early event in pulmonary fibrosis, regardless of etiology. When the damage on epithelial cells and basement membranes is too severe to be repaired, the migration of fibroblasts into alveolar space leads to intraalveolar fibrosis.

proliferating mesenchymal and inflammatory cells from the alveolar airspace or alveolar wall.[98] Secondary, excessive apoptosis may cause diseases. Repeated inhalation of agonistic anti-Fas antibody into adult mice causes epithelial cell apoptosis and lung inflammation, which subsequently leads to pulmonary fibrosis.[99]

Bleomycin rapidly produces extensive DNA damage in the lung.[100] In vitro, bleomycin can induce apoptosis.[101] The persistence of DNA damage rather than the initial level of strand scission is associated with pulmonary fibrosis.[100] Biphasic DNA fragmentation is observed by the gel electrophoretic results of DNA extracted from lung tissues in mice with bleomycin-induced pulmonary fibrosis. Electron microscopic findings show the characteristic features of apoptosis in bronchiolar and alveolar epithelial cells.[102] Therefore, DNA damage and the apoptosis of epithelial cells may be associated with pulmonary fibrosis. There is DNA damage or apoptosis in bronchiolar and alveolar epithelial cells in IPF using an in situ DNA nick end-labeling method.[103] DNA damage and apoptosis in lung epithelial cells have been reported in acute lung injury[104] and diffuse alveolar damage,[105] as well as IPF. Although the mechanism of how epithelial cell apoptosis leads to pulmonary fibrosis is not clear, it probably has an important role in the pathogenesis of pulmonary fibrosis.

The tumor suppressor p53 protein is a transcription factor, which plays a central role in the cellular response to DNA damage, resulting in either G1 arrest or apoptosis.[106] $p21^{Waf1/Cip1/Sdi1}$ (p21) is induced in wild-type p53-containing cells following exposure to DNA-damaging agents. p21 inhibits cyclin-CDK complex kinase activity and is a critical downstream effector in the p53-specific pathway of growth control in mammalian cells.[107] p53 and p21 are expressed in the hyperplastic bronchial and alveolar epithelial cells of lung tissue from patients with IPF using immunohistochemistry.[103] The association between p53 expression and pulmonary fibrosis has also been demonstrated in mice with bleomycin-induced pulmonary fibrosis.[108,109] These results suggest that p53 and p21 are upregulated in association with chronic DNA damage, resulting in either G1 arrest or apoptosis in lung epithelial cells. In contrast, p53 knockout mice shows more severe inflammation and fibrosis in bleomycin-induced pulmonary fibrosis in mice.[110] The expression and balance between pro- and anti-apoptotic molecules may be critical in lung injury associated with p53 upregulation.

The Fas-Fas ligand (FasL) pathway is a representative system of apoptosis- signaling receptor molecules. Bleomycin-induced pulmonary fibrosis is an animal model for lung injury and fibrosis. In this model, FasL mRNA is upregulated in infiltrating lymphocytes, and Fas is upregulated in bronchiolar and alveolar epithelial cells, in which excessive apoptosis is detected.[102] The repeated inhalation of anti-Fas antibody mimicking Fas-FasL cross-linking induces excessive apoptosis of epithelial cells and inflammation, which resulted in pulmonary fibrosis in mice.[111] The neutralization of FasL by Fas-Ig fusion protein or neutralizing anti-FasL antibody could prevent the development of this model, and Fas- or FasL-deficient mice are resistant to the induction of fibrosis in this model.[112] Furthermore, Fas ligation not only induces apoptosis but also induces IL-8 expression via NF-κB activation in bronchiolar epithelial cells in vitro.[113] The involvement of the Fas-FasL pathway in IPF has been demonstrated.[114] Soluble FasL levels in BALF were significantly increased in ARDS[115] and in the active phase of IPF.[116] TGF-β, which has multiple effects on the process of fibrogenesis, can induce apoptosis in lung epithelial cell,[117] and also Fas-mediated apoptosis on lung epithelial cells in vitro and in vivo.[118] BALF obtained from patients with IPF induces apoptosis on lung epithelial cells, which is blocked by anti-Fas or by anti-TGF-β antibody in vitro.[118] These results may indicate that excessive apoptosis induced by the Fas-FasL pathway is not only critical for the development of pulmonary fibrosis, but is also of possible therapeutic use for molecules that prevent the Fas-FasL pathway in patients with lung injury and pulmonary fibrosis. Protecting epithelial cells from injury may be a more effective therapeutic target than inhibiting fibroblast proliferation following epithelial cell damage (Fig. 3).

Angiotensin converting enzyme (ACE) levels in BALF and serum are increased in fibrosing lung diseases, including sarcoidosis, IPF, asbestosis, silicosis and ARDS. Angiotensin II concentrations increase during radiation-induced pulmonary fibrosis.[119] Angiotensin II induces human lung fibroblast proliferation in vitro via activation of the angiotensin type I receptor and the autocrine action of TGF-β.[120] Angiotensin II receptor antagonist or ACE inhibitors attenuate cardiac and renal fibrosis in animal models. The ACE inhibitor captopril ameliorates pulmonary fibrosis induced by monocrotaline in rats.[121] Captopril inhibits the accumulation of collagens and mast cells in the irradiated rat lung, and also inhibits fibroblast proliferation in the presence of bFGF in vitro.[122,123] Fas-induced apoptosis of human lung epithelial cells in culture is potently inhibited by captopril at concentrations readily attained in vivo.[123] The inhibitory actions of captopril on pulmonary fibrosis may be due to the prevention of lung epithelial cell apoptosis.

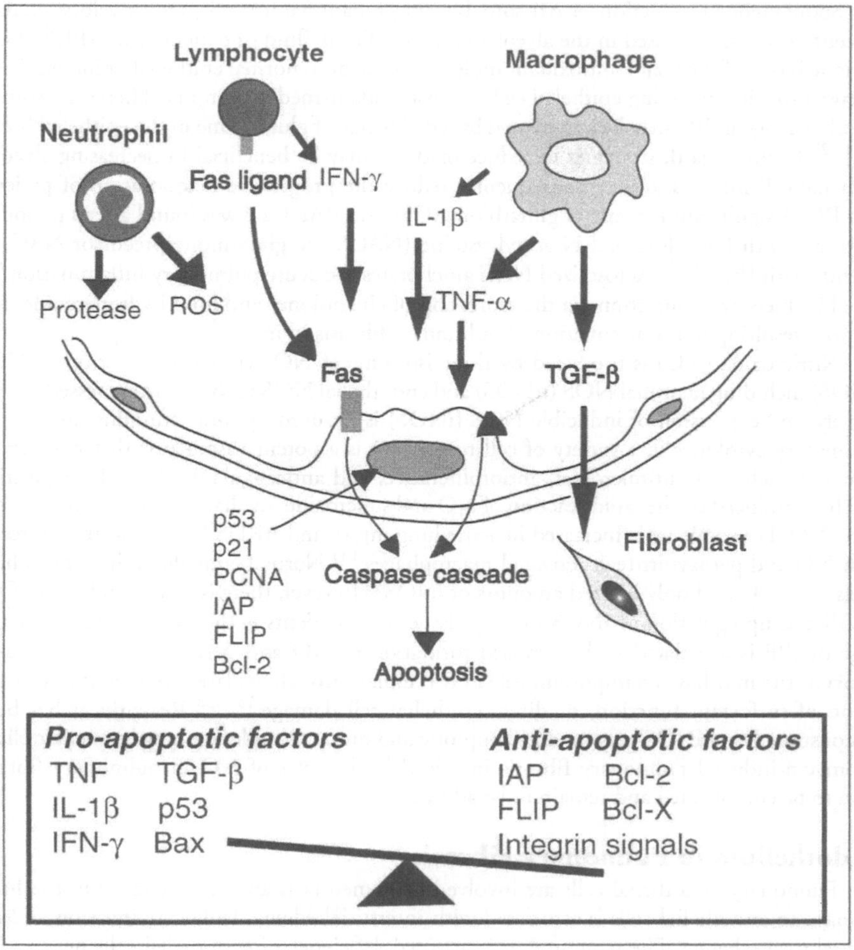

Figure 3. Lung epithelial cell apoptosis. Apoptosis plays a major role in homeostasis to maintain a balance between cell proliferation and cell death. There is increasing attention in the importance of epithelial cell death. The imbalance between pro- and anti-apoptotic factors may regulate the course of pulmonary fibrosis.

Oxidant-Mediated Lung Injury in Pulmonary Fibrosis

Inflammatory cells sequestrating alveolar walls can release oxidants capable of directly damaging lung epithelial cells. In patients with acute lung injury caused by viral infection, shock, severe trauma, or drugs, it is often necessary to administer oxygen. Pulmonary lesions seen in patients dying from ARDS are not only caused by the initial disease, but also by the toxic effects of oxygen therapy. It has been reported that a damaged lung is more susceptible to oxygen than an undamaged one.[124] Experimental animal studies with the intratracheal instillation of proteases and oxidants, phorbol myristate acetate, formylated norleu-leu-phe, or immune complexes, have demonstrated that an oxidant burden in the lower respiratory tract results in severe lung damage that includes epithelial cell death.[125-127] These changes are followed by severe interstitial fibrosis.

Spontaneous production of oxidants by lung inflammatory cells, and myeloperoxidase concentration are increased in the alveolar epithelial lining fluid of patients with IPF.[128] Glutathione is one of the major antioxidant molecules present in normal epithelial lining fluid and it plays a role in protecting epithelial cells against oxidant-mediated injury. Therefore, epithelial cell damage in IPF may be augmented by a deficiency of glutathione in the epithelial lining fluid.[129] In this regard, strategies to reduce oxidants may be beneficial in decreasing alveolar epithelial cell injury and may consequently reduce the progressive deterioration of patients with IPF. A significant increase of glutathione (GSH) in BAL fluid was found after a period of treatment with high-dose oral N-acetylcysteine (NAC), the glutathione precursor NAC, in patients with IPF.[130,131] Aerosolized NAC ameliorates the acute pulmonary inflammation induced by bleomycin injection via the repression of chemokines and lipid hydroperoxide production, resulting in the attenuation of pulmonary fibrosis in mice.[132]

Nitric oxide (NO) is produced by three isoforms of NO synthases. Constitutive NOS (cNOS) including neuronal NOS (nNOS) and endothelial NOS (eNOS) are expressed constitutively, and expression of inducible NOS (iNOS) is induced by some stimulus, such as inflammatory cytokines in a variety of cell types. NO is a potent vasodilator that has various functions including antimicrobial, antiproliferative, and anticoagulant effects. Peroxynitrite, which is produced by the rapid reaction of NO and superoxide, mediates the cytotoxic effect of nitric NO. Peroxynitrite is increased in acute lung injury and ARDS,[133,134] and the source of both NO and peroxynitrite is activated macrophages.[135] Normal epithelial cells express high levels of cNOS and only limited amounts of iNOS. However, there is downregulation of the cNOS and upregulation of iNOS in lung tissue from patients with IPF and that the active stage of IPF is associated with increased formation of NO and peroxynitrite.[136] Increased peroxynitrite may have an important role in IPF, either through oxidation of glutathione, inhibition of surfactant function, or direct epithelial cell damage.[137,138] Recently, it has been demonstrated that iNOS plays as anti-apoptotic and protective roles on lung epithelialc ells in bleomycin-induced pulmonary fibrosis in mice.[110] The roles of NO in pulmonary fibrosis seem to be complicated and remain to be addressed.

Endothelium in Pulmonary Fibrosis

Pulmonary endothelial cells are involved in homeostasis and gas exchange in the lung. Damage to endothelial cells is associated with interstitial edema, leukocyte invasion and decreased gas exchange. The structural or functional deficiency of endothelial cells may lead to malfunction or cell death of type II epithelial cells and may promote pulmonary fibrosis.[139] Ultrastructural findings show that injured epithelial and endothelial cells are found in early fibrosing alveolitis.[140] Wendt et.al demonstrated that alveolar type II cells release a peptide(s) that protects endothelial cells from the apoptosis induced by TNF-α.[141] Alveolar epithelial cells may regulate endothelial cell apoptosis during injury and repair.

Pulmonary endothelial cells are activated in a variety of inflammatory lung diseases, and their dysfunction may be important in pulmonary fibrosis. Endothelial cells express inflammatory and fibrogenic cytokines, which have an important role in inflammatory infiltration, cellular growth and matrix synthesis.[141] Apoptosis occurrs within the intra-alveolar granulation tissue during lung injury and BALF from these patients induces apoptosis in mesenchymal cells.[98] Endothelial cell proliferation and angiogenesis are essential in intra-alveolar granulation of injured lungs.

The existence of neovascularization in bleomycin-induced pulmonary fibrosis has been identified.[142] Keane et al demonstrated that the CXC chemokine, IFN-γ-inducible protein-10 (IP-10) was inversely correlated to total lung collagen and a greater angiogenic response in mice with bleomycin-induced pulmonary fibrosis compared with control lung tissue.[143] A significant increase in macrophage inflammatory protein (MIP)-2 is correlated with a significant increase in angiogenesis in mice with bleomycin-induced pulmonary fibrosis compared with the controls. Moreover, angiogenesis and pulmonary fibrosis are significantly attenuated with anti-MIP-2 antibody.[144] In addition to the fibroproliferative effect, the angiogenic effect of bFGF may play a role in pulmonary fibrosis.[142] These results suggest that antiangiogenic therapy could be another possible strategy in the search for a treatment for pulmonary fibrosis.

Epithelial-Fibroblast Interaction in Pulmonary Fibrosis

Severe injury and insufficient repair of lung epithelial cells impair normal epithelial-fibroblast interaction, which leads to pulmonary fibrosis. If epithelial cell repair does not proceed smoothly and completely, fibroblasts will proliferate, eventually leading to pulmonary fibrosis. Mouse lung explants with severe epithelial damage induced by prior hyperoxic lung injury exhibit marked fibroblast proliferation and collagen deposition in culture, whereas less severely injured explants do not.[145] Normal repair of the epithelial layer occurs through the proliferation and differentiation of type II alveolar epithelial cells. This process is affected by factors produced by lung fibroblasts.[146,147] The cuboidal epithelium of the fibrotic human lung is composed of both proliferating and dying cells, and that apoptotic and necrotic epithelial cells are observed in proximity to α-actin-positive interstitial cells.[148] Abnormal fibroblast phenotypes isolated from the fibrotic human lung produce factors capable of inducing apoptosis and necrosis of alveolar epithelial cells in vitro.[149] Neither inflammation nor fibrosis correlates with survival, and the only pathological data that showed a significant correlation with mortality are numbers of areas with fibroblastic foci.[150] Studies on the repopulation of denuded tracheal explants by epithelial cells show that the denuded tracheal implants are rapidly replaced by fibroblasts, unless enough epithelial cells are introduced into the lumen to control fibroblast proliferation.[151] Alternatively, epithelial cells may control fibroblasts by releasing cytokines that downregulate fibroblast activity. These abnormal epithelial-mesenchymal interactions contribute to the pathogenesis and exacerbation of fibrotic lung disease by preventing normal epithelial repair and progression of abnormal fibroblast proliferation. Lovastatin is an HMG-CoA reductase inhibitor widely used in the treatment of hypercholesteremia. Lovastatin induces apoptosis in normal and fibrotic fibroblasts both in vitro and in vivo, and dramatically reduces granulation tissue formation in a guinea pig wound chamber model with ultrastructural evidence of fibroblast apoptosis.[152] To induce apoptosis of fibroblasts may be one possible approach in the attempt to block fibroblast proliferation and collagen synthesis.

The persistence of fibrin and insufficient function of plasmin promote fibrosis. Fibrin forms within the interstitium and alveolar spaces during acute and chronic inflammatory lung diseases. Fibrin is generated from plasma that leaks from the vasculature to the overlying lung epithelial cells due to a deficiency in normal alveolar fibrinolytic activity.[153] Tissue factor and plasminogen activator inhibitors (PAI) are both strongly upregulated in alveolar epithelial cells of IPF.[153,154] Adenovirus-mediated transfer of urokinase-type plasminogen activator genes successfully upregulates fibrinolysis both in vitro and in vivo, and attenuates the development of bleomycin-induced pulmonary fibrosis.[155] Mice with targeted deletion of PAI-1 genes develop less intra-alveolar fibrin or fibrosis in bleomycin-induced pulmonary fibrosis.[156] Therefore, upregulation of fibrinolytic activity may be beneficial for fibroproliferative lung injury.

Matrix Remodeling in Pulmonary Fibrosis

The extracellular matrix in lung tissue is composed of collagen (especially types I and III), elastin, proteoglycans, fibronectin and other assorted proteins. Insufficient production and degradation of the ECM in tissue remodeling after injury may lead to the accumulation of ECM in the alveolar walls and spaces in pulmonary fibrosis. Various lung cells, such as macrophages, endothelial cells, alveolar and bronchiolar epithelial cells, and fibroblasts, can produce ECM components following lung injury in human diseases and animal models. It was demonstrated that the phenotype of the altered fibroblasts was one of the causes of exaggerated ECM production in fibrosing lung diseases.[157] ECM has many biological effects and modulates inflammatory and fibrotic processes. ECM is chemotactic to many types of cells, and activates immune cells to produce inflammatory factors, such as prostaglandins, superoxide and cytokines that enhance inflammation and fibrosis.[158,159] Cell adhesion to ECM activates transcription factors, such as AP-1 and NF-κB, which bind to promoter regions of genes that encode for proinflammatory mediators such as IL-1 and TNF-α.[160] ECM-derived signals may regulate cell proliferation and differentiation, as well as apoptosis.[161]

Degradation of ECM is controlled primarily by matrix metalloproteinases (MMPs). The levels of expression, the activation of proenzyme, and interactions between tissue inhibitors of MMPs (TIMPs) and MMPs are closely associated with the metabolism of ECM. The MMPs are a family of zinc-dependent, secreted enzymes comprising at least 18 members, which include the collagenases, gelatinases, stromelysins, macrophage metalloelastase, matrilysin and membrane type MMPs. MMPs are secreted in a latent form, whereby they are cleaved and activated by other enzymes and autocatalytic cleavage. Catalytic activity is specifically inhibited by TIMPs, which bind to the MMP active site and also to latent forms of MMPs. The TIMP family consists of TIMP-1, -2, -3 and -4.

The occurrence of intra-alveolar fibrosis in the pathogenesis of IPF suggests that there are defects in the alveolar epithelial lining and basement membrane through which mesenchymal cells such as fibroblasts and inflammatory cells migrate from the interstitium into the alveolar space.[162] Fibronectin and collagens-I, III and IV are detected in the intra-alveolar fibrosis in addition to the interstitial fibrosis in IPF.[163] MMP-1, MMP-2, MMP-9 and TIMP-2 are detected in the regenerated epithelial cells covering the intra-alveolar fibrosis. TIMP-2 in myofibroblasts in IPF may contribute to stable ECM deposition and irreversible pulmonary structural remodeling.[163] MMP-9 expression is increased in alveolar macrophages and treatment with steroids and immunosuppressive agents normalizes the MMP-9 expression in patients with IPF.[164] TGF-β inhibits the expression of MMP-1 and MMP-2 in fibroblasts, stimulates the synthesis and secretion of TIMP, and attenuates collagen degradation.[165] The differential expression of MMP and TIMP genes is observed in mice with bleomycin-induced pulmonary fibrosis,[166] and activation of MMP-2 may lead to the elongation and migration of alveolar epithelial cells in the repair process of this model.[167] A regulated balance of active MMPs and TIMPs is maintained during normal tissue metabolism, and accordingly an imbalance between MMPs and TIMPs could play a critical role in the pathogenesis of pulmonary fibrosis.

Conclusions

Recent advances with regard to the molecular mechanisms of pulmonary fibrosis concerning endothelial and epithelial cell injury, inflammatory reactions, fibroblast proliferation, collagen deposition and lung repair result in increasing recognition of its complexity. The major targets of therapy have focused on inflammatory cells, and this has led to the use of anti-inflammatory agents. However, conventional therapies, such as corticosteroids and cytotoxic agents, have been reported to be only minimally effective. In animal models of pulmonary fibrosis or human diseases such as IPF, various inflammatory mediators and death factors induce epithelial cell damage. The survival and recovery of epithelial cells, and the prevention of fibroblast proliferation and ECM deposition appear to be the key in the prognosis of patients. Protecting parenchymal cells from injury and maintaining their function may be an effective therapeutic strategy against pulmonary fibrosis.

Acknowledgements

The language editing of this manuscript was performed by Miss K. Miller (Royal English Language Center, Fukuoka, Japan).

References

1. Crystal RG, Bitterman PB, Mossman B et al. Future research directions in idiopathic pulmonary fibrosis: Summary of a national heart, lung, and blood institute working group. Am J Respir Crit Care Med 2002; 166:236-246.
2. Pantelidis P, Fanning GC, Wells AU et al. Analysis of tumor necrosis factor-alpha, lymphotoxin-alpha, tumor necrosis factor receptor II, and interleukin-6 polymorphisms in patients with idiopathic pulmonary fibrosis. Am J Respir Crit Care Med 2001; 163:1432-1436.
3. Thomas AQ, Lane K, Phillips 3rd J et al. Heterozygosity for a surfactant protein C gene mutation associated with usual interstitial pneumonitis and cellular nonspecific interstitial pneumonitis in one kindred. Am J Respir Crit Care Med 2002; 165:1322-1328.
4. Fukuda Y, Ishizaki M, Masuda Y et al. The role of intraalveolar fibrosis in the process of pulmonary structural remodeling in patients with diffuse alveolar damage. Am J Pathol 1987; 126:171-182.
5. Watters LC, Schwarz MI, Cherniack RM et al. Idiopathic pulmonary fibrosis. Pretreatment bronchoalveolar lavage cellular constituents and their relationships with lung histopathology and clinical response to therapy. Am Rev Respir Dis1987; 135:696-704.
6. Hallgren R, Bjermer L, Lundgren R et al. The eosinophil component of the alveolitis in idiopathic pulmonary fibrosis. Signs of eosinophil activation in the lung are related to impaired lung function. Am Rev Respir Dis 1989; 139:373-377.
7. Mori H, Kawada K, Zhang P et al. Bleomycin-induced pulmonary fibrosis in genetically mast cell-deficient WBB6F1-W/Wv mice and mechanism of the suppressive effect of tranilast, an antiallergic drug inhibiting mediator release from mast cells, on fibrosis. Int Arch Allergy Appl Immunol 1991; 95:195-201.
8. Thrall RS, Phan SH, McCormick JR et al. The development of bleomycin-induced pulmonary fibrosis in neutrophil-depleted and complement-depleted rats. Am J Pathol 1981; 105:76-81.
9. Schrier DJ, Kunkel RG, Phan SH. The role of strain variation in murine bleomycin-induced pulmonary fibrosis. Am Rev Respir Dis 1983; 127:63-66.
10. Piguet PF, Collart MA, Grau GE et al. Tumor necrosis factor/cachectin plays a key role in bleomycin-induced pneumopathy and fibrosis. J Exp Med 1989; 170:655-663.
11. Okazaki T, Nakao A, Nakano H et al. Impairment of bleomycin-induced lung fibrosis in CD28-deficient mice. J Immunol 2001; 167:1977-1981.
12. Martinet Y, Rom WN, Grotendorst GR et al. Exaggerated spontaneous release of platelet-derived growth factor by alveolar macrophages from patients with idiopathic pulmonary fibrosis. N Engl J Med 1987; 317:202-209.
13. Carre PC, King Jr TE, Mortensen R et al. Cryptogenic organizing pneumonia: increased expression of interleukin-8 and fibronectin genes by alveolar macrophages. Am J Respir Cell Mol Biol 1994; 10:100-105.
14. Meyrick B, Hoffman LH, Brigham KL. Chemotaxis of granulocytes across bovine pulmonary artery intimal explants without endothelial cell injury. Tissue Cell 1984; 16:1-16.
15. Milks LC, Brontoli MJ, Cramer EB. Epithelial permeability and the transepithelial migration of human neutrophils. J Cell Biol 1983; 96:1241-1247.
16. Parsons PE, Sugahara K, Cott GR et al. The effect of neutrophil migration and prolonged neutrophil contact on epithelial permeability. Am J Pathol 1987; 129:302-312.
17. Nakao A, Hasegawa Y, Tsuchiya Y et al. Expression of cell adhesion molecules in the lungs of patients with idiopathic pulmonary fibrosis. Chest 1995; 108:233-239.
18. Piguet PF, Rosen H, Vesin C et al. Effective treatment of the pulmonary fibrosis elicited in mice by bleomycin or silica with anti-CD-11 antibodies. Am Rev Respir Dis 1993; 147:435-441.
19. Azuma A, Takahashi S, Nose M et al. Role of E-selectin in bleomycin induced lung fibrosis in mice. Thorax 2000; 55:147-152.
20. Hynes RO. Integrins: versatility, modulation, and signaling in cell adhesion. Cell 1992; 69:11-25.

21. Piguet PF, Tacchini-Cottier F, Vesin C. Administration of anti-TNF-alpha or anti-CD11a antibodies to normal adult mice decreases lung and bone collagen content: evidence for an effect on platelet consumption. Am J Respir Cell Mol Biol 1995; 12:227-231.
22. Munger JS, Huang X, Kawakatsu H et al. The integrin alpha v beta 6 binds and activates latent TGF beta 1: A mechanism for regulating pulmonary inflammation and fibrosis. Cell 1999; 96:319-328.
23. Luster AD. Chemokines—Chemotactic cytokines that mediate inflammation. N Engl J Med 1998; 338:436-45.
24. Schwaltz MI. Internal Medicine. In: WN Kelly ed. Philadelphia, PA: J.P. Lippincott Co, 1989:1902-1905.
25. Suga M, Iyonaga K, Ichiyasu H et al. Clinical significance of MCP-1 levels in BALF and serum in patients with interstitial lung diseases. Eur Respir J 1999; 14:376-382.
26. Smith RE, Strieter RM, Phan SH et al. TNF and IL-6 mediate MIP-1alpha expression in bleomycin-induced lung injury. J Leukoc Biol 1998; 64:528-536.
27. Jiang Y, Beller DI, Frendl G et al. Monocyte chemoattractant protein-1 regulates adhesion molecule expression and cytokine production in human monocytes. J Immunol 1992; 148:2423-2428.
28. Standiford TJ, Kunkel SL, Lukacs NW et al. Macrophage inflammatory protein-1 alpha mediates lung leukocyte recruitment, lung capillary leak, and early mortality in murine endotoxemia. J Immunol 1995; 155:1515-1524.
29. Moore BB, Paine 3rd R, Christensen PJ et al. Protection from pulmonary fibrosis in the absence of CCR2 signaling. J Immunol 2001; 167:4368-4377.
30. Smith RE, Strieter RM, Phan SH et al. Production and function of murine macrophage inflammatory protein-1 alpha in bleomycin-induced lung injury. J Immunol 1994; 153:4704-4712.
31. Ziegenhagen MW, Zabel P, Zissel G et al. Serum level of interleukin 8 is elevated in idiopathic pulmonary fibrosis and indicates disease activity. Am J Respir Crit Care Med 1998; 157:762-768.
32. Carre PC, Mortenson RL, King Jr TE et al. Increased expression of the interleukin-8 gene by alveolar macrophages in idiopathic pulmonary fibrosis. A potential mechanism for the recruitment and activation of neutrophils in lung fibrosis. J Clin Invest 1991; 88:1802-1810.
33. Kodama N, Yamaguchi E, Hizawa N et al. Expression of RANTES by bronchoalveolar lavage cells in nonsmoking patients with interstitial lung diseases. Am J Respir Cell Mol Biol 1998; 18:526-531.
34. Tokuda A, Itakura M, Onai N et al. Pivotal role of CCR1-positive leukocytes in bleomycin-induced lung fibrosis in mice. J Immunol 2000; 164:2745-2751.
35. Piguet PF, Ribaux C, Karpuz V et al. Expression and localization of tumor necrosis factor-alpha and its mRNA in idiopathic pulmonary fibrosis. Am J Pathol 1993; 143:651-655.
36. Miyazaki Y, Araki K, Vesin C et al. Expression of a tumor necrosis factor-alpha transgene in murine lung causes lymphocytic and fibrosing alveolitis. A mouse model of progressive pulmonary fibrosis. J Clin Invest1995; 96:250-259.
37. Piguet PF, Collart MA, Grau GE et al. Requirement of tumour necrosis factor for development of silica-induced pulmonary fibrosis. Nature 1990; 344:245-247.
38. Ohtsuka Y, Munakata M, Ukita H et al. Increased susceptibility to silicosis and TNF-alpha production in C57BL/6J mice. Am J Respir Crit Care Med 1995; 152:2144-2149.
39. Liu JY, Brass DM, Hoyle GW et al. TNF-alpha receptor knockout mice are protected from the fibroproliferative effects of inhaled asbestos fibers. Am J Pathol 1998; 153:1839-1847.
40. Ortiz LA, Lasky J, Lungarella G et al. Upregulation of the p75 but not the p55 TNF-alpha receptor mRNA after silica and bleomycin exposure and protection from lung injury in double receptor knockout mice. Am J Respir Cell Mol Biol 1999; 20:825-833.
41. Raines EW, Dower SK, Ross R. Interleukin-1 mitogenic activity for fibroblasts and smooth muscle cells is due to PDGF-AA. Science 1989; 243:393-396.
42. Goldring MB, Birkhead J, Sandell LJ et al. Interleukin 1 suppresses expression of cartilage-specific types II and IX collagens and increases types I and III collagens in human chondrocytes. J Clin Invest 1988; 82:2026-2037.
43. Dayer JM, de Rochemonteix B, Burrus B et al. Human recombinant interleukin 1 stimulates collagenase and prostaglandin E2 production by human synovial cells. J Clin Invest 1986; 77:645-648.

44. Smith DR, Kunkel SL, Standiford TJ et al. Increased interleukin-1 receptor antagonist in idiopathic pulmonary fibrosis. A compartmental analysis. Am J Respir Crit Care Med 1995; 151:1965-1973.
45. Kolb M, Margetts PJ, Anthony DC et al. Transient expression of IL-1beta induces acute lung injury and chronic repair leading to pulmonary fibrosis. J Clin Invest 2001; 107:1529-1536.
46. Sempowski GD, Beckmann MP, Derdak S et al. Subsets of murine lung fibroblasts express membrane-bound and soluble IL-4 receptors. Role of IL-4 in enhancing fibroblast proliferation and collagen synthesis. J Immunol 1994; 152:3606-3614.
47. Gharaee-Kermani M, Phan SH. Lung interleukin-5 expression in murine bleomycin-induced pulmonary fibrosis. Am J Respir Cell Mol Biol 1997; 16:438-447.
48. Wallace WA, Ramage EA, Lamb D et al. A type 2 (Th2-like) pattern of immune response predominates in the pulmonary interstitium of patients with cryptogenic fibrosing alveolitis (CFA). Clin Exp Immunol 1995; 101:436-441.
49. Walker C, Bauer W, Braun RK et al. Activated T cells and cytokines in bronchoalveolar lavages from patients with various lung diseases associated with eosinophilia. Am J Respir Crit Care Med 1994; 150:1038-1048.
50. Duncan MR, Berman B. Gamma interferon is the lymphokine and beta interferon the monokine responsible for inhibition of fibroblast collagen production and late but not early fibroblast proliferation. J Exp Med 1985; 162:516-527.
51. Granstein RD, Murphy GF, Margolis RJ et al. Gamma-interferon inhibits collagen synthesis in vivo in the mouse. J Clin Invest 1987; 79:1254-1258.
52. Gurujeyalakshmi G, Giri SN. Molecular mechanisms of antifibrotic effect of interferon gamma in bleomycin-mouse model of lung fibrosis: downregulation of TGF-beta and procollagen I and III gene expression. Exp Lung Res 1995; 21:791-808.
53. Ziesche R, Hofbauer E, Wittmann K et al. A preliminary study of long-term treatment with interferon gamma-1b and low-dose prednisolone in patients with idiopathic pulmonary fibrosis. N Engl J Med 1999; 341:1264-1269.
54. Smith RE, Strieter RM, Phan SH et al. TNF and IL-6 mediate MIP-1alpha expression in bleomycin-induced lung injury. J Leukoc Biol 1998; 64:528-536.
55. Yoshida M, Sakuma J, Hayashi S et al. A histologically distinctive interstitial pneumonia induced by overexpression of the interleukin 6, transforming growth factor beta 1, or platelet-derived growth factor B gene. Proc Natl Acad Sci USA 1995; 92:9570-9574.
56. Arai T, Abe K, Matsuoka H et al. Introduction of the interleukin-10 gene into mice inhibited bleomycin-induced lung injury in vivo. Am J Physiol Lung Cell Mol Physiol 2000; 278:914-922.
57. Huaux F, Louahed J, Hudspith B et al. Role of interleukin-10 in the lung response to silica in mice. Am J Respir Cell Mol Biol 1998; 18:51-59.
58. Keane MP, Belperio JA, Burdick MD et al. IL-12 attenuates bleomycin-induced pulmonary fibrosis. Am J Physiol Lung Cell Mol Physiol 2001; 281:L92-97.
59. Maeyama T, Kuwano K, Kawasaki M et al. Attenuation of bleomycin-induced pneumopathy in mice by monoclonal antibody to interleukin-12. Am J Physiol Lung Cell Mol Physiol 2001; 280:1128-1137.
60. Saleh D, Furukawa K, Tsao MS et al. Elevated expression of endothelin-1 and endothelin-converting enzyme-1 in idiopathic pulmonary fibrosis: possible involvement of proinflammatory cytokines. Am J Respir Cell Mol Biol 1997; 16:187-193.
61. Mutsaers SE, Foster ML, Chambers RC et al. Increased endothelin-1 and its localization during the development of bleomycin-induced pulmonary fibrosis in rats. Am J Respir Cell Mol Biol 1998; 18:611-619.
62. Park SH, Saleh D, Giaid A et al. Increased endothelin-1 in bleomycin-induced pulmonary fibrosis and the effect of an endothelin receptor antagonist. Am J Respir Crit Care Med 1997; 156:600-608.
63. Coker RK, Laurent GJ, Shahzeidi S et al. Transforming growth factors-beta 1, -beta 2, and -beta 3 stimulate fibroblast procollagen production in vitro but are differentially expressed during bleomycin-induced lung fibrosis. Am J Pathol 1997; 150:981-991.
64. Khalil N, O'Connor RN, Unruh HW et al. Increased production and immunohistochemical localization of transforming growth factor-beta in idiopathic pulmonary fibrosis. Am J Respir Cell Mol Biol 1991; 5:155-162.

65. Raghow B, Irish P, Kang AH. Coordinate regulation of transforming growth factor beta gene expression and cell proliferation in hamster lungs undergoing bleomycin-induced pulmonary fibrosis. J Clin Invest 1989; 84:1836-1842.
66. Sime PJ, Xing Z, Graham FL et al. Adenovector-mediated gene transfer of active transforming growth factor-beta1 induces prolonged severe fibrosis in rat lung. J Clin Invest 1997; 100:768-776.
67. Giri SN, Hyde DM, Hollinger MA. Effect of antibody to transforming growth factor beta on bleomycin induced accumulation of lung collagen in mice. Thorax 1993; 48:959-966.
68. Wang Q, Wang Y, Hyde DM et al. Reduction of bleomycin induced lung fibrosis by transforming growth factor beta soluble receptor in hamsters. Thorax 1999; 54:805-812.
69. Kolb M, Margetts PJ, Galt T et al. Transient transgene expression of decorin in the lung reduces the fibrotic response to bleomycin. Am J Respir Crit Care Med 2001; 163:770-777.
70. Heldin CH, Miyazono K, ten Dijke P. TGF-beta signalling from cell membrane to nucleus through SMAD proteins. Nature 1997; 390:465-471.
71. Zhao J, Shi W, Wang YL et al. Smad3 deficiency attenuates bleomycin-induced pulmonary fibrosis in mice. Am J Physiol Lung Cell Mol Physiol 2002; 282:585-593.
72. Nakao A, Afrakhte M, Moren A et al. Identification of Smad7, a TGFbeta-inducible antagonist of TGF-beta signalling. Nature 1997; 389:631-635.
73. Hayashi H, Abdollah S, Qiu Y et al. The MAD-related protein Smad7 associates with the TGFbeta receptor and functions as an antagonist of TGFbeta signaling. Cell 1997; 89:1165-1173.
74. Nakao A, Fujii M, Matsumura R et al. Transient gene transfer and expression of Smad7 prevents bleomycin-induced lung fibrosis in mice. J Clin Invest 1999; 104:5-11.
75. Antoniades HN, Bravo MA, Avila RE et al. Platelet-derived growth factor in idiopathic pulmonary fibrosis. J Clin Invest 1990; 86:1055-1064.
76. Martinet Y, Rom WN, Grotendorst GR et al. Exaggerated spontaneous release of platelet-derived growth factor by alveolar macrophages from patients with idiopathic pulmonary fibrosis. N Engl J Med 1987; 317:202-209.
77. Liu JY, Morris GF, Lei WH et al. Rapid activation of PDGF-A and -B expression at sites of lung injury in asbestos-exposed rats. Am J Respir Cell Mol Biol 1997; 17:129-140.
78. Hoyle GW, Li J, Finkelstein JB et al. Emphysematous lesions, inflammation, and fibrosis in the lungs of transgenic mice overexpressing platelet-derived growth factor. Am J Pathol 1999; 154:1763-1775.
79. Yi ES, Lee H, Yin S et al. Platelet-derived growth factor causes pulmonary cell proliferation and collagen deposition in vivo. Am J Pathol 1996; 149:539-548.
80. Yoshida M, Sakuma-Mochizuki J, Abe K et al. In vivo gene transfer of an extracellular domain of platelet-derived growth factor beta receptor by the HVJ-liposome method ameliorates bleomycin-induced pulmonary fibrosis. Biochem Biophys Res Commun 1999; 265:503-508.
81. Rom WN, Basset P, Fells GA et al. Alveolar macrophages release an insulin-like growth factor I-type molecule. J Clin Invest 1988; 82:1685-1693.
82. Maeda A, Hiyama K, Yamakido H et al. Increased expression of platelet-derived growth factor A and insulin-like growth factor-I in BAL cells during the development of bleomycin-induced pulmonary fibrosis in mice. Chest 1996; 109:780-786.
83. Bitterman PB, Adelberg S, Crystal RG. Mechanisms of pulmonary fibrosis. Spontaneous release of the alveolar macrophage-derived growth factor in the interstitial lung disorders. J Clin Invest 1983; 72:1801-1813.
84. Henke C, Marineili W, Jessurun J et al. Macrophage production of basic fibroblast growth factor in the fibroproliferative disorder of alveolar fibrosis after lung injury. Am J Pathol 1993; 143:1189-1199.
85. Inoue Y, King Jr TE, Tinkle SS et al. Human mast cell basic fibroblast growth factor in pulmonary fibrotic disorders. Am J Pathol 1996; 149:2037-2054.
86. Liu JY, Morris GF, Lei WH et al. Up-regulated expression of transforming growth factor-alpha in the bronchiolar-alveolar duct regions of asbestos-exposed rats. Am J Pathol 1996; 149:205-217.
87. Madtes DK, Busby HK, Strandjord TP et al. Expression of transforming growth factor-alpha and epidermal growth factor receptor is increased following bleomycin-induced lung injury in rats. Am J Respir Cell Mol Biol 1994; 11:540-551.

88. Korfhagen TR, Swantz RJ, Wert SE et al. Respiratory epithelial cell expression of human transforming growth factor-alpha induces lung fibrosis in transgenic mice. J Clin Invest 1994; 93:1691-1699.
89. Sugahara K, Iyama K, Kuroda MJ et al. Double intratracheal instillation of keratinocyte growth factor prevents bleomycin-induced lung fibrosis in rats. J Pathol 1998; 186:90-98.
90. Sakai T, Satoh K, Matsushima K et al. Hepatocyte growth factor in bronchoalveolar lavage fluids and cells in patients with inflammatory chest diseases of the lower respiratory tract: detection by RIA and in situ hybridization. Am J Respir Cell Mol Biol 1997; 16:388-397.
91. Maeda J, Ueki N, Hada T et al. Elevated serum hepatocyte growth factor/scatter factor levels in inflammatory lung disease. Am J Respir Crit Care Med 1995; 152:1587-1591.
92. Shiratori M, Michalopoulos G, Shinozuka H et al. Hepatocyte growth factor stimulates DNA synthesis in alveolar epithelial type II cells in vitro. Am J Respir Cell Mol Biol 1995; 12:171-180.
93. Yaekashiwa M, Nakayama S, Ohnuma K et al. Simultaneous or delayed administration of hepatocyte growth factor equally represses the fibrotic changes in murine lung injury induced by bleomycin. A morphologic study 1997; 156:1937-1944.
94. Nagahori T, Dohi M, Matsumoto K et al. Interferon-gamma upregulates the c-Met/hepatocyte growth factor receptor expression in alveolar epithelial cells. Am J Respir Cell Mol Biol 1999; 21:490-497.
95. Mason RJ, Shannon JM. Alveolar type II cells. In: Crystal RG, ed. The lung: scientific foundations. Philadelphia: Lippincot-Raven Publishers, 1997.
96. Coalson JJ. The ultrastructure of human fibrosing alveolitis. Virchows Arch A Pathol Anat Histol 1982; 395:181-199.
97. Barbas-Filho JV, Ferreira MA, Sesso A et al. Evidence of type II pneumocyte apoptosis in the pathogenesis of idiopathic pulmonary fibrosis (IFP)/usual interstitial pneumonia (UIP). J Clin Pathol 2001; 54:132-138.
98. Polunovsky VA, Chen B, Henke C et al. Role of mesenchymal cell death in lung remodeling after injury. J Clin Invest 1993; 92:388-397.
99. Hagimoto N, Kuwano K, Miyazaki H et al. Induction of apoptosis and pulmonary fibrosis in mice in response to ligation of Fas antigen. Am J Respir Cell Mol Biol 1997; 17:272-278.
100. Harrison Jr JH, Hoyt DG, Lazo JS. Acute pulmonary toxicity of bleomycin: DNA scission and matrix protein mRNA levels in bleomycin-sensitive and -resistant strains of mice. Mol Pharmacol 1989; 36:231-238.
101. Tounekti O, Pron G, Belehradek Jr J et al. Bleomycin, an apoptosis-mimetic drug that induces two types of cell death depending on the number of molecules internalized. Cancer Res 1993; 53:5462-5469.
102. Hagimoto N, Kuwano K, Nomoto Y et al. Apoptosis and expression of Fas/Fas ligand mRNA in bleomycin-induced pulmonary fibrosis in mice. Am J Respir Cell Mol Biol 1997; 16:91-101.
103. Kuwano K, Kunitake R, Kawasaki M et al. P21Waf1/Cip1/Sdi1 and p53 expression in association with DNA strand breaks in idiopathic pulmonary fibrosis. Am J Respir Crit Care Med 1996; 154:477-483.
104. Bardales RH, Xie SS, Schaefer RF et al. Apoptosis is a major pathway responsible for the resolution of type II pneumocytes in acute lung injury. Am J Pathol 1996; 149:845-852.
105. Guinee Jr D, Fleming M, Hayashi T et al. Association of p53 and WAF1 expression with apoptosis in diffuse alveolar damage. Am J Pathol 1996; 149:531-538.
106. Nelson WG, Kastan MB. DNA strand breaks: the DNA template alterations that trigger p53-dependent DNA damage response pathways. Mol Cell Biol 1994; 14:1815-1823.
107. El-Deiry WS, Harper JW, O'Connor PM et al. WAF1/CIP1 is induced in p53-mediated G1 arrest and apoptosis. Cancer Res 1994; 54:1169-1174.
108. Mishra A, Doyle NA, Martin WJ. Bleomycin-mediated pulmonary toxicity: evidence for a p53-mediated response. Am J Respir Cell Mol Biol 2000; 22:543-549.
109. Kuwano K, Hagimoto N, Tanaka T et al. Expression of apoptosis-regulatory genes in epithelial cells in pulmonary fibrosis in mice. J Pathol 2000; 190:221-229.
110. Davis DW, Weidner DA, Holian A et al. Nitric oxide–dependent activation of p53 suppresses bleomycin-induced apoptosis in the lung. J Exp Med 2000; 192:857-870.

111. Hagimoto N, Kuwano K, Miyazaki H et al. Induction of apoptosis and pulmonary fibrosis in mice in response to ligation of Fas antigen. Am J Respir Cell Mol Biol 1997; 17:272-278.
112. Kuwano K, Hagimoto N, Kawasaki M et al. Essential roles of the Fas-Fas ligand pathway in the development of pulmonary fibrosis. J Clin Invest 1999; 104:13-19.
113. Hagimoto N, Kuwano K, Kawasaki M et al. Induction of interleukin-8 secretion and apoptosis in bronchiolar epithelial cells by Fas ligation. Am J Respir Cell Mol Biol 1999; 21:436-445.
114. Kuwano K, Miyazaki H, Hagimoto N et al. The involvement of Fas-Fas ligand pathway in fibrosing lung diseases. Am J Respir Cell Mol Biol 1999; 20:53-60.
115. Matute-Bello G, Liles WC, Steinberg KP et al. Soluble Fas ligand induces epithelial cell apoptosis in humans with acute lung injury (ARDS). J Immunol 1999; 163:2217-2225.
116. Kuwano K, Kawasaki M, Maeyama T et al. Soluble form of fas and fas ligand in BAL fluid from patients with pulmonary fibrosis and bronchiolitis obliterans organizing pneumonia. Chest 2000; 118:451-458.
117. Yanagisawa K, Osada H, Masuda A et al. Induction of apoptosis by Smad3 and down-regulation of Smad3 expression in response to TGF-beta in human normal lung epithelial cells. Oncogene 1998; 17:1743-1747.
118. Hagimoto N, Kuwano K, Inoshima I et al. TGF-beta 1 as an enhancer of Fas-mediated apoptosis of lung epithelial cells. J Immunol 2002; 168:6470-6478.
119. Ward WF, Solliday NH, Molteni A et al. Radiation injury in rat lung. II. Angiotensin-converting enzyme activity. Radiat Res 1983; 96:294-300.
120. Marshall RP, McAnulty RJ, Laurent GJ. Angiotensin II is mitogenic for human lung fibroblasts via activation of the type 1 receptor. Am J Respir Crit Care Med 2000; 160:1999-2004.
121. Molteni A, Ward WF, Ts'ao CH et al. Monocrotaline-induced pulmonary fibrosis in rats: amelioration by captopril and penicillamine. Proc Soc Exp Biol Med 1985; 180:112-120.
122. Ward WF, Molteni A, Ts'ao CH et al. Captopril reduces collagen and mast cell accumulation in irradiated rat lung. Int J Radiat Oncol Biol Phys 1990; 19:1405-1409.
123. Uhal BD, Gidea C, Bargout R et al. Captopril inhibits apoptosis in human lung epithelial cells: a potential antifibrotic mechanism. Am J Physiol 1998; 275:L1013-1017.
124. Witschi HR, Haschek WM, Klein-Szanto AJ et al. Potentiation of diffuse lung damage by oxygen: determining variables. Am Rev Respir Dis 1981; 123:98-103.
125. Johnson KJ, Fantone 3rd JC , Kaplan J et al. In vivo damage of rat lungs by oxygen metabolites. J Clin Invest 1981; 67:983-993.
126. Schraufstatter IU, Revak SD, Cochrane CG. Proteases and oxidants in experimental pulmonary inflammatory injury. J Clin Invest 1984; 73:1175-1184.
127. Johnson KJ, Ward PA. Acute and progressive lung injury after contact with phorbol myristate acetate. Am J Pathol 1982; 107:29-35.
128. Cantin AM, North SL, Fells GA et al. Oxidant-mediated epithelial cell injury in idiopathic pulmonary fibrosis. J Clin Invest 1987; 79:1665-1673.
129. Cantin AM, Hubbard RC, Crystal RG. Glutathione deficiency in the epithelial lining fluid of the lower respiratory tract in idiopathic pulmonary fibrosis.Am Rev Respir Dis 1989; 139:370-372.
130. Behr J, Maier K, Degenkolb B et al. Antioxidative and clinical effects of high-dose N-acetylcysteine in fibrosing alveolitis. Adjunctive therapy to maintenance immunosuppression. Am J Respir Crit Care Med 1997; 156:1897-1901.
131. Meyer A, Buhl R, Magnussen H. The effect of oral N-acetylcysteine on lung glutathione levels in idiopathic pulmonary fibrosis. Eur Respir J 1994; 7:431-436.
132. Hagiwara SI, Ishii Y, Kitamura S. Aerosolized administration of N-acetylcysteine attenuates lung fibrosis induced by bleomycin in mice. Am J Respir Crit Care Med 2000; 162:225-231.
133. Kooy NW, Royall JA, Ye YZ et al. Evidence for in vivo peroxynitrite production in human acute lung injury. Am J Respir Crit Care Med 1995; 151:1250-1254.
134. Haddad IY, Pataki G, Hu P et al. Quantitation of nitrotyrosine levels in lung sections of patients and animals with acute lung injury. J Clin Invest.1994; 94:2407-2413.
135. Ischiropoulos H, Zhu L, Beckman JS. Peroxynitrite formation from macrophage-derived nitric oxide. Arch Biochem Biophys 1992; 298:446-451.
136. Saleh D, Barnes PJ, Giaid A. Increased production of the potent oxidant peroxynitrite in the lungs of patients with idiopathic pulmonary fibrosis. Am J Respir Crit Care Med 1997; 155:1763-1769.

137. Behr J, Degenkolb B, Maier K et al. Increased oxidation of extracellular glutathione by bronchoalveolar inflammatory cells in diffuse fibrosing alveolitis. Eur Respir J.1995; 8:1286-1292.
138. Haddad IY, Ischiropoulos H, Holm BA et al. Mechanisms of peroxynitrite-induced injury to pulmonary surfactants. Am J Physiol 1993; 265:555-564.
139. Pober JS, Cotran RS. The role of endothelial cells in inflammation. Transplantation 1990; 50:537-544.
140. Harrison NK, Myers AR, Corrin B et al. Structural features of interstitial lung disease in systemic sclerosis. Am Rev Respir Dis 1991; 144:706-713.
141. Wendt CH, Polunovsky VA, Peterson MS et al. Alveolar epithelial cells regulate the induction of endothelial cell apoptosis. Am J Physiol 1994; 267:C893-900.
142. Peao MN, Aguas AP, de Sa CM et al. Neoformation of blood vessels in association with rat lung fibrosis induced by bleomycin. Anat Rec 1994; 238:57-67.
143. Keane MP, Belperio JA, Arenberg DA et al. IFN-gamma-inducible protein-10 attenuates bleomycin-induced pulmonary fibrosis via inhibition of angiogenesis. J Immunol1999; 163:5686-5692.
144. Keane MP, Belperio JA, Moore TA et al. Neutralization of the CXC chemokine, macrophage inflammatory protein-2, attenuates bleomycin-induced pulmonary fibrosis. J Immunol 1999; 162:5511-5518.
145. Adamson IY, Young L, Bowden DH. Relationship of alveolar epithelial injury and repair to the induction of pulmonary fibrosis. Am J Pathol 1988; 130:377-383.
146. Mason RJ, Leslie CC, McCormick-Shannon K et al. Hepatocyte growth factor is a growth factor for rat alveolar type II cells. Am J Respir Cell Mol Biol 1994; 11:561-567.
147. Panos RJ, Rubin JS, Csaky KG et al. Keratinocyte growth factor and hepatocyte growth factor/scatter factor are heparin-binding growth factors for alveolar type II cells in fibroblast-conditioned medium. J Clin Invest 1993; 92:969-977.
148. Uhal BD, Joshi I, Hughes WF et al. Alveolar epithelial cell death adjacent to underlying myofibroblasts in advanced fibrotic human lung. Am J Physiol 1998; 275:1192-1199.
149. Uhal BD, Joshi I, True AL et al. Fibroblasts isolated after fibrotic lung injury induce apoptosis of alveolar epithelial cells in vitro. Am J Physiol 1995; 269:819-828.
150. King Jr TE, Schwarz MI, Brown K et al. Idiopathic pulmonary fibrosis: relationship between histopathologic features and mortality. Am J Respir Crit Care Med 2001; 164:1025-1032.
151. Terzaghi M, Nettesheim P, Williams ML. Repopulation of denuded tracheal grafts with normal, preneoplastic, and neoplastic epithelial cell populations. Cancer Res 1978; 38:4546-4553.
152. Tan A, Levrey H, Dahm C et al. Lovastatin induces fibroblast apoptosis in vitro and in vivo. A possible therapy for fibroproliferative disorders. Am J Respir Crit Care Med 1999; 159:220-227.
153. Kotani I, Sato A, Hayakawa H et al. Increased procoagulant and antifibrinolytic activities in the lungs with idiopathic pulmonary fibrosis. Thromb Res 1995; 77:493-504.
154. Imokawa S, Sato A, Hayakawa H et al. Tissue factor expression and fibrin deposition in the lungs of patients with idiopathic pulmonary fibrosis and systemic sclerosis. Am J Respir Crit Care Med 1997; 156:631-636.
155. Sisson TH, Hattori N, Xu Y et al. Treatment of bleomycin-induced pulmonary fibrosis by transfer of urokinase-type plasminogen activator genes. Hum Gene Ther 1999; 10:2315-2323.
156. Eitzman DT, McCoy RD, Zheng X et al. Bleomycin-induced pulmonary fibrosis in transgenic mice that either lack or overexpress the murine plasminogen activator inhibitor-1 gene. J Clin Invest 1996; 97:232-237.
157. McDonald JA, Broekelmann TJ, Matheke ML et al. A monoclonal antibody to the carboxyterminal domain of procollagen type I visualizes collagen-synthesizing fibroblasts. Detection of an altered fibroblast phenotype in lungs of patients with pulmonary fibrosis. J Clin Invest 1986; 78:1237-1244.
158. Roman J. Extracellular matrix and lung inflammation. Immunol Res 1996; 15:163-178.
159. Gudewicz PW, Frewin MB, Heinel LA et al. Priming of human monocyte superoxide production and arachidonic acid metabolism by adherence to collagen- and basement membrane-coated surfaces. J Leukoc Biol 1994; 55:423-429.
160. Juliano RL, Haskill S. Signal transduction from the extracellular matrix. J Cell Biol 1993; 120:577-585.
161. Ruoslahti E, Reed JC. Anchorage dependence, integrins, and apoptosis. Cell 1994; 77:477-478.

162. Basset F, Ferrans VJ, Soler P et al. Intraluminal fibrosis in interstitial lung disorders. Am J Pathol 1986; 122:443-461.
163. Fukuda Y, Ishizaki M, Kudoh S et al. Localization of matrix metalloproteinases-1, -2, and -9 and tissue inhibitor of metalloproteinase-2 in interstitial lung diseases. Lab Invest 1998; 78:687-698.
164. Lemjabbar H, Gosset P, Lechapt-Zalcman E et al. Overexpression of alveolar macrophage gelatinase B (MMP-9) in patients with idiopathic pulmonary fibrosis: effects of steroid and immunosuppressive treatment. Am J Respir Cell Mol Biol 1999; 20:903-913.
165. Wright JK, Cawston TE, Hazleman BL. Transforming growth factor beta stimulates the production of the tissue inhibitor of metalloproteinases (TIMP) by human synovial and skin fibroblasts. Biochim Biophys Acta 1991; 1094:207-210.
166. Swiderski RE, Dencoff JE, Floerchinger CS et al. Differential expression of extracellular matrix remodeling genes in a murine model of bleomycin-induced pulmonary fibrosis. Am J Pathol 1998; 152:821-828.
167. Kunugi S, Fukuda Y, Ishizaki M et al. Role of MMP-2 in alveolar epithelial cell repair after bleomycin administration in rabbits. Lab Invest 2001; 81:1309-1318.

CHAPTER 15

Silica-Induced Inflammatory Mediators and Pulmonary Fibrosis

Andrea K. Hubbard, Sarah Mowbray, Michael Thibodeau and Charles Giardina

Abstract

Silicosis continues to be a lung disease with significant morbidity and mortality. Although silica-induced lung injury and cell activation and/or death have been investigated over the past several years, basic research continues to reveal the cell: cell and cell: mediator interactions critical to these events. This chapter will emphasize the production and participation of several inflammatory cytokines, mediators and cell processes in the development of silica-induced lung injury and fibrosis. Mediators to be discussed will include TNFα, IFNγ, IL-1β, IL-12, IL-18, IL-9, TGFβ, MMPs/TIMPs, ROS/RNS, caspases and Fas/FasL in the processes of cell activation, cell proliferation and cell death. The mediator networks and apoptotic pathways elicited by this inorganic particle are complex, driven by many cell types and affect numerous cell functions. Understanding these interactions will help in developing strategies for therapeutic intervention at different stages of the disease.

Silica and Silicosis

In the United States alone, silicosis was listed as a primary or contributing cause of death in 4313 individuals from 1979-1990.[1] With the significant occupational health risk posed by silica exposure,[2] research endeavors to explore the cellular and inflammatory mechanisms underlying silica induced lung injury are critical.

Silica-Induced Lung Injury (Fig. 1)

Much of the research that describes human silicosis is the result of investigations in animal models of silica-induced lung injury. In instances of low levels of silica exposure, there can be sufficient clearance to result in minimal toxicity and resolution of cell injury. At higher levels of exposure, there is crystal deposition at the alveolar duct bifurcations, phagocytosis of particles[3] and a rapid influx of neutrophils (PMNs) and monocytes into the alveoli.[4,5] Phagocytosis by alveolar inflammatory cells may result in cell activation or cell death due to apoptosis[6] or necrosis-the latter resulting in the release of lysosomal enzymes[4] and lactate dehydrogenase (LDH).[5] In addition, there is the generation of inflammatory cytokines, reactive oxygen species (ROS)[7,8] and reactive nitrogen species (RNS)[9,10] by pulmonary macrophages, lymphocytes and bronchiolar and alveolar epithelial cells eliciting focal damage to type I epithelial cells.[11] In response to this damage, silica exposure is documented to cause proliferation of type

Fibrogenesis: Cellular and Molecular Basis, edited by Mohammed S. Razzaque.

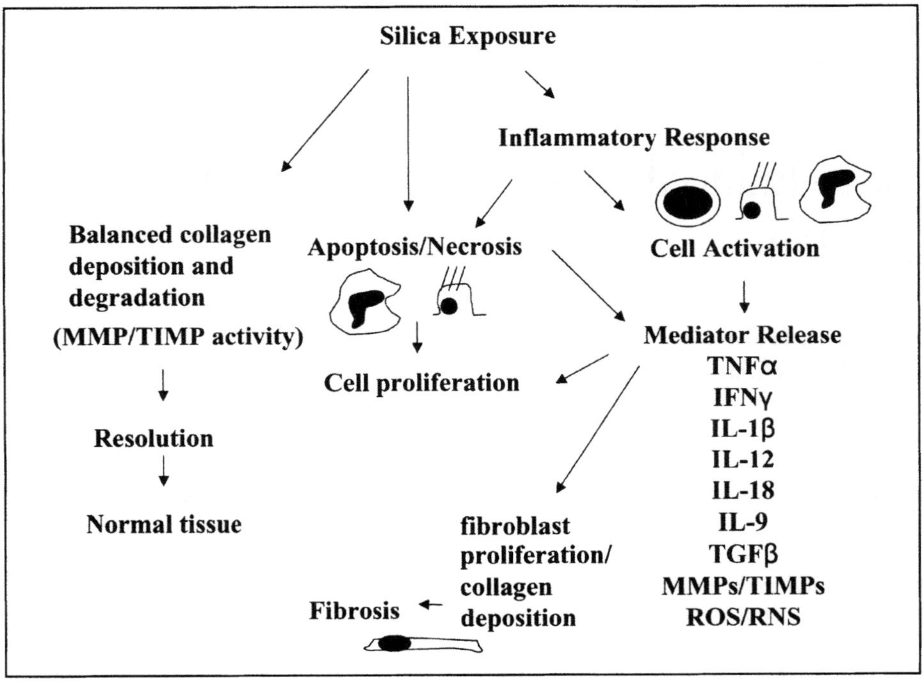

Figure 1. Silica-induced cell injury and resulting mediator release. At levels of low silica exposure, there may be little cell damage and ultimate resolution of injury with return to normal structure. Alternately, at higher levels of exposure, there may be cell death by either apoptotic and/or necrotic pathways with compensatory cell proliferation. With the initiation of a leukocyte inflammatory response, there is resulting activation of multiple cell types including macrophages, lymphocytes, and epithelial cells (bronchiolar, alveolar) with the production of numerous mediators leading to the dysregulation of collagen production by fibroblasts and ensuing fibrosis.

II epithelial cells at low doses and cell death at higher doses.[12] Silica induced activation of pulmonary cells, and the subsequent release of inflammatory mediators and fibrogenic factors,[13] appears to lead to pulmonary fibrosis.[14] As a reflection of this fibrotic response, animals exposed to silica also evidence impaired lung function (e.g., decreased lung compliance, vital capacity and diffusion capacity).[15]

Silica-Induced Mediators in Pulmonary Fibrosis

Most pulmonary fibrosis begins in the alveolus and develops in definable stages over time with epithelial cell injury and alveolar inflammation, organization of the ensuing alveolar exudates and incorporation of the alveolar fibroproliferative process into alveolar walls. Repeating cycles of these events can lead to the characteristic signature of distorted and dysfunctional lung parenchyma.[16,17] Integral in the understanding of the cellular and molecular basis of silica induced pulmonary fibrosis is the examination of silica induced inflammatory mediators in this process.

Tumor Necrosis Factor α (TNFα)

TNFα primarily produced by activated macrophages is a cytokine with numerous pro-inflammatory effects (e.g., macrophage activation, adhesion molecule expression, lymphocyte development, etc).[18] Many reports in the literature have documented the induction of

TNFα following silica exposure in vitro or in vivo. Driscoll et al.[19] demonstrated that silica exposure in vivo primed alveolar macrophages (AM) to release both interleukin-1 (IL-1) and TNFα following in vitro exposure to lipopolysaccharide (LPS). In the murine macrophage-like cell line RAW264.7, silica elicited a significant increase in TNFα secretion after 48 hrs of culture.[20] Intratracheal instillation of silica crystals into mice also elicited an increase in TNFα mRNA synthesis in lung tissue.[21] There appears to be a direct relationship between TNFα expression and silica induced pulmonary fibrosis. For example, Piguet et al confirmed that silica -induced collagen deposition could be attenuated by an anti-TNFα antibody, but significantly increased by continuous infusion of mouse recombinant TNF. This same group noted that infusion of a recombinant soluble TNF receptor prevented silica-induced fibrosis evidenced by decreased lung hydroxyproline.[22] In silica instilled mice there were significantly elevated levels of TNFα released from lavaged cells 28 days after exposure.[23] Similarly, Gossart et al[24] noted in rats increased TNFα in the bronchoalveolar lavage (BAL) and from lavaged AM three days after silica instillation. Rats exposed to silica aerosol increased TNFα production by BAL cells after 30 days of exposure with continued increases throughout the remaining 86 days. In addition, silica has been found to increase the expression of mRNA of both TNFα and the p75 TNFα receptor (R), but not the p55 TNFα receptor.[25] These authors also noted in double TNFR knockout mice that although silica could increase TNFα expression in these animals, they were protected from the fibrogenic effects of the particle.[26] In contrast, in another mouse model of silica induced lung damage, Huaux et al[27] noted increased TNFα mRNA in lung tissue without concomitant increases in TNFα protein.

Interferon γ (IFNγ)

IFNγ is a cytokine produced by the TH1 CD4⁺ lymphocyte population and by natural killer (NK) cells, which participates in the activation of macrophages.[28] Garn et al[29] reported a threefold increase in IFNγ mRNA in enlarged thoracic lymph nodes of rats exposed to silica by aerosol. In addition, in mice exposed to silica by aerosol, there was an increase in IFNγ mRNA in lung tissue and increased IFNγ in BAL cells (e.g., $CD4^+$ lymphocytes, $CD8^+$ lymphocytes, NK cells and γδ TCR lymphocytes). These authors speculated that IFNγ might contribute to silica-induced fibrosis through the activation of lung macrophages and concomitant TNFα production.[30] They supported their hypothesis by examining the silicotic response in IFNγKO mice. These animals evidenced much less fibrotic disease in response to silica aerosol as measured by histopathological analysis and by reduced lung collagen content.[31]

Interleukin-1β (IL-1β)

IL-1β is released by activated macrophages and participates in the activation of $CD4^+$ lymphocytes[32] and has been implicated in collagen production.[33] Approximately 10 years ago, release of IL-1β by silica activated resident peritoneal cells was first reported.[34] Since then other reports have documented the expression and participation of IL-1β in silica induced lung injury. Mice exposed to silica by inhalation increased production of both TNFα and IL-1β in alveolar mononuclear cells, in lung silicotic nodules and in lymphoid tissue.[35] In rats exposed to silica by inhalation, increased IL-1β was detected in BAL cells.[25] Mice genetically deficient in IL-1β (knockout) evidenced significantly fewer silicotic lesions 12 weeks after silica inhalation exposure.[36] These studies would suggest that silica activated pulmonary macrophages release IL-1β that in turn stimulates collagen synthesis.

Interleukin-12 (IL-12) / Interleukin-18 (IL-18)

IL-12 is a cytokine produced by macrophages, monocytes, dendritic cells and B cells and reported to induce IFNγ production by TH1 lymphocytes and NK cells.[37] Interleukin 18 produced by monocytes and macrophages is a weak inducer of IFNγ directly, but a potent inducer of IL-12, which in turn can induce IFNγ.[38] In mice exposed to silica by aerosol, there was increased mRNA expression for both the IL-12 p40 subunit and IL-18 in lung tissue. IL-12 mRNA was increased early (2 wks) and late (16 wks) whereas IL-18 was increased late during the fibroproliferative phase.[31] Garn et al[29] noted increased IL-12, but not IL-18 mRNA in the enlarged lymph nodes of silicotic rats and Huaux et al[39] reported increased IL-12 p40 subunit protein and mRNA in mice instilled with silica. Through their indirect activation of macrophages by augmenting IFNγ production, these mediators would also most likely contribute to silica induced collagen deposition.

Interleukin 9 (IL-9)

IL-9 has been implicated in the regulation of lung diseases including silica induced pulmonary fibrosis. This interleukin is produced primarily by TH2 lymphocytes[40] and participates as a growth factor for T cells and mast cells. Using Tg5 transgenic mice, which over express IL-9, Arras et al[41] noted an inhibition of silica-induced lung fibrosis with an accompanying decrease in TH2 responses, but an accumulation of B lymphocytes in the lung parenchyma. Thus, the role of TH1 and/or TH2 cells in this particle-induced fibrotic response is not resolved with each cell type indirectly modulating the response.

Transforming Growth Factor β (TGFβ)

TGFβ1 is expressed by bronchiolar alveolar epithelial cells as well as by mesenchymal cells and lung macrophages and its expression is accompanied by collagen and fibronectin production.[42] The ability of silica to increase TGFβ1 has been demonstrated in animal models and with human cells. Rats exposed to alumina silica refractory fiber by either instillation or inhalation increased expression of TNFα, IL-6 and TGFβ1 mRNA in lung tissue.[43] The interaction between TNFα and TGFβ1 is also evidenced by the observation that TNFα receptor KO mice do not produce TGFβ1 in response to fibers.[44] However, overexpression of TGFβ1 in mice through an adenovirus vector was sufficient to initiate fibroproliferative lung disease in nonparticle exposed TNFR KO animals.[45] Other studies, however, suggest that silica can decrease expression of TGFβ1. Investigators evaluated the effect of silica in vitro on a human lung fibroblast cell line (WI-1003) and found that silica internalization directly stimulated collagen synthesis, but elicited less TGFβ1 than nontreated cells. Silica also antagonized TGFβ1 activities though down regulation of this cytokine.[46] It is clear that the precise role of TGFβ1 in silica-induced fibrosis remains to be determined.

Matrix Metalloproteinases (MMPs)

MMPs are a large group of zinc-dependent endopeptidases, distinguished from other peptidases by their requirement for metal ions to degrade components of the extracellular matrix (ECM). MMPs are typically secreted in the pro-enzyme form and require cleavage of the signal peptide for activity. Additionally, the activity of MMPs is tightly regulated at the transcriptional level by the NFκB and AP-1 transcription factors.[47] MMP activity is also influenced by natural tissue inhibitors, tissue inhibitors of matrix metalloproteinases (TIMPs), which bind to the active site of MMPs and render them enzymatically inactive. MMP secretion is modulated by a variety of cytokines and growth factors including IL-1, TNF-α, and TGF-β.[48] MMPs are sub-classified according to substrate specificity and amino acid sequences to include collagenases (MMP-1,-8,-13,-18), stromelysins (MMP-3,-10,-11), gelatinases (MMP-2,-9), macrophage metalloelastase (MMP-12), matrilysin (MMP-7,-26), and the membrane type MMPs (MTMMPs) (MMP-14-17,-23-25).[49] These enzymes are made by a wide variety of cell types including, leukocytes, epithelial cells, endothelial cells, fibroblasts, osteoclasts, and keratinocytes.[50-58] Although MMP activity is integral to normal processes such as wound healing and embryogenesis, unregulated MMP activity has been implicated in a variety of diseases including rheumatoid arthritis,[59] asthma,[60] multiple sclerosis[61] and pulmonary fibrosis.[62,63]

Since MMPs are the major enzyme group responsible for the degradation of extracellular matrix, including collagen, it is postulated that MMP activity (or its dysregulation) may play a role in the pathogenesis of the fibrotic response. In support of this theory, the gene expression of MMP-1,-2,-7, and -9 were found to be upregulated in the lungs of patients suffering from idiopathic pulmonary fibrosis (IPF).[62] In a rat model of experimental lung silicosis, the expression of MMPs-2, -9 and -13 and TIMPs-1 and -2 differed during the progression of silicosis.[63] In early silicotic nodules (day 15), immunohistochemical staining and in situ hybridization for MMP-2, -9 and -13 and TIMP-1 and -2 was intense. In contrast, at 60 days post intratracheal instillation, lower MMP expression was evident with only a slight reduction in TIMP expression . These results suggest that the change in MMP/TIMP balance seen over the course of a fibrotic response is key to the regulation of ECM remodeling associated with silicosis. As discussed above, TNFα is a cytokine found to play a major role in the cellular response to silica exposure, as well as in the development of fibrosis.[21] In a mouse model of silicosis, mice deficient in either TNF receptor p55 or p75 exhibited significantly less collagen accumulation than wild type animals with an accompanying decrease in TIMP-1, but not MMP-13.[64] Thus the absence of TNFα receptor signaling may promote matrix degradation by decreasing overexpression of TIMP1. The precise role for MMPs in silicosis has yet to be defined adequately, but studies described suggest the contribution or dysregulation of MMP activity participates in silica induced lung injury and fibrosis.

ROS/NOS

The generation of ROS following silica exposure may be derived directly from the surface of the crystal[65,66] or indirectly through silica activation of inflammatory cells.[67] For the former, two distinct kinds of surface centers—silica based surface radicals and poorly coordinated iron ions—can generate $O_2^{-\bullet}$ and $HO^\bullet$ in aqueous solution.[68] Cell generated radicals have also been documented in vitro and in vivo. Gossart et al[24] noted that AM lavaged from rats evidenced increased zymosan or phorbol ester-triggered free radical production (i.e., chemiluminescence) as early as 1 day after silica exposure. Vallyathan et al[69] demonstrated that silica exposure of rats elicited significant pulmonary oxidative stress as well as increased lipid peroxidation and increased levels of antioxidant enzymes as early as 2 days post silica instillation. The relationship between generation of ROS and pulmonary fibrosis is documented by the observation that antioxidants reduce the silicotic fibrosis. Pretreatment of rats with a free radical scavenger (N-ter-butyl-α-phenylnitrone) prior to silica exposure reversed pathologic changes in the lung and decreased AM ROS and TNFα expression.[24] In addition, treatment of rat AM with the antioxidant, NAC, decreased silica induced cell injury[70] and the synthesis of several cytokines including TNFα.[71]

Production of RNS may also contribute to silica induced lung injury and fibrosis. Rats exposed to silica through intratracheal injection evidenced inducible nitric oxide synthetase (iNOS) expression in lungs and an elevated level of NO in pulmonary lesions.[10] In another study, cells lavaged from rats demonstrated increased iNOS mRNA 24 hrs after silica instillation.[9] Silica exposure of mice genetically deficient in iNOS elicited significantly fewer histopathologic changes including silicotic lesions.[36]

FasL, Caspase Activation and Apoptosis

Although silica has been documented to cause necrotic cell death, substantial evidence is emerging in vitro and in vivo that silica also induces apoptosis. Apoptosis is an active form of cell death requiring coordinated cellular and molecular activities. Apoptosis may contribute to silicotic inflammation culminating in pulmonary fibrosis. The induction of macrophage apoptosis by crystalline silica is demonstrable by cell morphology, caspase activation, and DNA fragmentation.[34,72] The DNA fragmentation induced by silica is evidenced by the formation of low molecular weight oligonucleosomal fragments as well as by increased formation of 3'-hydroxyl ends detected by nuclear TUNEL staining.[72-74]

Recently, the ability of a particle to induce alveolar macrophage apoptosis in vitro was associated with the particle's ability to produce pulmonary fibrosis. Under these conditions, the more fibrogenic crystalline α-quartz silica caused greater apoptosis than the less fibrogenic particles, amorphous silica or titanium dioxide.[75] However, this is not to say that poorly fibrogenic particles do not cause macrophage apoptosis, since higher concentrations of either of the less fibrogenic particles induced an apoptotic phenotype.[75] In addition to the induction of apoptosis in vitro, crystalline silica also causes higher levels of apoptosis in the lung. Rats administered a single intratracheal instillation of α-quartz have apoptotic cell morphology discernible in lung histology sections and bronchoalveolar lavage cells at 10-days post-exposure.[6] DNA extracts from these bronchoalveolar lavage cells display the paradigmatic DNA laddering of apoptosis. Similarly, mice after a 10-day inhalation exposure to α-quartz have increased TUNEL-staining in lung histology sections at 1 and 6-weeks post-exposure.[36]

Although the molecular mechanisms leading to macrophage apoptosis by silica remain uncertain, there is an apparent regulatory role by both ROS and RNS. In vitro exposure of rat bronchoalveolar macrophages to α-quartz leads to the generation of ROS that precedes the activation of caspases-9 and -3, poly(ADP-ribose) polymerase (PARP) cleavage, and TUNEL-staining.[73] Additionally, nitric oxide (NO) is important to silica-induced apoptosis since prevention of NO formation by the nitric oxide synthetase inhibitor L-NAME ameliorates α-quartz-induced apoptosis of the mouse macrophage cell line IC-21.[36] In addition, mice lacking inducible nitric oxide synthetase (iNOS) not only have decreased TUNEL-staining in lung histology sections after inhalation of α-quartz, but also have reduced inflammation compared to wild-type controls.[36]

Particular importance has recently been attributed to Fas ligand (FasL), which appears to be indispensable for apoptosis and lung injury elicited by silica. BALB.*gld* mice deficient in FasL are particularly resistant to the development of silica-induced inflammation, TNF-α production, and fibrosis.[76] Similar resistance to pulmonary injury occurs in wild type mice administered neutralizing antibody against FasL after intratracheal instillation of α-quartz.[76] The Fas-FasL pathway is already well established as an important molecular mechanism in the pathogenesis of pulmonary fibrosis.[77-79] In the mouse model of silica-induced pulmonary injury, the macrophage is presumed to be the cell type responsible for FasL mediated lung injury and inflammation. Wild-type mice reconstituted with BALB.*gld* FasL deficient bone marrow fail to develop silica-induced inflammation, indicating bone marrow-derived cells, not lung parenchyma, initiate the pulmonary injury.[76] Likewise, FasL has increased expression in silicotic mouse lungs, and the majority of FasL-expressing cells stain for the F4/80 marker of murine macrophages.[76] Although FasL is important to apoptosis, increased FasL activity could also potentially contribute to inflammation and fibrosis because trimerization of the Fas receptor can activate the transcription factor NF-κB, which may lead to the downstream elaboration of pro-inflammatory mediators.[77,80-82]

Further strengthening the link between apoptosis and silica-induced injury are recent observations that the prevention of apoptosis by the administration of caspase inhibitors can abrogate the development of pulmonary inflammation and fibrosis.[83,84] Caspases are a family of cysteine-containing aspartate-specific proteases that cleave after aspartic acid residues of certain protein substrates. Although caspase-independent apoptosis is reported, these proteases are a usual component of apoptosis, such as that induced by silica. Caspases-1, -2, -3, -6, and -9 are activated in silica treated macrophages.[73,84] Macrophage apoptosis as measured by cell morphology or DNA fragmentation can be partially abolished by the irreversible pancaspase inhibitor N-benzyloxycarbonyl-Val-Ala-Asp-fluoromethylketone (Z-VAD-FMK).[72] In addition, the caspase-3 inhibitor N-benzyloxycarbonyl-Asp-Glu-Val-Asp-fluoromethyl ketone (Z-DEVD-FMK) also prevents silica-induced apoptosis of macrophages.[72,73] Caspase inhibition studies have since been expanded to a mouse model of silica-induced lung injury in vivo.

Mice administered an intratracheal instillation of α-quartz will accumulate neutrophils in bronchoalveolar lavages and deposit collagen at sites of silicotic inflammation. Such changes are significantly decreased by the administration of the pancaspase inhibitor Z-VAD-FMK.[84] These findings corroborate reports of others that caspase inhibition ameliorates parenchymal injury, inflammation, and fibrosis occurring with stimuli as diverse as bleomycin and ischemia-reperfusion.[83,86]

How caspase activation and apoptosis may influence pulmonary inflammation and fibrosis remains unknown. It is accepted as dogma that apoptotic activation of caspases leads to the degradation of homeostatic and structural proteins such as PARP, lamins, fodrin, cytokeratin 18, and transglutaminase II.[87] However, multiple lines of evidence also show a requirement for caspase activity in the proteolytic maturation of several pro-inflammatory cytokines. For example, caspase-1 contributes to the maturation of IL-1β and IL-18; whereas the effector caspases-7 and -3 can process endothelial monocyte-activating polypeptide II (EMAP II) and IL-16 to their active forms.[88-92] Both IL-1β and IL-18 have already been suggested to play a role in mouse models of silicosis.[30,34,93-95] As more caspase substrates become elucidated, a greater understanding for the role of apoptosis to the silicotic induction of inflammation and fibrosis will surely become apparent.

Conclusions

This Chapter has emphasized the production and participation of several inflammatory mediators and cell processes in the development of silica-induced lung injury and fibrosis (Table 1). Several key points are apparent. The mediator networks elicited by this inorganic particle

Table 1. The role of silica-induced mediators in pulmonary fibrosis

Mediator	Source	Function	Role in Silica-Induced Fibrosis
TNFα	Macrophages	Pleotrophic Pro-inflammogen	+++
IFNγ	TH1 CD4+ lymphocytes NK cells	Activates macrophages	++
IL-1β	Macrophages	Activates macrophages Stimulates collagen synthesis	++
IL-12	Macrophages; B cells	Induces IFNγ	+
IL-18	Macrophages	Induces IL-12	+
IL-9	TH2 CD4+ lymphocytes	Growth factor for lymphocytes	↓
TGFβ	Macrophages Epithelial cells	Modulates collagen, fibronectin synthesis	↓↑
ROS	Macrophages PMNs	Enhances cytokine expression; Lipid peroxidation	+++
RNS	Epithelial cells		+++
MMPs/TIMPs	Leukocytes Epithelial cells Endothelial cells	Modulates collagen deposition	↓↑
Caspase activation	Macrophages Epithelial cells	Apoptosis Cytokine activation	+++
FasL	Leukocytes Epithelial cells	Apoptosis	+++

are complex, driven by many cell types and affecting numerous cell functions to include cell activation, cell proliferation and cell death (i.e., necrosis, apoptosis). In addition, silica-induced mediator production is often redundant with tremendous interaction between one mediator and another through synergy and/or antagonism. This is illustrated by the following examples: ROS increasing TNFα expression, IL-12 increasing IFNγ and IL-18, in turn, increasing IL-12, the interaction of TNFα with TGFβ, the induction of MMPs by IL-1, TNFα and TGFβ and the activation of IL-1 and IL-18 by caspase 1. Clearly, future research will continue to elucidate this intricate, potent and interacting network of mediators key to silica induced lung injury and fibrosis. It is through the exploration of these interactions that future therapeutic intervention at different stages of disease may result.

Acknowledgements

Portions of this work were supported by NIEHS R15 ES09433, NHLBI F32 HL10360 and the University of Connecticut Research Foundation.

References

1. National Institute for Occupational Safety and Health. Division of Respiratory Disease Studies. Work Related Lung Disease Surveillance report. DHHS publication no. (NIOSH) 94-120, 1994.
2. Ad Hoc Committee of the Scientific Assembly on Environmental and Occupational Health. Adverse effects of crystalline silica exposure. Am J Respir Crit Care Med 1997; 155:761-765.
3. Brody AR, Roe MW, Evans JN et al. Deposition and translocation of inhaled silica in rats. Quantification of particle distribution, macrophage participation and function. Lab Invest 1982; 47:533-542.
4. Callis AH, Sohnle PG, Mandel GS et al. Kinetics of inflammatory and fibrotic changes in a mouse model of silicosis. J Lab Clin Med 1985; 105:547-533.
5. Driscoll KE, Maurer JK, Lindenschmidt RC et al. Respiratory tract responses to dust: Relationships between dust burden, lung injury, alveolar macrophage fibronectin release and the development of pulmonary fibrosis. Toxicol Appl Pharmacol 1990; 106:88-101.
6. Leigh J, Wang H, Bonin A et al. Silica-induced apoptosis in alveolar and granulomatous cells in vivo. Environ Health Perspect 1997; 105:1241-5.
7. Vallyathan V, Mega JF, Shi X et al. Enhanced generation of free radicals from phagocytes induced by mineral dusts. Am J Respir Cell Mol Biol 1992; 6:404-413.
8. Schapira RM, Ghio AJ, Effros RM et al. Hydroxyl radical production and lung injury in the rat following silica or titanium dioxide instillation in vivo. Am J Respir Cell Mol Biol 1995; 12:220-226.
9. Blackford JA, Antonini JM, Castranova V et al. Intratracheal instillation of silica up regulates inducible nitric oxide synthase gene expression and increases nitric oxide production in alveolar macrophages and neutrophils. Am J Respir Cell Mol Biol 1994; 11:426-431.
10. Setoguchi K, Takeya M, Akaike T et al. Expression of inducible nitric oxide synthase and its involvement in pulmonary granulomatous inflammation in rats. Am J Pathol 1996; 149:2005-2022.
11. Adamson IYR, Bowden DH. Role of polymorphonuclear leukocytes in silica-induced pulmonary fibrosis. Am J Pathol 1984; 117:37-43.
12. Lesur O, Cantin AM, Tanswell AK et al. Silica exposure induces cytotoxicity and proliferative acitivity of type II pneumocytes. Exp Lung Res 1992; 18:173-190.
13. Adamson IYR., Letourneau, HL, Bowden DH. Comparison of alveolar and interstitial macrophages in fibroblast stimulation after silica and long or short asbestos. Lab Inves 1991; 64:339-344.
14. Reiser KM, Hascher WM, Hesterberg TW et al. Experimental silicosis. II. Long term effects of intratacheally instilled quartz on collagen metabolism and morphologic characteristics of rat lungs. Am J Pathol 1983; 110:30-41.
15. Begin R, Dufresne A, Cantin A et al. Quartz exposure, retention and early silicosis in sheep. Exp Lung Res 1989; 15:409-428.
16. Kuhn C 3rd, Boldt J, King TE Jr et al. An immunohistochemical study of architectural remodeling and connective tissue synthesis in pulmonary fibrosis. Am Rev Respir Dis 1989; 140:1693-703.
17. Crouch E. Pathobiology of pulmonary fibrosis. Am J Physiol 1990; 259:L159-84.

18. Roitt I, Brostoff J, Male D. Immunology. 6th ed. New York: Harcourt Publishers Limited, 2001:119-129.
19. Driscoll KE, Lindenschmidt RC, Maurer JK et al. Pulmonary response to silica or TiO2: Inflammatory cells, alveolar macrophage-derived cytokines and histopath. Am J Respir Cell Mol Biol 1990; 2:381-390.
20. Claudio E, Segade F, Wrobel K et al. Activation of murine macrophages by silica particles in vitro is a process independent of silica-induced cell death. Am J Respir Cell Mol Biol 1995; 13:547-554.
21. Piguet PF, Collart MA, Grau GF et al. Requirement of tumor necrosis factor for development of silica-induced pulmonary fibrosis. Nature 1990; 344:245-7.
22. Piguet PF, Vesin C. Treatment by human recombinant soluble TNF receptor of pulmonary fibrosis induced by bleo or silica in mice. Eur Respir J 1994; 7:515-518.
23. Ohtsuka Y, Munakata M, Ukita H et al. Increased susceptibility to silicosis and TNF production in C57Bl/6 mice. Am J Respir Crit Care Med 1995; 152:2144-2149.
24. Gossart S, Cambon C, Orfila C et al. Reactive oxygen intermediates as regulators of TNF production in rat lung inflammation induced by silica. J Immunol 1996; 156:1540-1548.
25. Porter DW, Ye J, Mas J et al. Time course of pulmonary response of rats to inhalation of crystalline silica: NF-kappa B activation, inflammation, cytokine production, and damage. Inhal Toxicol 2002; 14:349-67.
26. Ortiz LA, Lasky J, Lungarella G et al. Upregulation of the p75 but not the p55 TNF-alpha receptor mRNA afte silica and bleomycin exposure and protection from lung injury in double recpetor kncokout mice. Am J Respir Cell Mol Biol 1999; 20:825-33.
27. Huaux F, Arraas M, Vink A et al. Soluble tumor necrosis factr (TNF) receptors p55 and p75 and interluekin-10 downregulate TNF-alpha activity during the lung response to silica particles in NMR1 mice. Am J Respir Cell Mol Biol 1999; 21:137-45.
28. Roitt I, Brostoff J, Male D. Immunology. 6th ed. New York (NY): Harcourt Publishers Limited, 2001:147-162.
29. Garn H, Friedetzky A, Kirchner A et al. Experimental silicosis: a shift to a preferential IFN-gamma-based Th1 response in thoracic lymph nodes. Am J Physiol Lung Cell Mol Physiol 2000; 278:L1221-30.
30. Davis GS, Pfeiffer LM, Hemenway DR. Interferon-gamma production by specific lung lymphocyte phenotypes in silicosis in mice. Am J Respir Cell Mol Biol 2000; 22:491-501.
31. Davis GS, Holmes CE, Pfeiffer LM et al. Lymphocytes, lymphokines, and silicosis. J Environ Pathol Toxicol Oncol 2001; 20:53-65.
32. Roitt I, Brostoff J, Male D. Immunology. 6th ed. New York: Harcourt Publishers Limited, 2001:105-118.
33. Zhang Y, Lee TC, Guillemin B et al. Enhanced IL-1 beta and tumor necrosis factor-alpha release and messenger RNA expression in macrophages from idiopathic pulmonary fibrosis or after asbestos exposure. J Immunol 1993; 150:4188-96.
34. Sarih M, Souvannavong V, Brown SC et al. Silica induces apoptosis in macrophages and the release of interleukin-1 alpha and interleukin-1 beta. J Leukoc Biol 1993; 54:407-13.
35. Davis GS, Pfeiffer LM, Hemenway DR. Persistent overexpression of interleukin-1beta and tumor necrosis factor-alpha in murine silicosis. J Environ Pathol Toxicol Oncol 1998; 17:99-114.
36. Srivastava KD, Rom WN, Jagirdar J et al. Crucial role of interleukin-1beta and nitric oxide synthase in silica-induced inflammation and apoptosis in mice. Am J Respir Crit Care Med 2002; 165:527-33.
37. Gately MK, Renzetti LM, Magram J et al. The interleukin-12/interleukin-12-receptor system: role in normal and pathologic immune responses. Annu Rev Immunol 1998; 16:495-521.
38. Dinarello CA. IL-18: A TH1-inducing, proinflammatory cytokine and new member of the IL-1 family. J Allergy Clin Immunol 1999; 103(1 Pt 1):11-24.
39. Huaux F, Lardot C, Arras M et al. Lung fibrosis induced by silica particles in NMRI mice is associated with an upregulation of the p40 subunit of interleukin-12 and Th-2 manifestations. Am J Respir Cell Mol Biol 1999; 20:561-72.
40. Renauld JC, Goethals A, Houssiau F et al. Human P40/IL-9. Expression in activated CD4+ T cells, genomic organization, and comparison with the mouse gene. J Immunol 1990; 144:4235-41.

41. Arras M, Huaux F, Vink A et al. Interleukin-9 reduces lung fibrosis and type 2 immune polarization induced by silica particles in a murine model. Am J Respir Cell Mol Biol 2001; 24:368-75.
42. Broekelmann TJ, Limper AH, Colby TV et al. Transforming growth factor beta 1 is present at sites of extracellular matrix gene expression in human pulmonary fibrosis. Proc Natl Acad Sci USA 1991; 88:6642-6.
43. Morimoto Y, Tsuda T, Yamato H et al. Comparison of gene expression of cytokines mRNA in lungs of rats induced by intratracheal instillation and inhalation of mineral fibers. Inhal Toxicol 2001; 13:589-601.
44. Liu JY, Brody AR. Increased TFG-beta1 in the lungs of asbestos-expsoed rats and mice: Reduced expression in TNF-alpha receptor knockout mice. J Environ Pathol Toxicol Oncol 2001; 20:97-108.
45. Liu JY, Sime PJ, Wu T et al. Transforming growth factor-beta(1) overexpression in tumor necrosis factor-alpha receptor knockout mice induces fibroproliferative lung disease. Am J Respir Cell Mol Biol 2001; 25:3-7.
46. Baroni T, Bodo M, D'Alessandro A et al. Silica and its antagonistic effects on transforming growth factor-beta in lung fibroblast extracellular matrix production. J Invest Med 2001; 49:146-56.
47. Himelstein BP, Lee EJ, Sato H et al. Transcriptional activation of the matrix metalloproteinase-9 gene in an H-ras and v-myc transformed rat embryo cell line. Oncogene 1997; 14:1995-8.
48. Moore BA, Aznavoorian S, Engler JA et al. Induction of collagenase-3 (MMP-13) in rheumatoid arthritis synovial fibroblasts. Biochim Biophys Acta 2000; 1502:307-18.
49. Parks WC, Shapiro SD. Matrix metalloproteinases in lung biology. Respir Res 2001; 2:10-19.
50. Trocme C, Gaudin P, Berthier S et al. Human B lymphocytes synthesize the 92-kDa gelatinase, matrix metalloproteinase-9. J Biol Chem 1998; 273:20677-84.
51. Pagenstecher A, Stalder AK, Kincaid CL et al. Differential expression of matrix metalloproteinase and tissue inhibitor of matrix metalloproteinase genes in the mouse central nervous system in normal and inflammatory states. Am J Pathol 1998; 152:729-41.
52. Leppert D, Waubant E, Galardy R et al. T cell gelatinases mediate basement membrane transmigration in vitro. J Immunol 1995; 154:4379-89.
53. Hibbs MS. Expression of 92 kDa phagocyte gelatinase by inflammatory and connective tissue cells. Matrix Suppl 1992; 1:51-7.
54. Pardo A, Ridge K, Uhal B et al. Lung alveolar epithelial cells synthesize interstitial collagenase and gelatinases A and B in vitro. Int J Biochem Cell Biol 1997; 29:901-10.
55. Harkness KA, Adamson P, Sussman JD et al. Dexamethasone regulation of matrix metalloproteinase expression in CNS vascular endothelium. Brain 2000; 123:698-709.
56. Sato T, del Carmen Ovejero M, Hou P et al. Identification of the membrane-type matrix metalloproteinase MT1-MMP in osteoclasts. J Cell Sci 1997; 110:589-96.
57. Larjava H, Lyons JG, Salo T et al. Anti-integrin antibodies induce type IV collagenase expression in keratinocytes. J Cell Physiol 1993; 157:190-200.
58. Bachmeier BE, Boukamp P, Lichtinghagen R et al. Matrix metalloproteinases-2,-3,-7,-9 and-10, but not MMP-11, are differentially expressed in normal, benign tumorigenic and malignant human keratinocyte cell lines. Biol Chem 2000; 381:497-507.
59. Yoshihara Y, Nakamura H, Obata K et al. Matrix metalloproteinases and tissue inhibitors of metalloproteinases in synovial fluids from patients with rheumatoid arthritis or osteoarthritis. Ann Rheum Dis 2000; 59:455-61.
60. Mautino G, Oliver N, Chanez P et al. Increased release of matrix metalloproteinase-9 in bronchoalveolar lavage fluid and by alveolar macrophages of asthmatics. Am J Respir Cell Mol Biol 1997; 17:583-91.
61. Leppert D, Ford J, Stabler G et al. Matrix metalloproteinase-9 (gelatinase B) is selectively elevated in CSF during relapses and stable phases of multiple sclerosis. Brain 1998; 121:2327-34.
62. Zuo F, Kaminski N, Eugui E et al. Gene expression analysis reveals matrilysin as a key regulator of pulmonary fibrosis in mice and humans. Proc Natl Acad Sci USA 2002; 99:6292-6297.
63. Perez-Ramos J, de Lourdes Segura-Valdez M, Vanda B et al. Matrix metalloproteinases 2, 9, and 13, and tissue inhibitors of metalloproteinases 1 and 2 in experimental lung silicosis. Am J Respir Crit Care Med 1999; 160:1274-1282.

64. Ortiz L, Lasky J, Gozal E et al. Tumor necrosis factor receptor deficiency alters matrix metalloproteinase 13/tissue inhibitor of metalloproteinase 1 expression in murine silicosis. Am J Respir Crit Care Med 2001; 163:244-52.
65. Shi X, Dalal NS, Vallyathan V. ESR evidence for the hydroxyl radical formation in aqueous suspension of quartz particles and its possible significance to lipid peroxidation in silicosis. J Toxicol Environ Hlth 1998; 25:237-245.
66. Daniel LN, Mao Y, Wang TCL et al. DNA strand breakage, thymine glycol production and hydroxyl radical generation induced by different samples of crystalline silica in vitro. Environ Res 1995; 71:60-73.
67. Antonini JM, Van Dyke K, Ye Z et al. Introduction of luminol-dependent chemiluminescence as a method to study silica inflammation in the tissue and phagocytic cells of rat lung. Environ Health Perspect 1994; 102:37-42.
68. Fubini B, Bolis V, Giamello E. The surface chemistry of crushed quartz dusts is relation to its pathogenicity. Inorg Chim Acta, Bioinorg Chem 1987; 138:193-197.
69. Vallyathan V, Leonard S, Kuppusamy P et al. Oxidative stress in silicosis: evidence for the enhanced clearance of free radicals from whole lungs. Mol Cell Biochem 1997; 168:125-32.
70. Zhang Z, Shen H-M, Zhang Q-F et al. Critical role of GSH in silica-induced oxidative stress, cytotoxicity, and genotoxicity in alveolar macrophages. Am J Physiol 1999; 277(4 Pt 1):L743-8.
71. Barrett EG, Johnston C, Oberdorster G et al. Antioxidant treatment attenuates cytokine and chemokine levels in murine macrophages following silica exposure. Toxicol Appl Pharmacol 1999; 158:211-20.
72. Iyer R, Holian A. Involvement of the ICE family of proteases in silica-induced apoptosis in human alveolar macrophages. Am J Physiol 1997; 273(4 Pt 1):L760-7.
73. Shen HM, Zhang Z, Zhang QF et al. Reactive oxygen species and caspase activation mediate silica-induced apoptosis in alveolar macrophages. Am J Physiol Lung Cell Mol Physiol 2001; 280:L10-7.
74. Lecoeur H. Nuclear apoptosis detection by flow cytometry: influence of endogenous endonucleases. Exp Cell Res. 2002; 277:1-14.
75. Iyer R, Hamilton RF, Li L et al. Silica-induced apoptosis mediated via scavenger receptor in human alveolar macrophages. Toxicol Appl Pharmacol 1996; 141:84-92.
76. Borges VM, Falcao H, Leite-Junior JH et al. Fas ligand triggers pulmonary silicosis. J Exp Med 2001; 194:155-64.
77. Chapman HA. A Fas pathway to pulmonary fibrosis. J Clin Invest 1999; 104:1-2.
Hagimoto N, Kuwano K, Miyazaki H et al. Induction of apoptosis and pulmonary fibrosis in mice in response to ligation of Fas antigen. Am J Respir Cell Mol Biol 1997; 17:272-278.
78. Kuwano K, Hagimoto N, Kawasaki M et al. Essential roles of the Fas-Fas ligand pathway in the development of pulmonary fibrosis. J Clin Invest 1999; 104:13-9.
79. Hagimoto N, Kuwano K, Kawasaki M et al. Induction of interleukin-8 secretion and apoptosis in bronchiolar epithelial cells by Fas ligation. Am J Respir Cell Mol Biol 1999; 21:436-45.
80. Schaub FJ, Han DK, Liles WC et al. Fas/FADD-mediated activation of a specific program of inflammatory gene expression in vascular smooth muscle cells. Nat Med 2000; 6:790-6.
81. Manos EJ, Jones DA. Assessment of tumor necrosis factor receptor and Fas signaling pathways by transcriptional profiling. Cancer Res 2001; 61:433-8.
82. Kuwano K, Hara N. Signal transduction pathways of apoptosis and inflammation induced by the tumor necrosis factor receptor family. Am J Respir Cell Mol Biol 2000; 22:147-9.
83. Kuwano K, Kunitake R, Maeyama T et al. Attenuation of bleomycin-induced pneumopathy in mice by a caspase inhibitor. Am J Physiol Lung Cell Mol Physiol 2001; 280:L316-25.
84. Borges VM, Lopes MF, Falcao H et al. Apoptosis underlies immunopathogenic mechanisms in acute silicosis. Am J Respir Cell Mol Biol 2002; 27:78-84.
85. Chao SK, Hamilton RF, Pfau JC et al. Cell surface regulation of silica-induced apoptosis by the SR-A scavenger receptor in a murine lung macrophage cell line (MH-S). Toxicol Appl Pharmacol 2001; 174:10-6.
86. Daemen MA, van 't Veer C, Denecker G, Heemskerk VH, Wolfs TG, Clauss M, Vandenabeele P, Buurman WA. Inhibition of apoptosis induced by ischemia-reperfusion prevents inflammation. J Clin Invest. 1999;104:541-9.

87. Cohen GM. Caspases: the executioners of apoptosis. Biochem J 1997; 326:1-16.
88. Fantuzzi G, Dinarello CA. Interleukin-18 and interleukin-1 beta: two cytokine substrates for ICE (caspase-1). J Clin Immunol 1999; 19:1-11.
89. Chang HY, Yang X. Proteases for cell suicide: Functions and regulation of caspases. Microbiol Mol Biol Rev 2000; 64:821-46.
90. Knies UE, Behrensdorf HA, Mitchell CA et al. Regulation of endothelial monocyte-activating polypeptide II release by apoptosis. Proc Natl Acad Sci USA 1998; 95:12322-7.
91. Behrensdorf HA, van de Craen M, Knies UE et al. The endothelial monocyte-activating polypeptide II (EMAP II) is a substrate for caspase-7. FEBS Lett 2000; 466:143-7.
92. Sciaky D, Brazer W, Center DM et al. Cultured human fibroblasts express constitutive IL-16 mRNA: cytokine induction of active IL-16 protein synthesis through a caspase-3-dependent mechanism. J Immunol 2000; 164:3806-14.
93. Orfila C, Lepert JC, Gossart S et al. Immunocytochemical characterization of lung macrophage surface phenotypes and expression of cytokines in acute experimental silicosis in mice. Histochem J 1998; 30:857-67.
94. Driscoll KE, Lindenschmidt RC, Maurer JK et al. Pulmonary response to silica or titanium dioxide: inflammatory cells, alveolar macrophage-derived cytokines, and histopathology. Am J Respir Cell Mol Biol 1990; 2:381-90.
95. Oghiso Y, Kubota Y. Interleukin 1 production and accessory cell function of rat alveolar macrophages exposed to mineral dust particles. Microbiol Immunol 1987; 31:275-87.

INDEX

D

E

F

G

H

I

J

K

L

M

N

O

P

V

W

MIX
Papier aus verantwortungsvollen Quellen
Paper from responsible sources
FSC® C105338
FSC
www.fsc.org

If you have any concerns about our products,
you can contact us on
ProductSafety@springernature.com

In case Publisher is established outside the EU,
the EU authorized representative is:
Springer Nature Customer Service Center GmbH
Europaplatz 3, 69115 Heidelberg, Germany

Printed by Libri Plureos GmbH
in Hamburg, Germany